The Decision Problem
Solvable Classes of
Quantificational Formulas

The Decision Problem

Solvable Classes of Quantificational Formulas

Burton Dreben

Harvard University

Warren D. Goldfarb

Harvard University

1979

Addison-Wesley Publishing Company, Inc.
Advanced Book Program
Reading, Massachusetts

London·Amsterdam·Don Mills, Ontario·Sydney·Tokyo

Library of Congress Cataloging in Publication Data

Dreben, Burton.
    The decision problem.

    Bibliography: p.
    Includes indexes.
        1. Predicate calculus. 2. Unsolvability (Mathemat-
ical logic) I. Goldfarb, Warren D., joint author.
II. Title.
QA9.35.D73   511'.3   79-18458
ISBN 0-201-02540-X

ABCDEFGHIJ-HA-79

American Mathematical Society (MOS) Subject Classification Scheme (1970):
  02B10, 02G05, 68A40

Manufactured in the United States of America

To Judge Dreben
 and
 Ella *and* Norman Goldfarb

Contents

Preface

Hilbert called the decision problem "the fundamental problem of mathematical logic," and Herbrand wrote that its positive solution "would furnish a general method for mathematics that would permit mathematical logic to play the same role with respect to ordinary mathematics that analytic geometry plays vis-à-vis Euclidean." A noble hope, reminiscent perhaps of Leibniz's vision of a *calculus ratiocinator*. But such dreams are not to be. Gödel's Incompleteness Theorem of 1931—as Herbrand was among the very first to recognize—made enormously implausible a positive solution for the decision problem, and Church in 1936 showed such a solution impossible. Still, various special cases of the decision problem do have positive solutions; that is, decision procedures exist for various classes of quantificational formulas. The aim of the present work is to give, by building on insights of Skolem and Herbrand, a systematic account of these solvable classes and to suggest an approach towards a general theory of solvability.

BURTON DREBEN
WARREN D. GOLDFARB

Acknowledgments

This book has been long in the making, and the debts incurred are numerous and extensive.

Stål Aanderaa's role has been unique. His name is attached to many of the most important and difficult results in the subject and in the book.

The small number of references to John Denton belies how much we owe to him. Only a detailed acquaintance with his doctoral dissertation of 1963 and his unpublished manuscripts of the mid-60s would convey the depth and pervasiveness of his contributions to this book. We are also obliged to him for several recent remarks that have improved what we had thought was the final version.

Andrew Gleason's unpublished generalization of a combinatorial lemma of Gödel was essential to the development of Chapter 3.

Harry Lewis has worked closely with us for the past eight years. The frequent occurrence of his name in the text barely hints at the immense impact he has had and the benefits we have gained.

W. V. Quine is our teacher. Through the years he has, of course, made many wise suggestions, but our true debt to him is inestimable. The standards of lucidity and penetration that are coextensive with his name have been both our inspiration and our despair.

Hartley Rogers has been constantly encouraging, and indeed called the attention of our publisher to Goldfarb's doctoral dissertation (1975), the immediate source of this book.

Gerald Sacks, too, in his inimitable way, has been encouraging, particularly in urging that no book should be flawless.

We are deeply indebted to Joseph Ullian. He has read with extraordinary care and understanding several versions of this and related works, and his numerous comments have always been of enormous value.

Alonzo Church, Charles Parsons, T. M. Scanlon, and Hao Wang have each had a much greater influence on this book than they imagine.

For a variety of comments, criticisms, and helpful advice we are grateful to Peter Andrews, George Boolos, Bradford Dunham, Herbert Enderton, Richard Fridshal, Michael Friedman, James Geiser, Stephen Gerig, Richard Goldberg, Wilbur D. Hart III, Peter Hylton, Daniel Isaacson, David Isles, Louis Jaeckel, A. S. Kahr, D. A. Martin, John Miller, Rohit Parikh, Michael

Rabin, Thomas G. Ricketts, Henry Rosovsky, Naomi Scheman, Kurt Schütte, Richard Shore, Leslie Tharp, A. Thomas Tymoczko, Dirk van Dalen, Jean van Heijenoort, and Joseph Wholey.

For not abandoning hope we thank Rogers Albritton, Stanley Cavell, Roderick Firth, Nelson Goodman, Robert Nozick, Hilary Putnam, John Rawls, Israel Scheffler, and Morton White.

The Society of Fellows at Harvard harbored one of us from 1952 to 1955 and the other from 1971 to 1975. Our appreciation for the intellectual stimulation and financial support it provided knows no bounds. Thanks are also due to the John Simon Guggenheim Memorial Foundation and to the National Science Foundation.

Much of the manuscript was written at the summer home of Judith and Richard Wurtman in Falmouth, Massachusetts. We thank them.

Finally, the older author is grateful to Florence and Robert Dreben, to Shalom Spiegel and his daughter, and to Elizabeth and Jon for their unflagging support, forbearance, and affection.

The Decision Problem
Solvable Classes of
Quantificational Formulas

Introduction

A class of quantificational formulas is *solvable* if and only if there is a decision procedure for satisfiability, that is, an effective procedure that determines of every formula in the class whether it has a model. In this book we treat questions of solvability by examining the Herbrand expansions of quantificational formulas. (Herbrand expansions are described on page 3.) As a result of this examination, a number of syntactically specified classes of formulas will be shown solvable.‡ The study of solvable classes has been bedeviled since its inception by overly syntactic and *ad hoc* arguments. It is our hope, however, to elicit the mathematical structures that underlie solvability. Our intent is not merely to compile a list of solvable and unsolvable classes. Rather, we aim at illuminating those features of expansions that give rise to decision procedures.

To make the definition of "solvability" precise, an explication of the notion of effectiveness must be adopted. We identify effectiveness with recursiveness or Turing computability. Thus solvability requires only that some recursive decision procedure exist; whether or not we know how to specify such a procedure is immaterial.‖ But, with one minor exception (Chapter 7, §3), our solvability proofs are all constructive: they actually provide decision procedures for the classes considered. The exact bounds of the notion of effectiveness are not relevant here; the decision procedures provided are clearly effective in any plausible intuitive sense of the word. In fact, prior to the first general explications of "effective" in the 1930s, it was only by actually exhibiting clearly effective decision procedures that classes could be shown solvable. Proofs of this type were devised for the class of formulas containing just monadic predicate letters (Löwenheim,

---

‡ Except when specifically noted, when we speak of classes of formulas in this Introduction we speak of classes of formulas of pure quantification theory that contain neither free variables nor function signs nor constants nor the identity sign.

‖ This has an unintuitive consequence: every finite class of formulas, and every class effectively reducible to a finite class of formulas, is solvable.

Burton Dreben and Warren D. Goldfarb, The Decision Problem: Solvable Classes of Quantificational Formulas

ISBN 0-201-02540-X

1915; Skolem, 1919; Behmann, 1922) and for the following prefix-classes:

(1) the class of prenex formulas with prefixes ∃···∃∀···∀ (Bernays and Schönfinkel, 1928);

(2) the class of prenex formulas with prefixes ∃···∃∀∀∃···∃ (Gödel, 1932; Kalmár, 1933; Schütte, 1934).

Of course, before the notion of effectiveness was made mathematically precise there could be no question of proving unsolvability. Instead, classes were shown to be reduction classes, that is, the decision problem for all of quantification theory was effectively reduced to the decision problem for these classes. Among these classes were the class of formulas containing only dyadic predicate letters (Löwenheim, 1915) and the class of prenex formulas with prefixes ∀···∀∃···∃ (Skolem, 1920). The unsolvability of reduction classes followed once Church (1936) and Turing (1937) proved the unsolvability of quantification theory. Turing's proof was quite direct. He encoded Turing machines by quantificational formulas, so that the unsolvability of quantification theory was immediately implied by the undecidability of the halting problem for Turing machines. Both reduction methods and encoding methods have continued to be used in showing unsolvability. (In the end, however, the distinction between these two methods is more heuristic than mathematical. In many cases they are in essence the same. See Lewis, 1974, 1979.) Of the following minimal prefix-classes, (3) was first shown unsolvable by reduction and (4) by encoding the "domino problem":

(3) the class of prenex formulas with prefixes ∀∀∀∃ (Surányi, 1950);

(4) the class of prenex formulas with prefixes ∀∃∀ (Kahr, Moore, and Wang, 1961).

We have mentioned results (1)–(4) because collectively they settle the decision problem for classes of prenex formulas specified solely by prefix: only (1) and (2) and their subclasses are solvable. (But note that if we extend quantification theory to include a sign for the identity relation, then there remains a prefix-class whose status is not settled, namely (2). Indeed, the status of the class of prenex formulas containing the identity sign with prefixes ∀∀∃ is not settled. We discuss this problem at length in Chapter 8.)

Classes (3) and (4) show that unsolvability arises even when quantificational structure is quite simple. They are examples of a phenomenon that pervades our subject: even apparently quite stringent syntactic restrictions yield unsolvability. Thus, rather detailed means of specifying classes must be employed for solvability to occur. In this book we go beyond specification by prefix in two ways: by restricting the truth-functional form of the

ISBN 0-201-02540-X

matrix, and by restricting the sequences of variables that may occur as arguments of atomic subformulas. Obviously, specifications may be refined *ad infinitum*; and since the specifications we consider are of necessity less natural ways of sorting quantificational formulas than by prefix alone, the enterprise may look like a pointless taxonomy. However, as we have already said, our interest is not in individual results, but in the picture given by the results taken as a whole, together with the methods used to obtain them. Unsolvability proofs ordinarily turn on a quite straightforward interplay between the syntax of formulas and reductions or encodings. Similarly, our treatment of a variety of solvable classes within a unified framework exhibits how their solvability depends on their syntactic specifications.

Our tool for exploring this dependence is something we call Herbrand expansions. We investigate the relations between the syntactic form of a quantificational formula and structural properties of its expansions, and also the ways these structural properties can permit decision procedures.

We now briefly describe Herbrand expansions. (Here we limit ourselves to prenex formulas F; complete and precise definitions are contained in Chapter 0.) With every F we effectively associate a (usually infinite) set of terms, called the *domain* of F, and a (usually infinite) number of quantifier-free instances of F, called *Herbrand instances*. The Herbrand instances are not all the instances of F. Although any term in the domain may be substituted for the universal variables of F, the substituents in any Herbrand instance for the existential variables are terms uniquely determined by the substituents for the universal variables. The *expansion* $E(F, \omega)$ is the set of all Herbrand instances of F. Moreover, $E(F, \omega)$ can be naturally viewed as the union of finite expansions $E(F, p)$, $p > 0$; each $E(F, p)$ is a finite subset of $E(F, \omega)$ in which all terms are of some limited complexity. The Expansion Theorem (Chapter 0, §3) asserts the equivalence of three conditions: the satisfiability of F, the truth-functional satisfiability of each $E(F, p)$, and the truth-functional satisfiability of the infinite expansion $E(F, \omega)$.

Various versions of the Expansion Theorem were used quite early to obtain solvability proofs. It is often simpler to find effective conditions equivalent to the truth-functional satisfiability of the expansion $E(F, \omega)$ than to the existence of a model for F. Thus, in their solvability proofs for prefix-class (2) above, Gödel, Kalmár, and Schütte formulate an effective syntactic condition on the matrix of a formula F. They then show, for any F in this class, what amounts to the following (cf. Chapter 7, §1): if there is a truth-assignment verifying $E(F, \omega)$, then the matrix of F satisfies the condition, and if the matrix satisfies the condition, then a truth-assignment verifying $E(F, \omega)$ may be inductively constructed.

However, a more intrinsic use of expansions may be made. Since each $E(F, p)$ is a finite set of quantifier-free formulas, there is an effective

ISBN 0-201-02540-X

method that determines, given any F and any p, whether $E(F, p)$ is truth-functionally satisfiable. Hence the solvability of a class $\mathfrak{C}$ follows if there is an algorithm for calculating, given any F in $\mathfrak{C}$, a number p such that the truth-functional satisfiability of $E(F, p)$ yields that of $E(F, \omega)$. For in this case the truth-functional satisfiability of $E(F, p)$ is equivalent to the satisfiability of F, and so a decision procedure for $\mathfrak{C}$ phrased entirely in terms of expansions is obtained. Thus we may concentrate solely on the expansions of F, and try to pass from a truth-assignment verifying some finite expansion $E(F, p)$ to one verifying $E(F, \omega)$. This approach to the decision problem was originated by Skolem (1928) and Herbrand (1930, 1931), and was applied by Church, in both a survey paper (1951) and his book (1956), to yield a number of classically known results. Although Church's proof sketches are quite intricate, a basic idea runs throughout. In each case a truth-assignment verifying a suitable finite expansion $E(F, p)$ is extended to one verifying $E(F, \omega)$ by a periodic iteration of values. The study of this basic idea by Dreben in the 1950s resulted in an unpublished monograph (1959) and a paper (1961); further study resulted in papers by Dreben, Kahr, and Wang (1962), Dreben and Denton (1963), and Goldberg (1963), as well as the more extensive dissertations of Denton (1963) and Goldfarb (1975). The present book, in turn, builds on that work.

The Expansion Theorem focuses our attention on the structure of expansions, which can be viewed as having two central features:

1.   All Herbrand instances of a formula F share the same truth-functional structure. That is, since all Herbrand instances of F are quantifier-free instances of F, they differ only in the arguments of their atomic subformulas: the placement of the truth-functional connectives and the predicate letters is the same in each. This common structure depends only on the internal truth-functional structure of F itself, for the same structure is shared by the matrix of F as well as by the Herbrand instances. Quantificational aspects of F play no role here.

2.   The Herbrand instances of a formula are not independent; two Herbrand instances may have atomic subformulas in common. For the moment let us say that any two such Herbrand instances are *connected*. (This informal notion of connection will be replaced in Chapter 1 by the precise notion of *coinstantiation*.) Naturally, they may be connected in different ways, depending on which atomic subformulas of one are identical to which of the other. The connections occurring in the expansion of F reflect quantificational aspects of F. For the existence of a connection—say, where the ith atomic subformula of the Herbrand instance H_1 is identical to the jth atomic subformula of the Herbrand instance H_2—depends not just on the fact that the ith and jth atomic subformulas of the matrix of F have the

ISBN 0-201-02540-X

same predicate letter, but also on the variables that are the arguments of these atomic subformulas. (A precise account of this dependency is not possible without an examination of the rules for constructing Herbrand instances, that is, the rules determining the substituents in an Herbrand instance for the existential variables.)

Connections among Herbrand instances are the source of the difficulty in showing solvability, for they offer the main obstacle to the construction of truth-assignments verifying $E(F, \omega)$. Clearly, if H_1 and H_2 are not connected and $\mathfrak{A}_1$ and $\mathfrak{A}_2$ are truth-assignments to H_1 and H_2, then there always is a truth-assignment $\mathfrak{A}$ that agrees with $\mathfrak{A}_1$ on H_1 and with $\mathfrak{A}_2$ on H_2. Not so if H_1 and H_2 are connected: if $\mathfrak{A}_1$ and $\mathfrak{A}_2$ assign opposite values to a common atomic subformula of H_1 and H_2, then there is no such $\mathfrak{A}$. Hence, were there no connections between Herbrand instances, $E(F, \omega)$ would be truth-functionally satisfiable provided that each individual Herbrand instance were truth-functionally satisfiable. Moreover, truth-functional unsatisfiability of an expansion $E(F, \omega)$ may result not just from connections between two Herbrand instances but from large patterns of connection, that is, connections among the members of sets $\{H_1,...,H_n\}$ of Herbrand instances. The unsolvability of quantification theory shows that there is no effective way, given an arbitrary formula F, to limit the size of the patterns that might cause $E(F, \omega)$ to be unsatisfiable. However, the syntactic restrictions on formulas in our solvable classes enable us to obtain the needed effective control on these patterns in the expansion. We mention the two main methods by which this is done.

The simpler method is truth-functional, and applies to classes of formulas specified by the truth-functional form of the matrix. Although the patterns of connection in the expansions of formulas F in these classes may be as intricate as in the general case, if the truth-functional form of the matrix of F—and hence that of each Herbrand instance of F—is "weak" enough, then the possible truth-functional effects of these patterns are limited. That is, by elementary truth-functional arguments we can show how to calculate from F a number n such that at most n connections between Herbrand instances could be involved in causing the expansion to be truth-functionally unsatisfiable. Thus we need only consider the truth-functional satisfiability of sets containing at most n Herbrand instances; and this yields solvability. In Chapter 1 we show by this method the solvability of several classes. The proofs there depend only on the truth-functional form of a formula and not on its quantificational structure.

The second method handles the far more interesting solvable classes specified by quantificational restrictions: restrictions on prefix and on atomic subformulas. The connections between the Herbrand instances of a formula F subject to such restrictions are repetitive or periodic in various

ISBN 0-201-02540-X

respects. Hence, we shall be able to construct, given any truth-assignment $\mathfrak{A}$ on a large enough finite expansion $E(F, p)$—where p is calculated in an effective manner from F—a truth-assignment $\mathfrak{B}$ whose behavior on $E(F, \omega)$ repeatedly mimics the behavior of $\mathfrak{A}$ on $E(F, p)$. More precisely, we shall so construct $\mathfrak{B}$ that, for each Herbrand instance H of F, the action of $\mathfrak{B}$ on H is identical to the action of $\mathfrak{A}$ on some Herbrand instance $\Psi(H)$ that lies in $E(F, p)$. Thus, if $\mathfrak{A}$ verifies $E(F, p)$, then $\mathfrak{B}$ verifies $E(F, \omega)$. Since $\mathfrak{A}$ is arbitrary, we may then infer the truth-functional satisfiability of $E(F, \omega)$ from that of $E(F, p)$. To exploit those properties of expansions that permit the construction of a suitable $\mathfrak{B}$, we use some (very simple) algebraic machinery; in particular we use mappings from atomic formulas to atomic formulas. The desired $\mathfrak{B}$ is defined in terms of such a mapping Γ by letting $\mathfrak{B}$ assign to each atomic formula A what $\mathfrak{A}$ assigns to its image $\Gamma(A)$. Our machinery enables us to carry out the constructions on an abstract level. Since we are concerned solely with *comparing* the action of truth-assignments on different Herbrand instances, we may ignore the actual values that these instances take, as well as the truth-functional form of the matrix of the formula F. In addition, we need not analyze ever larger patterns of connections; instead we focus on the atomic formulas themselves and on the ways in which they may occur multiply in expansions.

This method, called the *amenability method*, is introduced in Chapter 2. Our first two applications of it are to well-known classes: the class of formulas containing just monadic predicate letters, and the Ackermann Class, that is, the class of prenex formulas with prefixes $\forall y \exists x_1 \cdots \exists x_n$. Most of Chapter 2 is occupied with amenability arguments for a sequence of increasingly general classes of formulas. The syntactic specification of each successive class is reflected in increasingly complex patterns of connections, and these complexities require new subtleties in the construction of the mappings.

Solvability questions are not the only ones we treat by analyses of expansions. Chapter 3 takes up *finite controllability*: a class is finitely controllable if and only if each formula in it that has a model at all has a finite model. Many, though not all, solvable classes are also finitely controllable (among them the prefix-classes (1) and (2) mentioned on page 2). A formula F possesses a finite model provided there is a truth-assignment that verifies a set of instances of F obtained from $E(F, \omega)$ by replacing each term with a term in a finite set (this is the Finite Model Lemma of Chapter 3, §1). Thus our attention is still focused on the truth-functional satisfiability of a collection of (quantifier-free) instances of F: given any truth-assignment $\mathfrak{A}$ verifying the expansion $E(F, \omega)$, we seek a truth-assignment $\mathfrak{B}$ verifying a collection of instances of F that furnishes a finite model for F. Consequently, the machinery developed in earlier chapters may be applied. Chap-

ISBN 0-201-02540-X

ter 3 contains finite controllability proofs for all the solvable classes of Chapters 1 and 2; each proof exhibits strong analogies with its counterpart solvability proof. In particular, for the classes of Chapter 2, we prove finite controllability by establishing a property similar to amenability, a property we call *amiability*.

We have said that the amenability method ignores truth-functional features of the matrix of formulas. Indeed, at the end of Chapter 2 we show that the method does not apply to any solvable class whose specification essentially involves truth-functional restrictions. In Chapter 4 we present such a class, the Maslov Class. This class is specified by restrictions that are both quantificational and truth-functional in nature, and we show it solvable and finitely controllable by means of a hybrid method, combining techniques from amenability and amiability proofs with truth-functional arguments resembling those in Chapter 1.

Chapter 5 treats solvable classes that are not finitely controllable. We present three such: one amenable and two of the hybrid type like the Maslov Class. If a class is not finitely controllable, then a further question arises: is the class *docile*, that is, is there an effective method for deciding, given any F in the class, whether F has a finite model? Again by use of our basic machinery we answer this question affirmatively for the solvable classes of Chapter 5, as well as for an additional class, whose solvability is open.

Reduction proofs are the subject of Chapter 6. A *reduction* of a class of formulas to another class is an effective function that associates to each formula F in the first class a formula G in the second such that F has a model if and only if G has a model. Reductions can be put to work in two directions: reducing a class $\mathfrak{C}$ to a class already known to be solvable establishes the solvability of $\mathfrak{C}$; reducing a class already known to be unsolvable to a class $\mathfrak{C}'$ establishes the unsolvability of $\mathfrak{C}'$. We give a general procedure useful for reduction proofs in the first direction (the Splitting Lemma of §1), and with its aid obtain solvable extensions of the classes of Chapter 2. We also give two general procedures useful for reduction proofs in the second direction; these procedures exploit specific sorts of connections between Herbrand instances. Reduction proofs in both directions furnish information about the dividing line between solvable and unsolvable classes, particularly with regard to Skolem classes, that is, classes of prenex formulas with prefixes $\forall y_1 \cdots \forall y_m \exists x_1 \cdots \exists x_n$. Chapter 6 closes with our versions of two reductions known from the literature: one eliminates from a prenex formula any existential quantifiers at the beginning of the prefix; the other eliminates from a prenex formula with prefix $\forall y_1 \exists x \forall y_2$ all but dyadic predicate letters.

In Chapter 7 we return to the amenability method, and examine further

ISBN 0-201-02540-X

the link between amenability and repetitive properties of patterns of connections in Herbrand expansions. This examination has a surprising by-product: all the classes we showed solvable by the amenability method are reducible to the class of formulas containing just monadic predicate letters. We end this chapter with two results obtained by elementary analyses of connections: the solvability and finite controllability of the class of formulas containing at most two distinct atomic subformulas; and the solvability, for each m and n, of the class of formulas containing at most m universal quantifiers and n distinct atomic subformulas. The latter is our only non-constructive solvability proof. We reduce each such class to a finite class, but do not specify a decision procedure for each. Indeed, there can be no uniform specification of decision procedures for each of these classes; since the union of these classes is the class of all formulas, any such uniform specification would yield a decision procedure for all of quantification theory.

In Chapter 8 we consider quantification theory extended by a sign for the identity relation. Here a new source of difficulty in showing solvability appears. For if two terms are equated by a truth-assignment $\mathfrak{A}$, then $\mathfrak{A}$ must assign the same value to any pair of atomic formulas that differ solely by substitution of one of these terms for the other. (That is, $\mathfrak{A}$ must verify all instances of the law of substitutivity of identicals.) Thus Herbrand instances can fail to be independent not only by containing a common atomic subformula, but also by containing atomic subformulas differing by such substitutions. As a result, many solvable classes of pure quantification theory become unsolvable when identity is allowed; others remain solvable, but the proofs become more complex; moreover, an important open problem emerges.

We view this book as a prolegomenon to an abstract study of solvability and related notions, a study not concerned with particular classes. Is there an informative general criterion that distinguishes those syntactic restrictions that do from those syntactic restrictions that do not lead to solvable classes? We hope our examination of the structural properties of expansions provides the data and tools needed to attack this question, and points to the general concepts in terms of which an answer might be formulated.

ISBN 0-201-02540-X

Notation and the Expansion Theorem

1. Quantificational Language

The formal language in which we couch quantification theory has as primitive signs the quantifiers "∀" and "∃", the truth-functional connectives "∧", "∨", and "−", left and right parentheses, an infinite supply of variables, for each $k > 0$ an infinite supply of k-place predicate letters and k-place function signs, and an infinite supply of 0-place function signs, also called constants. (In Chapter 8 we shall include a sign "≡" for the identity predicate. The regular identity sign "=" is used only in the metalanguage.) We use certain latin letters, with and without subscripts, in a double sense (made clear by the context): both as syntactic variables and as particular object-language signs. "P", "Q", "R" range over predicate letters and are actual predicate letters; "x", "y", "z" range over variables and are object-language variables; "f" and "g" range over function signs and are function signs; and "c" ranges over constants and is a constant. However, the letters "v" and "w" are only syntactic variables ranging over object-language variables.

The *terms* of the language are constructed from variables and constants by iterated application of the function signs. More precisely,

(1) a *term of height* 1 is a variable or constant;

(2) a *term of height* $p + 1$, $p \geq 1$, is any string $f(t_1 \cdots t_k)$ where f is a k-place function sign, $k > 0$, and $t_1, \ldots, t_k$ are terms of height $\leq p$ of which at least one is of height p. The terms $t_1, \ldots, t_k$ are said to be the *arguments* of the term $f(t_1 \cdots t_k)$.

A term is then a term of some height ≥ 1; the height of a term t, often abbreviated $h(t)$, is a measure of the maximum number of nested function signs in t.

Burton Dreben and Warren D. Goldfarb, The Decision Problem: Solvable Classes of Quantificational Formulas

The *atomic formulas* of the language are just the strings $Pt_1\cdots t_k$, where P is any k-place predicate letter, $k > 0$, and $t_1,\ldots,t_k$ are any terms. The terms $t_1,\ldots,t_k$ are said to be the *arguments* of the atomic formula $Pt_1\cdots t_k$. A *formula* is either an atomic formula or else an expression constructed in the usual manner from atomic formulas by means of the truth-functional connectives and quantifiers; the notions of scope of a quantifier-occurrence or connective-occurrence in a formula, free and bound variable-occurrences, and so on, are defined in the ordinary way. A *signed atomic formula* is either an atomic formula or an atomic formula preceded by the negation sign "$-$"; it is said to be positively signed in the first case and negatively signed in the second. A formula is said to be in *conjunctive normal form* iff it is a conjunction of disjunctions of signed atomic formulas, and is said to be in *disjunctive normal form* iff it is a disjunction of conjunctions of signed atomic formulas.‡

We often take advantage of the associativity of conjunction and disjunction, and write conjunctions and disjunctions of formulas without parentheses. If $K = K_1 \wedge \cdots \wedge K_n$ is a conjunction, each K_i is said to be a *conjunct in K*; similarly, if $K = K_1 \vee \cdots \vee K_n$, each K_i is said to be a *disjunct in K*. We sometimes use $\vee$ and $\wedge$ for many-fold disjunction and conjunction. Thus $\bigvee_{1 \leq i \leq n} K_i$ is $K_1 \vee \cdots \vee K_n$. Often we omit the bounds on the index i, when these are clear from the context; sometimes we give these bounds in words following the formula.

The connectives "$\Rightarrow$" and "$\Leftrightarrow$" are employed as shorthand notation in the usual way: we write $F \Rightarrow G$ for the formula $-F \vee G$, and $F \Leftrightarrow G$ for the formula $(-F \vee G) \wedge (F \vee -G)$.

The *matrix* of a formula F is the formula F^M obtained from F by deleting each occurrence of a quantifier as well as the occurrence of the variable immediately to its right. Thus even nonprenex formulas have matrices.

We shall be primarily concerned with two sublanguages of our language. The first is that in which no function sign occurs. A formula F of this sublanguage is *rectified* iff no variable occurs both bound and free in F, and no variable is bound by two different quantifier-occurrences. By relettering bound variables we can of course obtain a rectified equivalent for any given formula. A *schema* is a formula of this sublanguage that is rectified and closed (contains no free variables). This nonstandard use of "schema" is a mere terminological convenience. Throughout this book we deal with decision problems for classes of schemata.

Let F be a schema. A variable v *governs* a variable w in F iff the scope of the quantifier-occurrence in F that binds v properly includes the scope of the quantifier-occurrence that binds w. A variable v is an *x-variable* of

---

‡ We use "iff" for "if and only if".

ISBN 0-201-02540-X

F iff the quantifier-occurrence binding v in F is either an $\exists$-occurrence within the scopes of an even number (possibly zero) of negation sign occurrences, or else an $\forall$-occurrence within the scopes of an odd number. Otherwise, a variable v of F is a *y-variable*. Thus the x-variables of F are precisely those which become existential variables in any prenex equivalent of F obtained by the usual prenexing rules, and the y-variables of F are those which become universal variables. We shall always so write schemata that each use of ''y'', subscripted or not, represents a y-variable, and each use of ''x'' an x-variable.‡

The second sublanguage we consider is that with neither quantifiers nor variables. Thus a term in this sublanguage is a variable-free term, and a formula either is an atomic formula containing only variable-free terms or else is constructed from such atomic formulas by means of the truth-functional connectives. Atomic formulas not containing variables are called *atoms*. A *signed atom* is a signed atomic formula containing no variables.

Components. We introduce certain functions for specifying atomic subformulas of a formula F. With each formula F we associate a positive integer $c(F)$, called the length of F, and for each i, $1 \le i \le c(F)$, an atomic subformula $\langle F \rangle_i$ of F, called the ith component of F. Intuitively, $c(F)$ is the number of occurrences of atomic formulas in F, and $\langle F \rangle_i$ is the atomic formula an occurrence of which is the ith atomic-formula-occurrence in F, counting from the left. More precisely,

(1) If F is an atomic formula, then $c(F) = 1$ and $\langle F \rangle_1 = F$;

(2) If F is $-G$, $\forall v G$, or $\exists v G$, then $c(F) = c(G)$ and $\langle F \rangle_i = \langle G \rangle_i$ for $1 \le i \le c(F)$;

(3) If F is $G_1 \wedge G_2$ or $G_1 \vee G_2$, then $c(F) = c(G_1) + c(G_2)$ and $\langle F \rangle_i = \langle G_1 \rangle_i$ for $1 \le i \le c(G_1)$, $\langle F \rangle_i = \langle G_2 \rangle_{i-c(G_1)}$ for $c(G_1) < i \le c(G_1) + c(G_2)$.

We also stipulate that if $i > c(F)$, then $\langle F \rangle_i = \langle F \rangle_{c(F)}$; thus we have defined a function that takes any formula F and any positive integer i to the atomic formula $\langle F \rangle_i$.

Note that $\langle F \rangle_i$ is an atomic formula, not an occurrence of an atomic formula. We may have $\langle F \rangle_i = \langle F \rangle_j$ even though $i \ne j$. Also note that the value of $\langle F \rangle_i$ does not depend on the presence or absence of truth-functional connectives, quantifiers, or parentheses. (In particular $\langle F \rangle_i = \langle F^M \rangle_i$ for every i, where F^M is the matrix of F.) For example, suppose F is

$$\forall y \exists x [Py \wedge (-Px \vee Qy \vee \exists z(-Py \wedge Qz))].$$

Then $c(F) = 5$, $\langle F \rangle_1 = Py = \langle F \rangle_4$, $\langle F \rangle_2 = Px$, $\langle F \rangle_3 = Qy$, and $\langle F \rangle_5 = Qz$.

---

‡ The following mnemonic is due to W. V. Quine: ''x'' for x-istential, ''y'' for you-niversal.

ISBN 0-201-02540-X

Instances.   Let F be a formula, let $v_1,...,v_k$ be distinct variables, and let $t_1,...,t_k$ be (not necessarily distinct) terms, variable-free or not. Then $F(v_1/t_1,...,v_k/t_k)$ is the result of replacing simultaneously for each i, $1 \leq i \leq k$, the variable v_i at all its free occurrences in F by t_i. The variable v_i need not actually occur free in F for this notation to be defined. If, for example, v_1 does not occur free in F, then $F(v_1/t_1,...,v_k/t_k)$ is just $F(v_2/t_2,...,v_k/t_k)$.

An *instance* of a formula F is any formula resulting from the matrix F^M of F by substituting, for each variable, some variable-free term. That is, an instance of F is any formula $I = F^M(v_1/t_1,...,v_k/t_k)$, where $v_1,...,v_k$ are all the (distinct) variables of F^M and $t_1,...,t_k$ are any (not necessarily distinct) variable-free terms. Thus I is a variable-free formula. The instance I is said to be an instance of F *over* D iff D is any set of terms including $t_1,...,t_k$. The *substituent in I for the variable* v_i is the term t_i.

2. Expansions and Domains

We now associate with each variable v of the language and each number $k \geq 0$ a k-place function sign called the *k-place indicial correlate of v*; we also pick out a special constant, that is, a 0-place function sign, $\dashv$. No distinct variables have indicial correlates in common, and $\dashv$ is not an indicial correlate of any variable. We often write an indicial correlate of a variable v as f_v, and the 0-place indicial correlate of v as c_v. An indicial correlate of any variable is called an *indicial function sign*.

Let F be a schema. The *functional form F^* of F* is the result of replacing in the matrix F^M of F each x-variable x of F by the term $f_x(y_{i_1} \cdots y_{i_r})$, where $r \geq 0$ is the number of y-variables governing x in F, f_x is the r-place indicial correlate of x, and $\langle y_{i_1},...,y_{i_r} \rangle$ is the sequence of y-variables of F that govern x, in left-to-right order of their first occurrence in F. Thus F^* is a quantifier-free formula whose free variables are precisely the y-variables of F. The term $f_x(y_{i_1} \cdots y_{i_r})$ is called the *indicial term supplanting x in F^**. If x is governed by no y-variable, then the indicial term supplanting x in F^* is just the 0-place indicial correlate of x.

EXAMPLES.   (1)   Let F be the schema

$$\exists x_1 \forall y_1 P y_1 x_1 \wedge \forall y_2 \forall y_3 - \forall x_2 (P x_2 y_2 \vee \exists y_4 P y_4 y_3).$$

Then F^* is

$$P y_1 c_{x_1} \wedge -(P f_{x_2}(y_2 y_3) y_2 \vee P y_4 y_3),$$

where c_{x_1} is the 0-place indicial correlate of x_1 and f_{x_2} is the 2-place indicial correlate of x_2. That is, $F^* = F^M(x_1/c_{x_1}, x_2/f_{x_2}(y_2 y_3))$.

ISBN 0-201-02540-X

(2) Let F be a *Skolem schema*, that is, a prenex schema with prefix $\forall y_1 \cdots \forall y_m \exists x_1 \cdots \exists x_n$. Then the functional form F^* is obtained from the matrix of F by replacing each x-variable x_k, $1 \le k \le n$, with the term $f_{x_k}(y_1 \cdots y_m)$, where f_{x_k} is the m-place indicial correlate of x_k.

Thus in constructing a functional form we replace each x-variable with an indicial term containing a function sign; the function sign can be viewed as representing a Skolem function that provides values for the x-variable. The use of functional forms is essentially our only foray into the full quantificational language specified above; except for our dealings with functional forms, we shall always be concerned either with formulas not containing function signs or with formulas not containing variables.

The *domain $D(F, p)$ of a schema F of order p*, $p \ge 1$, is the set of all variable-free terms of height $\le p$ generated from $\dashv$ and the indicial constants occurring in F^* by the k-place indicial function signs occurring in F^*, $k > 0$. Note that the domains are cumulative, that is, $D(F, p) \subseteq D(F, p+1)$ for each p. In Example (1), $D(F, 1) = \{c_{x_1}, \dashv\}$ and $D(F, 2) = \{c_{x_1}, \dashv, f_{x_2}(\dashv \dashv), f_{x_2}(c_{x_1} c_{x_1}), f_{x_2}(c_{x_1} \dashv), f_{x_2}(\dashv c_{x_1})\}$. In Example (2), $D(F, 1) = \{\dashv\}$ and $D(F, 2) = \{\dashv, f_{x_1}(\dashv \cdots \dashv), \ldots, f_{x_n}(\dashv \cdots \dashv)\}$. Domains of larger order rapidly become too big to give as examples.

The *domain $D(F, \omega)$* is the union of the domains $D(F, p)$. Thus $D(F, \omega)$ is infinite whenever F^* contains at least one r-place function sign for $r > 0$, that is, whenever there is an x-variable in F governed by at least one y-variable. We assume for each F a fixed ordering of the domain $D(F, \omega)$, an ordering in which terms of lesser height precede those of greater height. Thus it makes sense to speak of the earliest term in a domain $D(F, \omega)$ with such-and-such a property. If x_k is an x-variable of F, then an x_k-*term* in $D(F, \omega)$ is any term whose outermost function sign is the indicial correlate of x_k that occurs in F^*. Each term t in $D(F, \omega)$ except $\dashv$ is an x_k-term for a unique x-variable x_k of F; we let $\chi(t)$ be this unique x-variable. We also let $\chi(\dashv) = \dashv$; hence if F contains m x-variables, then $\chi(t)$ takes on $m + 1$ values as t ranges over $D(F, \omega)$. In Example (1) the term c_{x_1} is the only x_1-term in $D(F, \omega)$, and all other terms in $D(F, \omega)$ but $\dashv$ are x_2-terms. In Example (2), an x_k-term is any term $f_{x_k}(t_1 \cdots t_m)$ in $D(F, \omega)$, $1 \le k \le n$.

An *Herbrand instance of F of order p*, $p \ge 1$, is any instance of the functional form F^* over the domain $D(F, p)$. That is, an Herbrand instance of order p is any formula $H = F^*(y_1/t_1, \ldots, y_k/t_k)$, where $y_1, \ldots, y_k$ are all the y-variables of F (and hence all the free variables of F^*), and $t_1, \ldots, t_k$ are any (not necessarily distinct) terms in $D(F, p)$. An Herbrand instance of F (simpliciter) is of course an Herbrand instance of F of some order. Every Herbrand instance is a variable-free formula. Moreover, every Herbrand instance H of F is an instance of F fulfilling the following constraints:

ISBN 0-201-02540-X

(1) the substituents in H for the y-variables of F are in $D(F, \omega)$;

(2) if the substituents in H for the y-variables $y_{i_1},\ldots,y_{i_r}$ of F are $t_1,\ldots,t_r$, respectively, and $f_{x_k}(y_{i_1}\cdots y_{i_r})$ is the indicial term supplanting the x-variable x_k of F in F^*, then the term $f_{x_k}(t_1\cdots t_r)$ is the substituent in H for x_k.

In particular, the substituent in any Herbrand instance H of F for any x-variable x_k of F is an x_k-term. Note that if H is of order p, then terms of height $p + 1$ may occur in H: for if one of the t_i is of height exactly p, then the substituent $f_{x_k}(t_1\cdots t_r)$ for x_k is of height $p + 1$. Thus the Herbrand instances of order p of F are all the instances of F^* over $D(F, p)$, but are only certain of the instances of F over $D(F, p + 1)$.

Note too that for each term t in $D(F, \omega)$ except $\dashv$, t occurs as the substituent in at least one Herbrand instance of F for the x-variable $\chi(t)$. Indeed, t so occurs in every Herbrand instance H in which the substituents in H for the y-variables of F that govern $\chi(t)$ are just the arguments of t.

We use the letter "H", with or without subscripts, as a syntactic variable over the Herbrand instances of a schema. Although there is a natural ordering of Herbrand instances, the subscripts used on the letter "H" never refer to this ordering. Thus, if in some context dealing with a schema F we use "H_1" and "H_2", then these are meant to range independently over arbitrary Herbrand instances of F.

The *expansion* $E(F, p)$ *of* F *of order* p, $p \geq 1$, is the set of all Herbrand instances of F of order p. Hence $E(F, p)$ is a finite set of variable-free formulas, and $E(F, p) \subseteq E(F, p + 1)$ for each p. The *expansion* $E(F, \omega)$ of F is the union of all the expansions $E(F, p)$, and hence is the set of all Herbrand instances of F. In general $E(F, \omega)$ is infinite. The Expansion Theorem, which we state in the next section, connects expansions of a schema with the existence of models for the schema.

We now introduce some notation that will be used—starting in Chapter 2, §2—to depict relationships between terms in domains $D(F, \omega)$. Let F be a schema. With each variable v of F we associate a mapping $\hat{v}$ defined on (some) terms in $D(F, \omega)$: for each x_k-term t in $D(F, \omega)$, if there is a unique term s that is the substituent for v in every Herbrand instance of F in which t is the substituent for x_k, then $\hat{v}t = s$; otherwise, the mapping $\hat{v}$ is not defined on t, or, as we shall often say, $\hat{v}t$ is undefined.

Thus $\hat{v}t$ is undefined in just the following cases: (a) $t = \dashv$; (b) $t \neq \dashv$ but v is a y-variable not governing the x-variable $\chi(t)$; (c) $t \neq \dashv$ but v is an x-variable governed by at least one y-variable that does not govern $\chi(t)$.

EXAMPLES. (1) Let F be a Skolem schema $\forall y_1\cdots\forall y_m\exists x_1\cdots\exists x_n F^M$.

ISBN 0-201-02540-X

If $t = f_{x_k}(t_1 \cdots t_m)$, then $\hat{y}_i t = t_i$ and $\hat{x}_j t = f_{x_j}(t_1 \cdots t_m)$ for $1 \le i \le m$ and $1 \le j \le n$.

(2) Let F be a prenex schema whose y-variables are $y_1, \ldots, y_m$ and whose x-variables are $x_1, \ldots, x_n$, where both lists are in left-to-right order of occurrence in the prefix of F. For each x-variable x_k and each x_k-term t,

(i) if y_i governs x_k, then $\hat{y}_i t$ is the ith argument of t, $1 \le i \le m$;
(ii) if x_j is governed by the y-variables $y_1, \ldots, y_l$ and no other y-variables, and $y_1, \ldots, y_l$ all govern x_k, then $\hat{x}_j t = f_{x_j}(\hat{y}_1 t \cdots \hat{y}_l t)$, $1 \le j \le n$;
(iii) in all other cases, $\hat{v} t$ is undefined. End of Examples.

The following are some simple properties of the mappings $\hat{v}$: for each term $t \ne \mathbf{\dagger}$,

(1) if $\chi(t) = x_k$, then $\hat{x}_k t = t$;
(2) if v is an x-variable and $\hat{v} t$ is defined, then $\chi(\hat{v} t) = v$;
(3) if v governs an x-variable x_j that governs $\chi(t)$, then $\hat{v} t = \hat{v} \hat{x}_j t$;
(4) if t is the substituent in an Herbrand instance H of F for $\chi(t)$, and if $\langle F \rangle_i = P v_1 \cdots v_p$, then $\langle H \rangle_i = P \hat{v}_1 t \cdots \hat{v}_p t$, provided that each $\hat{v}_i t$ is defined.

If V is a set of variables of F and t is a term in $D(F, \omega)$, we also let $\hat{V}[t] = \{ \hat{v} t \mid v \in V \}$. This notation is always used with the presupposition that $\hat{v} t$ is defined for each v in V.

3. The Expansion Theorem

A *model* for a closed formula F consists of a nonempty universe U together with interpretations of the predicate letters and function signs of F—each k-place predicate letter is assigned a subset of the k-fold Cartesian product U^k of U with itself, and each k-place function sign is assigned a function from U^k to U (each 0-place function sign is assigned a member of U)—such that F is true in U under these interpretations. If F has a model, we often say that F is *satisfiable*; and if F has a model with universe U, we often say that F is *satisfiable over U*.

We have immediately from the construction of the functional form that any model for the universal closure of the functional form F^* of a schema F is a model for F: the interpretations of the function signs of F^* yield the values of the x-variables required for the truth of F. Conversely, assuming the axiom of choice, any model for F can be transformed into a model for F^* by adjoining appropriate interpretations of the indicial correlates occurring in F^*. The axiom of choice is used to obtain these interpretations if the given model for F is nondenumerable. However, without the axiom of

ISBN 0-201-02540-X

choice we can show that if F has a model, then there is some (denumerable) model for F^*. This proof goes by way of the Expansion Theorem below; indeed, the Expansion Theorem was used by Skolem to show the choice-free form of the Skolem–Löwenheim Theorem: a formula or denumerable set of formulas has a model only if it has a denumerable model.

Let S be a set of atomic formulas; let $\tilde{S}$ be the truth-functional closure of S, that is, the union of S and the set of all formulas constructed from members of S by the truth-functional connectives. A *truth-assignment on* $\tilde{S}$ is a function $\mathfrak{A}$ from $\tilde{S}$ into $\{0, 1\}$, a function uniquely determined by its values on S by means of the following inductive rules:

$$\mathfrak{A}(-F) = 1 \quad \text{iff} \quad \mathfrak{A}(F) = 0;$$

$$\mathfrak{A}(F \vee G) = 1 \quad \text{iff} \quad \mathfrak{A}(F) = 1 \text{ or } \mathfrak{A}(G) = 1;$$

$$\mathfrak{A}(F \wedge G) = 1 \quad \text{iff} \quad \mathfrak{A}(F) = 1 \text{ and } \mathfrak{A}(G) = 1.$$

A truth-assignment $\mathfrak{A}$ is said to *verify* a formula F if $\mathfrak{A}(F) = 1$ and to *falsify* F if $\mathfrak{A}(F) = 0$.

For the sake of brevity we extend our terminology thus: a truth-assignment on an arbitrary set E of quantifier-free formulas is just a truth-assignment on the set $\tilde{S}$, where S is the set of atomic subformulas of members of E. We shall ordinarily use "$\mathfrak{A}$" and "$\mathfrak{B}$" for truth-assignments on sets of formulas constructed from atoms (that is, variable-free formulas), and "$\mathfrak{T}$" for truth-assignments on sets of formulas containing variables.

A set E of quantifier-free formulas is *consistent* iff there is a truth-assignment $\mathfrak{A}$ on E such that $\mathfrak{A}$ verifies every member of E (or, as we shall say, verifies E). A formula F is consistent iff $\{F\}$ is consistent. Thus we use "consistent" in a semantic rather than syntactic sense; this property is more usually referred to as "truth-functional satisfiability".

There is, of course, a uniform effective method for deciding the consistency of any finite set of formulas. We need only check the finite number of different truth-assignments on the finite set of atomic formulas that actually occur in members of the set.

If F is a schema and $p > 0$, let $L(F, p)$ be the set of all atoms constructed from predicate letters of F and terms in the domain $D(F, p)$, and let $L(F, \omega)$ be the union of the $L(F, p)$. Thus $L(F, \omega)$ is usually infinite. We shall often take the consistency of an expansion $E(F, p)$ or $E(F, \omega)$ to imply the existence of a truth-assignment on $\tilde{L}(F, \omega)$ that verifies $E(F, p)$ or $E(F, \omega)$, even though not every member of $L(F, \omega)$ actually occurs in the expansion (and indeed, in the case of the finite expansion $E(F, p)$, only finitely many members of $L(F, \omega)$ occur). Clearly this is legitimate: if there is a truth-assignment $\mathfrak{A}$ verifying the expansion, and hence defined on all atoms

ISBN 0-201-02540-X

occurring in the expansion, then $\mathfrak{A}$ can be extended arbitrarily to the other atoms in $L(F, \omega)$.

Expansion Theorem.   Let F be a schema. Then the following are equivalent:

(1) F has a model;
(2) for each p the expansion $E(F, p)$ is consistent;
(3) the expansion $E(F, \omega)$ is consistent.

We shall not prove the Expansion Theorem; instead we make just a few observations. To obtain (2) from (1) the axiom of choice is not needed, since for each p the consistency of $E(F, p)$ does not require the infinite number of choices necessary to construct Skolem functions on the universe of the assumed model for F. To obtain (3) from (2) the Law of Infinite Conjunction is needed, that is, the inference to the consistency of an infinite set of quantifier-free formulas from the consistency of each finite subset. The Law of Infinite Conjunction is a special case of the König Infinity Lemma, but a special case that, unlike the general Lemma, does not require the axiom of dependent choice. Obtaining (1) from (3) is easy; any truth-assignment verifying $E(F, \omega)$ provides a model for the universal closure of F^*, and hence a model for F, over the universe $D(F, \omega)$; we look at this in a little more detail two paragraphs below.

Arguments for the Expansion Theorem, with various degrees of generality and precision, are contained in papers by Skolem (1922, 1928, 1929), Herbrand (1930), and Gödel (1930). It should be noted, however, that Herbrand's major purpose in his thesis (1930) was to argue not for the formulation of the Expansion Theorem that we have given, but rather for the equivalence of condition (2) above and the irrefutability of F in some standard axiomatic system of quantification theory. Similarly, Gödel's concern, as part of his completeness proof for quantification theory, was to obtain (2), then (3), and then (1), from the irrefutability of F. Skolem's papers, on the other hand, do not always make clear which connections among syntactic and semantic notions are being established. For further historical reflections, see the introduction to Herbrand's *Logical Writings* (1971), and the introductions to Skolem's 1928 paper and to Gödel's 1930 paper in van Heijenoort's sourcebook (1967); see also Goldfarb (1971, 1979).

If $\mathfrak{A}$ is a truth-assignment on $\bar{L}(F, \omega)$ that verifies the expansion $E(F, \omega)$, then $\mathfrak{A}$ provides a model for the universal closure of F^* thus: the universe of this model is simply the domain $D(F, \omega)$; each k-place predicate letter P of F is assigned the set of k-tuples $\langle t_1, \ldots, t_k \rangle$ from $D(F, \omega)$ such that $\mathfrak{A}$ verifies $P t_1 \cdots t_k$, and each k-place function sign f of F^* is assigned the function taking a k-tuple $\langle t_1, \ldots, t_k \rangle$ from $D(F, \omega)$ to the element $f(t_1 \cdots t_k)$ of $D(F, \omega)$. Thus the model obtained has built-in interpretations for the

ISBN 0-201-02540-X

function signs. By simply deleting these interpretations, retaining the interpretations of the predicate letters, we obtain a model for F over $D(F, \omega)$.

We cannot correspondingly obtain a model for F over the (finite) universe
$D(F, p)$ from a truth-assignment verifying the (finite) expansion $E(F, p)$.
For in general, Herbrand instances of F in $E(F, p)$ may contain terms of
height $p + 1$; the built-in interpretations of the function signs thus carry
$D(F, p)$ into $D(F, p + 1)$, that is, the universe $D(F, p)$ is not closed under
application of these interpreted function signs. (Of course, we cannot obtain
a model over $D(F, p + 1)$ either; for if $\mathfrak{A}$ verifies $E(F, p)$, then the truth
of F^*—under the interpretations defined from $\mathfrak{A}$ as in the preceding paragraph—is ensured only when the variables of F^* range over $D(F, p)$.)
Thus the expansion $E(F, p)$ is not to be construed as an expansion of F
over a finite universe, but rather as an initial segment of an expansion of F
over an infinite universe. A finite model for F is forthcoming if there is a
truth-assignment verifying not $E(F, \omega)$ but rather a set of instances of F
closely related to $E(F, \omega)$: we formulate a condition of this form in Chapter
3, §1. This condition then plays a central role in our proofs of finite controllability.

We close this section with two further terminological points. We say that
a formula F is *valid* iff F is true in every universe under every interpretation.
And we say that a formula F *logically implies* a formula G iff the formula
$F \Rightarrow G$ is valid.

4. Further Notation

We shall also employ some fairly standard mathematical notation. We
use angular brackets for finite sequences, that is, for ordered k-tuples,
$k > 0$; thus $\langle t_1, \ldots, t_k \rangle$ is the k-tuple whose ith member is t_i, $1 \le i \le k$.
(There should be no confusion between this usage and the component
notation defined earlier in §1.) Set inclusion (not necessarily proper) is
denoted by "$\subseteq$"; set union by "$\cup$"; and set intersection by "$\cap$". If S and
T are sets, then $|S|$ is the cardinality of S and $S - T$ is the set of elements
of S that are not elements of T. If S is a set and $n > 0$, then S^n is the n-fold
Cartesian product of S with itself. If η is a function and S is a set, then
$\eta[S]$ is the image of S under η; that is, $\eta[S] = \{\eta(a) \mid a \in S\}$. Finally, "$\square$"
indicates the end of a proof.

ISBN 0-201-02540-X

CHAPTER 1

Basic Considerations about Expansions

The Expansion Theorem transforms the question of the satisfiability of a quantificational schema F into the truth-functional question of the consistency of each expansion $E(F, p)$ or, equivalently, the consistency of the set $E(F, \omega)$ of all Herbrand instances of F. Thus the structure of these quantifier-free expansions comes to the center of our attention. Our proofs of solvability for classes $\mathfrak{C}$ of schemata usually take the following form: we show that the schemata in $\mathfrak{C}$ have well-behaved expansions, in the sense that there is a uniform method for calculating, for each F in $\mathfrak{C}$, a number p such that $E(F, \omega)$ is consistent if the finite expansion $E(F, p)$ is consistent.

The solvability of one class is immediate: the class of schemata F for which $E(F, \omega)$ is finite. For $E(F, \omega)$ is finite whenever no nonconstant function sign occurs in the functional form F^* of F, and then $E(F, \omega) = E(F, 1)$. Hence, from the definition of F^* we obtain

Class 1.1. Let F be a schema in which no y-variable governs an x-variable. Then F has a model iff $E(F, 1)$ is consistent. $\square$

The solvability of Class 1.1 is essentially a classical result, first obtained by Bernays and Schönfinkel (1928); for Class 1.1 quickly reduces, by application of the prenexing rules, to the Bernays–Schönfinkel Class: prenex schemata with prefixes $\exists x_1 \cdots \exists x_m \forall y_1 \cdots \forall y_n$. (Incidentally, this simple solvability result is stronger than might at first appear. As we shall see in Chapter 2, §2, the Monadic Class, that is, the class of schemata containing only 1-place predicate letters, reduces to Class 1.1.)

1. Classes Specified by Truth-Functional Form of Matrix

If a schema F is not in Class 1.1, then there are an infinite number of Herbrand instances of F. However, these Herbrand instances all resemble

Burton Dreben and Warren D. Goldfarb, The Decision Problem: Solvable Classes of Quantificational Formulas

ISBN 0-201-02540-X

each other in a crucial way: *all Herbrand instances of a schema F have the same truth-functional structure.* That is, any two Herbrand instances of F differ only in the arguments of their atomic subformulas; each predicate letter and each truth-functional connective occurs homologously in the two Herbrand instances. The truth-functional similarity of the Herbrand instances of a schema is basic to our enterprise of finding an expansion $E(F, p)$ whose consistency suffices for that of $E(F, \omega)$.

Of course, *all* instances of a schema F have the same truth-functional structure. Herbrand instances are further distinguished in that they are instances of the functional form F^*; hence in any Herbrand instance H the substituent for an x-variable x_k of F is that x_k-term in $D(F, \omega)$ whose arguments are the substituents in H for the y-variables governing x_k. This special property of Herbrand instances will play an important role later. First, however, we present several results based *solely* on truth-functional properties. We do not limit ourselves to Herbrand instances in our formulation of these results. Although in treating satisfiability we are ultimately concerned only with Herbrand instances, the proofs below are simplified by the use of arbitrary instances. In addition, the generality thereby gained will be needed to show that the classes soon to be shown solvable are also finitely controllable (see Chapter 3, §2).

Definition 1.1.   A schema F is *n-condensable, $n > 0$,* iff every inconsistent conjunction of instances of F contains an inconsistent subconjunction made up of n or fewer instances of F. A schema F is *condensable* iff it is *n*-condensable for some n.

The property of condensability is quite strong. Even such an obviously unsatisfiable schema as

$$\forall y_1 \forall y_2 \forall y_3 \forall y_4 [Py_1 \wedge (Py_2 \vee -Py_3) \wedge -Py_4],$$

whose sole Herbrand instance $[P\!\!+ \wedge (P\!\!+ \vee -P\!\!+) \wedge -P\!\!+]$ is inconsistent, is not condensable: for each $k > 0$ the conjunction

$$\bigwedge_{1 \leq i \leq k} [Pc_1 \wedge (Pc_{i+1} \vee -Pc_i) \wedge -Pc_{k+1}]$$

of instances, where the c_i are distinct constants, is inconsistent, yet this conjunction contains no inconsistent proper subconjunction. The schemata we show below to be condensable are specified by severe restrictions on the truth-functional forms of their matrices.

The Expansion Theorem tells us that an *n*-condensable schema is unsatisfiable just in case there is an inconsistent conjunction of n Herbrand instances of F. Hence our condensability proofs will yield solvability granted that there is an algorithm that determines, for each schema F and each number n, whether there is an inconsistent conjunction of n Herbrand instances of F. The existence of such an algorithm will be shown in §3 of

ISBN 0-201-02540-X

this chapter: the Compression Lemma asserts that if there is an inconsistent conjunction of n or fewer Herbrand instances of F, then there is such a conjunction composed entirely of Herbrand instances in $E(F, m \cdot n + 1)$, where m is the number of x-variables of F. Thus an n-condensable schema F containing m x-variables is satisfiable just in case $E(F, m \cdot n + 1)$ is consistent.

The solvability of our next class of schemata was originally shown by Herbrand (1930, 1931). The class following it, which is due to Krom (1967), extends it. Recall that a signed atomic formula is either an atomic formula or the negation of an atomic formula.

Class 1.2   (Herbrand Class).   Let F be a schema whose matrix is a conjunction of signed atomic formulas. Then F is 2-condensable.

Proof.   Any conjunction of instances of F is a conjunction of signed atoms. Hence such a conjunction K is inconsistent only if, for some atom A, both A and $-A$ are conjuncts in K. If there is such an A, then there are (not necessarily distinct) instances I_1 and I_2 of F, each of which is a subconjunction of K, such that A is a conjunct in I_1 and $-A$ is a conjunct in I_2. But then $I_1 \wedge I_2$ is inconsistent. $\square$

Class 1.3.   Let F be a schema whose matrix is a conjunction consisting of positively signed atomic formulas and of disjunctions of negatively signed atomic formulas. Then F is $(n + 1)$-condensable, where n is the largest number of signed atomic formulas in any conjunct in the matrix of F.

Proof.   A conjunction of instances of F is a conjunction of positively signed atoms and of disjunctions of negatively signed atoms. Such a conjunction K is inconsistent only if there are atoms $A_1, \ldots, A_j$, $j \leq n$, such that each of $A_1, \ldots, A_j$ is a conjunct in K, and the disjunction $-A_1 \vee \cdots \vee -A_j$ is also a conjunct in K. For if there were no such atoms, then the truth-assignment that verifies any atom A iff A is a (positively signed) conjunct in K would verify K.

If there are such atoms $A_1, \ldots, A_j$, then there are (not necessarily distinct) instances $I_1, \ldots, I_{j+1}$ of F, $j \leq n$, each of which is a subconjunction of K, such that A_i is a conjunct in I_i, $1 \leq i \leq j$, and $-A_1 \vee \cdots \vee -A_j$ is a conjunct in I_{j+1}. In this case, $I_1 \wedge \cdots \wedge I_{j+1}$ is inconsistent. $\square$

The same argument shows that if a schema F is in the "mirror-image" of Class 1.3, that is, the matrix of F is a conjunction consisting of negatively signed atomic formulas and of disjunctions of positively signed atomic formulas, then F is $(n + 1)$-condensable, where n is the largest number of signed atomic formulas in any conjunct in the matrix of F.

Our next condensability result comes from Harry Lewis and requires a somewhat more elaborate argument. A quantifier-free formula is in *conjunctive normal form* iff it is a conjunction of disjunctions of signed atomic formulas.

Class 1.4. Let F be a schema whose matrix is in conjunctive normal form and contains no more than two occurrences of each predicate letter. If n is the number of conjuncts in the matrix of F, then F is n-condensable.

Proof. Say that a set S of disjunctions of signed atomic formulas is *sparse* iff no predicate letter occurs more than twice in S. We prove the following assertion by induction on the cardinality of S:

(#) Any inconsistent conjunction of instances of the members of a sparse set S contains an inconsistent subconjunction consisting of at most one instance of each member of S.

Assertion (#) suffices for the n-condensability of any schema F in Class 1.4 whose matrix F^M contains n conjuncts. For the set of conjuncts in F^M is a sparse set S with n members; any conjunction K of instances of F is a conjunction of instances of the members of S; and any subconjunction of K consisting of at most one instance of each member of S is included in a subconjunction of K containing at most n instances of F.

If S contains only one member, then (#) holds vacuously, since no conjunction of instances of a disjunction of signed atomic formulas is inconsistent. Suppose S contains $n + 1$ members, and that (#) holds for all sparse sets of cardinality $\leq n$. Let K be an inconsistent conjunction of instances of members of S; we suppose K contains at least one instance of each member of S. We shall proceed by collapsing two members of S into one, by applying to K a truth-functional cut rule.

We pick a predicate letter P and two distinct members D_1 and D_2 of S with the following property: P occurs in one positively signed atomic formula in D_1 and in one negatively signed atomic formula in D_2. This can be done. If there were no such P, D_1, and D_2, then K would be consistent, since the following truth-assignment $\mathfrak{A}$ would verify K: $\mathfrak{A}$ verifies an atom A iff the predicate letter in A occurs in S in some positively signed atomic formula. Note that D_1 and D_2 contain the only (two) occurrences of P in S. We assume that the first disjuncts in D_1 and in D_2 are the signed atomic formulas that contain P.

Suppose first that D_1 and D_2 each contain only one disjunct, that is, D_1 is a positively signed atomic formula and D_2 is a negatively signed atomic formula. If there is an atom A such that A is an instance of D_1, $-A$ is an instance of D_2, and A and $-A$ are conjuncts in K, then clearly $A \wedge -A$ is the desired subconjunction of K. If there is no such A, then the subconjunction K' of K obtained by deleting all instances of D_1 and D_2 is inconsistent, since it does not contain P. Moreover, K' is a conjunction of instances of the sparse set $S - \{D_1, D_2\}$, which has $n - 1$ members. Hence (#) follows from the induction hypothesis.

Suppose that at least one of D_1, D_2 contains more than one disjunct. Let

ISBN 0-201-02540-X

D_3 be the disjunction of all signed atomic formulas in D_1 or D_2 save those two that contain P. Let $S' = (S - \{D_1, D_2\}) \cup \{D_3\}$. Then S' is a sparse set with n members, and P does not occur in S'. Let A be an atom containing the predicate letter P. Call a disjunction E of signed atoms an A-*companion* iff $A \vee E$ is an instance of D_1 that is a conjunct in K, and call E a $-A$-*companion* iff $-A \vee E$ is an instance of D_2 that is a conjunct in K. Let J be the conjunction formed thus:

(1) J contains every instance of a member of $S - \{D_1, D_2\}$ that is a conjunct in K;

(2) J contains every disjunction $E_1 \vee E_2$ such that, for some atom A containing P, E_1 is an A-companion and E_2 is a $-A$-companion. (Note that every such disjunction $E_1 \vee E_2$ is an instance of D_3.)

We claim that J is inconsistent. For suppose J consistent and let $\mathfrak{A}$ be a truth-assignment verifying it. Since $\mathfrak{A}$ verifies all disjunctions $E_1 \vee E_2$ formed in accordance with (2) above, the set $U_1 = \{A \mid A$ contains P and $\mathfrak{A}$ falsifies at least one A-companion$\}$ is disjoint from the set $U_2 = \{A \mid A$ contains P and $\mathfrak{A}$ falsifies at least one $-A$-companion$\}$. Hence we may so extend $\mathfrak{A}$ to atoms containing P that $\mathfrak{A}$ verifies every atom in U_1 and falsifies every atom in U_2. But this extension of $\mathfrak{A}$ verifies every instance $A \vee E_1$ of D_1 that is a conjunct in K, and verifies every instance $-A \vee E_2$ of D_2 that is a conjunct in K. Hence this extension of $\mathfrak{A}$ verifies K, contrary to the hypothesis that K is inconsistent.

Now J is a conjunction of instances of members of the sparse set S', and S' has n members. Hence, by the induction hypothesis, there is an inconsistent subconjunction J' of J containing no more than one instance of each member of S'. If J' contains no instances of D_3, then J' is itself the desired subconjunction of K. Otherwise $J' = J'' \wedge (E_1 \vee E_2)$, where J'' is some conjunction of at most one instance of each member of $S - \{D_2, D_2\}$, and where, for some atom A containing P, E_1 is an A-companion and E_2 is a $-A$-companion. Thus $A \vee E_1$ and $-A \vee E_2$ are conjuncts in K. But then $J'' \wedge (A \vee E_1) \wedge (-A \vee E_2)$ is an inconsistent subconjunction of K containing at most one instance of each member of S. We conclude that assertion (#) holds for S. □

Truth-functional arguments like that just given yield the condensability of schemata in several other classes similar to Class 1.4. We specify these classes in §1 of the Appendix. A pair of unsolvability results, again due to Harry Lewis, show that the specification of Class 1.4 cannot be liberalized in certain (obvious) ways. The following classes of schemata F are unsolvable: (i) each predicate letter occurs at most twice in F, but the matrix of F is not necessarily in conjunctive normal form; (ii) one predicate letter

ISBN 0-201-02540-X

occurs three times in F, all the other predicate letters occur twice, and the matrix of F is in conjunctive normal form.

We might also try to generalize the Herbrand Class. The matrix of an Herbrand schema can be viewed as a degenerate disjunctive normal form, that is, one with only one disjunct. However, if we allow two disjuncts, that is, matrices that are disjunctions of two conjunctions of signed atomic formulas, the resulting class is unsolvable (Orevkov, 1969; see also Lewis, 1979).

Finally, if we classify schemata in terms of conjunctive normal forms, we have the following: the class of schemata with matrices in conjunctive normal form with only one conjunct, that is, matrices that are disjunctions of signed atomic formulas, is solvable (indeed, every schema in this class has a model); but the class of schemata with matrices in conjunctive normal form with two conjuncts is unsolvable (Goldfarb, 1975; see also Lewis, 1979).

2. Multiple Occurrences of Atoms in Expansions

We now look more closely at the circumstances in which a conjunction of Herbrand instances is inconsistent. If each Herbrand instance of a schema F could be viewed as an independent copy of the matrix F^M of F, then no conjunction of Herbrand instances would be inconsistent, provided that F^M itself were consistent. However, even though each Herbrand instance has the same truth-functional structure as F^M, the Herbrand instances cannot be viewed as independent copies: the same atom may occur in different Herbrand instances or at different places in one Herbrand instance. Indeed, the consistency of F^M need not even yield the consistency of each Herbrand instance of F. For example, if F is $\forall y_1 \forall y_2 (Py_1 \wedge -Py_2)$, then F^M is consistent, yet the Herbrand instance $H = (P\mathord{\downarrow} \wedge -P\mathord{\downarrow})$ of F is not. Clearly this discrepancy arises from the fact that $\langle H \rangle_1 = \langle H \rangle_2$ whereas $\langle F \rangle_1 \neq \langle F \rangle_2$. (The component notation $\langle F \rangle_i$ is defined in Chapter 0, §1.)

Similarly, an inconsistency may ensue because an atom occurs in two distinct Herbrand instances. For example, if F is $\forall y \exists x (Py \wedge -Px)$, then each Herbrand instance of F is of the form $Pt \wedge -Pf_x(t)$ for some term t in $D(F, \omega)$. Thus, although each Herbrand instance of F is consistent, if H_1 is $[Pt \wedge -Pf_x(t)]$ and H_2 is $[Pf_x(t) \wedge -Pf_x(f_x(t))]$, then $H_1 \wedge H_2$ is inconsistent. Here the inconsistency is due to the fact that $\langle H_1 \rangle_2 = Pf_x(t) = \langle H_2 \rangle_1$. Of course, if the matrix of F has a more complicated form, then much more extensive connections among the Herbrand instances may be responsible for inconsistencies.

ISBN 0-201-02540-X

We are thus led to consider the multiple occurrences of atoms in Herbrand expansions. We begin by giving a precise explication of the intuitive notion of a double occurrence of an atom.

Definition 1.2. A *coinstantiation in the expansion of a schema F* is a quadruple $\langle H, i, H', j \rangle$ such that $\langle H \rangle_i = \langle H' \rangle_j$, where H and H' are (not necessarily distinct) Herbrand instances of F and i and j are (not necessarily distinct) numbers $\leq c(F)$. A *simultaneous coinstantiation* is a coinstantiation of the form $\langle H, i, H, j \rangle$, that is, one in which the two Herbrand instances are identical.

As we have just seen, it is the existence of coinstantiations in the expansion that makes the decision problem nontrivial. Ordinarily, there are an infinite number of coinstantiations in $E(F, \omega)$. To study these coinstantiations we must also consider the atomic subformulas of F underlying them.

Definition 1.3. Two atomic subformulas A and B of F *engender* a coinstantiation $\langle H, i, H', j \rangle$ iff $A = \langle F \rangle_i$ and $B = \langle F \rangle_j$. If A and B engender $\langle H, i, H', j \rangle$, then we say that A and B are *coinstantiated* in $\langle H, H' \rangle$. Moreover, A and B are *coinstantiable* iff they engender at least one coinstantiation, and are *simultaneously coinstantiable* iff they engender at least one simultaneous coinstantiation.

Note that coinstantiability is actually a 3-place relation: two atomic formulas are coinstantiable or not with respect to a given schema F in which they occur. For example, suppose $\langle F \rangle_1 = Pxy$ and $\langle F \rangle_2 = Pyy$, where y is a y-variable and x an x-variable. If y does not govern x in F, then $\langle F \rangle_1$ and $\langle F \rangle_2$ are coinstantiable; indeed, they are simultaneously coinstantiable, since $\langle H \rangle_1 = \langle H \rangle_2$ for every Herbrand instance H of F in which the substituent for y is identical with that for x. On the other hand, if y governs x in F, then $\langle F \rangle_1$ and $\langle F \rangle_2$ are not coinstantiable; if H is any Herbrand instance of F, the atom $\langle H \rangle_1$ has the form Pts, where t is an x-term in $D(F, \omega)$ of height strictly greater than that of s, while the atom $\langle H \rangle_2$ has the form Pss. Nevertheless, we usually talk of coinstantiability as a 2-place relation, for in any context we assume some fixed schema F.

Coinstantiability (with respect to any given schema F) is obviously reflexive. It is also symmetric. However, it is not transitive. For example, if x_1 and x_2 are distinct x-variables of F and y is a y-variable of F, then Px_1 and Py are coinstantiable, as are Px_2 and Py, but Px_1 and Px_2 are not, since in any Herbrand instance the substituent for x_1 is an x_1-term of $D(F, \omega)$ and that for Px_2 is an x_2-term.

Clearly two atomic subformulas of a schema F are coinstantiable only if they share the same predicate letter. If instead of restricting attention to Herbrand instances in our definition of coinstantiable we were to consider

ISBN 0-201-02540-X

arbitrary instances of F, then the sharing of the same predicate letter would also suffice: for example, if $\langle F \rangle_i$ and $\langle F \rangle_j$ both contain P, then $\langle I \rangle_i = \langle I \rangle_j$ for any instance I of F in which all variables are replaced by the same term. The restriction to Herbrand instances gives coinstantiability a more intricate structure, for then the rules determining the substituents for the x-variables in any Herbrand instance come into play. The examples in the preceding two paragraphs show that coinstantiability depends strongly on the details of quantificational structure: on which places in atomic formulas are occupied by x-variables, on which places by y-variables that govern or do not govern these x-variables, and so on. We give another example that highlights this.

EXAMPLE.   Let F be a *Skolem schema*, that is, a prenex schema with prefix $\forall y_1 \cdots \forall y_m \exists x_1 \cdots \exists x_n$. Then $\langle F \rangle_i$ and $\langle F \rangle_j$ are simultaneously coinstantiable iff one can be obtained from the other by replacing some or all occurrences of some or all y-variables by occurrences of other y-variables. Moreover, $\langle F \rangle_i$ and $\langle F \rangle_j$ are coinstantiable iff either they are simultaneously coinstantiable or else one of them, say $\langle F \rangle_i$, contains no x-variables and the other, $\langle F \rangle_j$, can be obtained from $\langle F \rangle_i$ by replacing occurrences of some variables by occurrences of other variables, subject to the following restriction: if some occurrences of a y-variable y_p are replaced by occurrences of an x-variable x_k, then all occurrences of y_p are replaced by occurrences of x_k. These criteria for coinstantiability rest on the fact that in any Herbrand instance of F the substituents for the x-variables are all distinct, and are of strictly greater height than all the substituents for the y-variables. Thus if $\langle F \rangle_i$ and $\langle F \rangle_j$ are coinstantiable and each contains at least one x-variable, then $\langle F \rangle_i$ and $\langle F \rangle_j$ are simultaneously coinstantiable. Indeed, $\langle F \rangle_i$ and $\langle F \rangle_j$ engender only simultaneous coinstantiations, since each term in $D(F, \omega)$ occurs as the substituent for an x-variable in at most one Herbrand instance of F.

Our criteria show that an atomic subformula $Py_p x_k$ is simultaneously coinstantiable only with atomic subformulas $Py_q x_k$, and is coinstantiable only with atomic subformulas $Py_q x_k$ and $Py_q y_r$ for $q \neq r$. An atomic subformula $Px_k x_l$ with $k \neq l$ is simultaneously coinstantiable only with itself, and is coinstantiable only with itself and with atomic subformulas $Py_q y_r$ for $q \neq r$. Further, $Py_1 y_1 y_2 x_1 x_2 y_3$ is coinstantiable but not simultaneously coinstantiable with $Py_2 y_2 y_2 y_3 y_1 y_4$; simultaneously coinstantiable with $Py_3 y_4 y_1 x_1 x_2 y_1$; and not coinstantiable at all with $Py_2 y_2 y_1 y_1 y_3 y_4$ or with $Py_1 y_1 x_1 y_2 y_3 y_4$.   End of Example.

Coinstantiability and simultaneous coinstantiability are effective notions, as we shall show in §3. Sometimes this fact alone can yield the solvability of a class of schemata. For example, let F be a schema in Class 1.2, the Herbrand Class. We saw in §1 that F is unsatisfiable iff there is an atom A

ISBN 0-201-02540-X

and Herbrand instances H_1 and H_2 of F such that A is a conjunct in H_1 and $-A$ is a conjunct in H_2. But this occurs just when there are numbers i and j such that $\langle F \rangle_i$ and $-\langle F \rangle_j$ are conjuncts in the matrix of F and $\langle F \rangle_i$ and $\langle F \rangle_j$ are coinstantiable.

In general, however, the existence or nonexistence of an inconsistent conjunction of Herbrand instances of a schema F is not determined solely by the coinstantiabilities between the atomic subformulas of F (and the truth-functional structure of F^M). For example, if F is

$$\forall y_1 \exists x \forall y_2 [Py_2y_2 \wedge (-Pxy_2 \vee -Py_1y_2)],$$

then $E(F, \omega)$ is consistent: it is verified by the truth-assignment that verifies an atom Pst iff $s = t$. However, if F is

$$\forall y_1 \exists x \forall y_2 [Py_2y_2 \wedge (-Pxy_2 \vee -Py_1y_1)],$$

then there are inconsistent conjunctions $H_1 \wedge H_2$ of Herbrand instances of F, namely, $H_1 = F^*(y_1/t, y_2/f_x(t))$ and $H_2 = F^*(y_1/t, y_2/t)$ for any term t in $D(F, \omega)$. But in both cases the coinstantiabilities between the atomic subformulas are the same: $\langle F \rangle_1$ and $\langle F \rangle_2$ are simultaneously coinstantiable, as are also $\langle F \rangle_1$ and $\langle F \rangle_3$, and $\langle F \rangle_2$ and $\langle F \rangle_3$ are coinstantiable but not simultaneously coinstantiable. The inconsistency in the second case results from the existence of a pair $\langle H_1, H_2 \rangle$ of Herbrand instances in which $\langle F \rangle_1$ and $\langle F \rangle_2$ are coinstantiated *and* $\langle F \rangle_3$ and $\langle F \rangle_1$ are also coinstantiated.

3. Formula-Mappings

The situation becomes far more complex when we focus on arbitrary conjunctions of Herbrand instances. The simplest way to handle these complexities is to use mappings of quantifier-free formulas that are determined by their values on atoms.

Definition 1.4. A *formula-mapping* for a schema F is a function Γ from and to the set $\tilde{L}(F, \omega)$ of all quantifier-free formulas constructed from the atoms in $L(F, \omega)$ such that

(1) Γ carries atoms to atoms;
(2) Γ commutes with the truth-functional connectives, that is, for all formulas J and K in $\tilde{L}(F, \omega)$,

$$\Gamma(-J) = -\Gamma(J),$$

$$\Gamma(J \vee K) = \Gamma(J) \vee \Gamma(K),$$

$$\Gamma(J \wedge K) = \Gamma(J) \wedge \Gamma(K).$$

Thus the values of Γ on $\tilde{L}(F, \omega)$ are determined by its values on $L(F, \omega)$,

ISBN 0-201-02540-X

that is, on the atoms. Moreover, Γ commutes with the component function: $\langle \Gamma(J) \rangle_i = \Gamma(\langle J \rangle_i)$ for each J in $\bar{L}(F, \omega)$ and each i.

Much of the structure of formulas in $\bar{L}(F, \omega)$ is invariant under formula-mappings Γ. First, a formula J and its image $\Gamma(J)$ share the same truth-functional structure, in the sense that the truth-functional connectives occur homologously in J and $\Gamma(J)$.‡ Second, because formula-mappings commute with the component function, if two formulas share an atom, then so do their images under Γ: if $\langle J \rangle_i = \langle K \rangle_j$, then $\langle \Gamma(J) \rangle_i = \langle \Gamma(K) \rangle_j$. Given these two facts, the following Lemma is not surprising.

Consistency Lemma.   A set S of formulas in $\bar{L}(F, \omega)$ is consistent if there is at least one formula-mapping Γ for which the image $\Gamma[S]$ of S is consistent.

Proof.   Suppose $\Gamma[S]$ consistent. Then there is a truth-assignment $\mathfrak{A}$ that verifies it. Let $\mathfrak{A}(\Gamma)$ be the truth-assignment that verifies an atom A iff $\mathfrak{A}$ verifies $\Gamma(A)$; that is, $\mathfrak{A}(\Gamma)$ is the composition of $\mathfrak{A}$ and Γ. Since Γ commutes with connectives, we have by induction on the construction of formulas that, for every J in $\bar{L}(F, \omega)$, $\mathfrak{A}(\Gamma)$ assigns to J what $\mathfrak{A}$ assigns to $\Gamma(J)$. In particular, then, $\mathfrak{A}(\Gamma)$ verifies every member of S; hence S is consistent.   □

The use of formula-mappings often enables us to avoid talking explicitly about coinstantiations. We can show the consistency of a given set S of Herbrand instances by constructing a formula-mapping that takes S to some set known to be consistent. Multiple occurrences of atoms in the members of S are automatically reflected in the possible values *any* formula-mapping can take on S.

For the remainder of this chapter we are interested especially in formula-mappings that take Herbrand instances to Herbrand instances. Call Γ *Herbrand instance preserving* on a set S of Herbrand instances iff $\Gamma[S]$ is a set of Herbrand instances.

If the coinstantiabilities between atomic subformulas of F are drastically constrained, then formula-mappings that are Herbrand instance preserving on all of $E(F, \omega)$ are easy to come by. For example, suppose there is at least one Herbrand instance H_0 of F such that $\langle H_0 \rangle_i = \langle H_0 \rangle_j$ whenever $\langle F \rangle_i$ and $\langle F \rangle_j$ are coinstantiable. Then there is a formula-mapping Γ that takes every Herbrand instance of F to H_0. Just let Γ be such that

$$\Gamma(\langle H \rangle_i) = \langle H_0 \rangle_i$$

for every H in $E(F, \omega)$ and every i. This mapping Γ is well-defined, since

---

‡ This is a weaker sense of identity of truth-functional structure than that used in §1, where we required predicate letters to occur homologously as well. The stronger identity property holds if in addition A and $\Gamma(A)$ share the same predicate letter, for each atom A. The formula-mappings we construct almost always obey this restriction.

ISBN 0-201-02540-X

if $\langle H \rangle_i = \langle H' \rangle_j$, then $\langle F \rangle_i$ and $\langle F \rangle_j$ are coinstantiable, and so $\Gamma(\langle H \rangle_i) = \langle H_0 \rangle_i = \langle H_0 \rangle_j = \Gamma(\langle H' \rangle_j)$. In this case, by the Consistency Lemma, the consistency of H_0 alone suffices for that of the entire expansion $E(F, \omega)$. Two simple solvable classes are immediate.

Class 1.5. Let F be a schema no distinct atomic subformulas of which are coinstantiable. Then F has a model just in case $E(F, 1)$ is consistent.

Proof. Let H_0 be any Herbrand instance in $E(F, 1)$. If $\langle F \rangle_i$ and $\langle F \rangle_j$ are coinstantiable, then $\langle F \rangle_i = \langle F \rangle_j$; hence $\langle H_0 \rangle_i = \langle H_0 \rangle_j$. $\square$

Class 1.5 includes, *inter alia*, every schema in which distinct atomic formulas contain distinct predicate letters.

Class 1.6. Let F be a Skolem schema such that any two atomic subformulas of F are coinstantiable only if they are simultaneously coinstantiable. Then F has a model just in case $E(F, 1)$ is consistent.

Proof. Let H_0 be the member of $E(F, 1)$ in which $+$ is the substituent for every y-variable of F. If $\langle F \rangle_i$ and $\langle F \rangle_j$ are coinstantiable, then they are simultaneously coinstantiable. Since the substituents in H_0 for all the y-variables are identical, by the criterion for simultaneous coinstantiability given in the Example of §2, we have $\langle H_0 \rangle_i = \langle H_0 \rangle_j$. $\square$

Class 1.6 includes each Skolem schema F such that no atomic subformula of F containing an x-variable shares a predicate letter with any atomic subformula lacking x-variables.

The main application we make of Herbrand instance preserving formula-mappings yields the Compression Lemma. Here we seek formula-mappings that are Herbrand instance preserving on *finite* sets of Herbrand instances. First, however, we need a few additional notions.

Definition 1.5. A *term-mapping* for a schema F is a function γ on and to the set $D(F, \omega)$ of terms.

Formula-mappings may be defined from term-mappings in a variety of ways. The simplest is to let the value of Γ on an atom $Pt_1 \cdots t_k$ be $P\gamma t_1 \cdots \gamma t_k$. In this case we say Γ is *γ-canonical* on $Pt_1 \cdots t_k$; and we call Γ the *γ-canonical formula-mapping* iff Γ is γ-canonical on every atom in $L(F, \omega)$.

For γ-canonical formula-mappings we may easily state a condition equivalent to Herbrand instance preservation. Let γ be a term-mapping and let Γ be the γ-canonical formula-mapping. We say γ is *homomorphic* on a term $t = f(t_1 \cdots t_n)$, $n \geq 0$, iff $\gamma t = f(\gamma t_1 \cdots \gamma t_n)$. (Thus if $n = 0$, so that t has height 1, γ is homomorphic on t iff $\gamma t = t$.) For any Herbrand instance H, $\Gamma(H)$ is itself an Herbrand instance provided that γ is homomorphic on every term $f_x(t_1 \cdots t_n)$ that occurs in H as a substituent for an x-variable. Indeed, if $H = F^*(y_1/s_1, \ldots, y_m/s_m)$, where $y_1, \ldots, y_m$ are all the y-variables of F, then $\Gamma(H) = F^*(y_1/\gamma s_1, \ldots, y_m/\gamma s_m)$. Moreover, this condition on γ is necessary as well: for otherwise a term $\gamma f_x(t_1 \cdots t_n)$ not identical with

ISBN 0-201-02540-X

$f_x(\gamma t_1 \cdots \gamma t_n)$ occurs in $\Gamma(H)$ as a substituent for an x-variable x, yet $\gamma t_1, \ldots, \gamma t_n$ are the substituents in $\Gamma(H)$ for the y-variables governing x. Hence the γ-canonical formula-mapping Γ will be Herbrand instance preserving on a set $S \subseteq E(F, \omega)$ iff γ is homomorphic on every term that occurs in at least one member of S as a substituent for an x-variable of F. Our proof of the Compression Lemma exploits the fact that if S is finite, then there is only a finite number of terms on which γ must be homomorphic, and so the range of γ can be bounded.

Herbrand Instance Preservation Lemma. Let F be a schema, let m be the number of x-variables of F, and let S be a set of n or fewer Hebrand instances of F. Then there is a formula-mapping Γ such that $\Gamma[S] \subseteq E(F, m \cdot n + 1)$.

Proof. Let T be the set of terms that occur in at least one member of S as a substituent for some x-variable of F. Let γ be the term-mapping that is homomorphic on every term in T and takes every term not in T to $\perp$. Then the γ-canonical formula-mapping Γ is Herbrand instance preserving on S. Moreover, T contains at most $m \cdot n$ members. But then, by the definition of homomorphic, the range of γ is contained in $D(F, m \cdot n + 1)$; hence $\Gamma[S] \subseteq E(F, m \cdot n + 1)$. $\square$

Compression Lemma. Let F be a schema, let m be the number of x-variables of F, and let n be a positive integer. There is an inconsistent conjunction of n or fewer Herbrand instances of F only if there is such a conjunction in $E(F, m \cdot n + 1)$.

Proof. Suppose $H_1 \wedge \cdots \wedge H_k$ is an inconsistent conjunction of Herbrand instances of F, where $k \leq n$. By the Herbrand Instance Preservation Lemma there is a formula-mapping Γ such that $\Gamma(H_i) \in E(F, m \cdot n + 1)$ for each $i \leq k$. By the Consistency Lemma, then $\Gamma(H_1) \wedge \cdots \wedge \Gamma(H_k)$ must be inconsistent. (Note that the images $\Gamma(H_i)$ need not be distinct.) $\square$

The Herbrand Instance Preservation Lemma also suffices for the decidability of the notions of coinstantiability and simultaneous coinstantiability.

Corollary. Let F be a schema containing m x-variables. Two atomic subformulas of F are coinstantiable only if they are coinstantiated in some pair of Herbrand instances in $E(F, 2m + 1)$, and are simultaneously coinstantiable only if they are coinstantiated in $\langle H, H \rangle$ for some H in $E(F, m + 1)$.

Proof. If $\langle F \rangle_i$ and $\langle F \rangle_j$ are coinstantiable, then by definition there are H and H' such that $\langle H \rangle_i = \langle H' \rangle_j$. By the Herbrand Instance Preservation Lemma there is a Γ such that $\Gamma(H)$ and $\Gamma(H')$ are in $E(F, 2m + 1)$. Moreover, $\langle F \rangle_i$ and $\langle F \rangle_j$ are coinstantiated in $\langle \Gamma(H), \Gamma(H') \rangle$, since $\langle \Gamma(H) \rangle_i = \langle \Gamma(H') \rangle_j$. The proof for simultaneous coinstantiability is analogous. $\square$

The decision procedures furnished by the Corollary are rather indirect. To check whether $\langle F \rangle_i$ and $\langle F \rangle_j$ are coinstantiable we have to search through

ISBN 0-201-02540-X

$E(F, 2m + 1)$. We shall sketch later a way of obtaining a more direct condition for coinstantiability in the course of establishing the solvability of the class of schemata containing two atomic subformulas (Chapter 7, §3). This condition is essentially syntactical: like that of the Example of §2, it is formulated in terms of the types of variables occurring in the atomic formulas in question, and of which of these variables govern which of the others.

Herbrand instance preserving formula-mappings are of limited use beyond this point. In general we could infer the consistency of an expansion $E(F, \omega)$ from that of a finite expansion $E(F, p)$ if there were a formula-mapping that takes $E(F, \omega)$ to $E(F, p)$. Unfortunately, there will ordinarily be no such formula-mapping. We shall dwell more on this in the next chapter, and develop a method for using formula-mappings that are not Herbrand instance preserving on all of $E(F, \omega)$. Naturally, the possibility of so doing depends on facts about the coinstantiations in $E(F, \omega)$. But instead of simple facts relating to a single Herbrand instance, like those about schemata in Classes 1.5 and 1.6, and instead of limitations to a bounded number of Herbrand instances obtained by truth-functional arguments, like those obtained for condensable schemata, we shall be interested in the interplay between coinstantiations and more-or-less periodic aspects of the behavior of truth-assignments on the expansion.

ISBN 0-201-02540-X

CHAPTER 2 _____

The Amenability Method

In this chapter we present a general method for constructing formula-mappings that yield solvability results. As we said at the very beginning of Chapter 1, we intend to find for each schema F in any given class under consideration an expansion $E(F, p)$ whose consistency suffices for that of $E(F, \omega)$. And as we saw, an expansion $E(F, p)$ will have this property whenever there is a formula-mapping that carries $E(F, \omega)$ to $E(F, p)$. However, the classes of schemata we now wish to treat do not allow the construction of such Herbrand instance preserving formula-mappings. More subtlety in the application of formula-mappings is required. Our method for doing this, which we call the *amenability* method, can be used to show the solvability of a large number of classes. Moreover, we shall see in the next chapter that a variant of the method provides finite controllability results as well.

1. Introduction of the Method

There is in general a conflict between requiring a formula-mapping Γ to be Herbrand instance preserving on all of $E(F, \omega)$ and requiring its image $\Gamma[E(F, \omega)]$ to be finite. This is immediate if, in addition, Γ is required to be γ-canonical for some term-mapping γ. For then Γ is Herbrand instance preserving on $E(F, \omega)$ only if γ is homomorphic on every term occurring in some Herbrand instance of F as a substituent for some x-variable of F, that is, only if γ is homomorphic on every term in $D(F, \omega)$ but ╪. But then the range of γ is infinite and so $\Gamma[E(F, \omega)]$ is infinite. However, even without the additional requirement of γ-canonicality, suitable Herbrand instance preserving formula-mappings will ordinarily not exist. To be sure, as we saw in the case of Classes 1.5 and 1.6, if the coinstantiabilities among atomic subformulas of F are drastically constrained, then there is no problem. Nevertheless, the following example shows that we rarely have to do

Burton Dreben and Warren D. Goldfarb, The Decision Problem: Solvable Classes of Quantificational Formulas

ISBN 0-201-02540-X

with such constrained coinstantiabilities. Suppose F has prefix $\forall y \exists x$ and that $\langle F \rangle_i = Py$ and $\langle F \rangle_j = Px$. Thus, for every Herbrand instance H of F, if $\langle H \rangle_i = Pt$, then $\langle H \rangle_j = Pf_x(t)$. Hence if Γ is Herbrand instance preserving on $E(F, \omega)$ and if $\Gamma(Pt) = Ps$, then $\Gamma(Pf_x(t))$ must be $Pf_x(s)$. Hence as the height of t increases so does the height of the term occurring in $\Gamma(Pf_x(t))$. But then $\Gamma[E(F, \omega)]$ must be infinite. More complex examples that preclude the existence of formula-mappings fulfilling the two requirements are easily devised: essentially, there is no such Γ whenever there is some "infinite pattern" of coinstantiations in $E(F, \omega)$ that is not mimicked in any finite set of Herbrand instances. This commonly happens when there are coinstantiable subformulas $\langle F \rangle_i$ and $\langle F \rangle_j$ such that $\langle F \rangle_i$ contains only y-variables and $\langle F \rangle_j$ contains x-variables governed by those y-variables.

Thus we wish to relax the requirement of Herbrand instance preservation. To this end we consider the effect of formula-mappings not only on formulas but also on truth-assignments.

Definition 2.1. Let F be a schema, $\mathfrak{A}$ a truth-assignment on $\hat{L}(F, \omega)$, and Γ a formula-mapping for F. The *truth-assignment induced by* $\mathfrak{A}$ *and* Γ is the assignment $\mathfrak{A}(\Gamma)$ on $\hat{L}(F, \omega)$ that assigns to an atom A the same value that $\mathfrak{A}$ assigns to $\Gamma(A)$. That is, for every atom A in $L(F, \omega)$,

$$\mathfrak{A}(\Gamma)(A) = \mathfrak{A}(\Gamma(A)).$$

(Induced truth-assignments are not new: the proof of the Consistency Lemma rests on the simple but important observation that for each—not necessarily atomic—formula J in $\hat{L}(F, \omega)$, $\mathfrak{A}(\Gamma)(J) = \mathfrak{A}(\Gamma(J))$.)

Now each formula-mapping Γ that carries $E(F, \omega)$ to $E(F, p)$ has the following property: if a truth-assignment $\mathfrak{A}$ verifies $E(F, p)$, then the induced truth-assignment $\mathfrak{A}(\Gamma)$ verifies $E(F, \omega)$. Thus, to look for an Herbrand instance preserving Γ is to look for a single formula-mapping that transforms every truth-assignment verifying $E(F, p)$ into one verifying $E(F, \omega)$. But the inference from the consistency of $E(F, p)$ to that of $E(F, \omega)$ is licensed by a weaker condition: for each $\mathfrak{A}$ verifying $E(F, p)$ there is a Γ such that $\mathfrak{A}(\Gamma)$ verifies $E(F, \omega)$. In other words, we permit different formula-mappings, which are normally not Herbrand instance preserving, for different truth-assignments. For, if the weaker condition holds, then the existence of any truth-assignment verifying $E(F, p)$ implies the existence of a truth-assignment verifying $E(F, \omega)$.

Given the definition of induced truth-assignment, our condition becomes: for every $\mathfrak{A}$ verifying $E(F, p)$ there is a Γ such that $\mathfrak{A}$ verifies $\Gamma[E(F, \omega)]$. And $\mathfrak{A}$ verifies $\Gamma[E(F, \omega)]$ whenever there is, for each H in $E(F, \omega)$, an H' in $E(F, p)$ such that $\Gamma(H)$ and H' are equivalent under $\mathfrak{A}$. Moreover, as we often said in Chapter 1, even though $\Gamma(H)$ need not be an Herbrand instance, it does have the same truth-functional structure as H and hence

ISBN 0-201-02540-X

the same truth-functional structure as *every* Herbrand instance of F. In particular, then, the desired equivalence of $\Gamma(H)$ and H' under $\mathfrak{A}$ holds if it holds component by component, that is, if

$$\mathfrak{A}(\langle\Gamma(H)\rangle_i) = \mathfrak{A}(\langle H'\rangle_i)$$

for every number i. Since formula-mappings commute with the component function, we may restate this thus:

$$\mathfrak{A}(\Gamma(\langle H\rangle_i)) = \mathfrak{A}(\langle H'\rangle_i)$$

for every i.

To simplify the final statement of our method, we introduce the notion of instance-mapping.

Definition 2.2. An *instance-mapping* for a schema F is a function Ψ from and to the set $E(F, \omega)$ of Herbrand instances of F. (An instance-mapping is *not* a formula-mapping.)

Central Fact. Let $\mathfrak{A}$ be a truth-assignment on $\hat{L}(F, \omega)$ and let Γ be a formula-mapping for F. The induced truth-assignment $\mathfrak{A}(\Gamma)$ verifies $E(F, \omega)$ if there is an instance-mapping Ψ for F such that

(1) $\mathfrak{A}$ verifies every Herbrand instance in the range of Ψ;

(2) $\mathfrak{A}(\Gamma(\langle H\rangle_i)) = \mathfrak{A}(\langle\Psi(H)\rangle_i)$ for every Herbrand instance H of F and every number i.

We call (2) the *central condition* on Γ and Ψ. We may view the central condition as requiring that the accompanying diagram commute for each i.

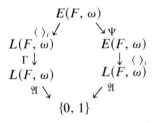

(In this diagram $\langle\ \rangle_i$ represents the function taking each formula to its ith component. Note that since formula-mappings commute with the component function, the labels $\langle\ \rangle_i$ and Γ on the arrows at the left may be interchanged, provided the labels $L(F, \omega)$ on the nodes are replaced by $\hat{L}(F, \omega)$.)

In our solvability proofs we can assume the given assignment $\mathfrak{A}$ verifies $E(F, p)$; hence (1) is satisfied if

$$\Psi[E(F, \omega)] \subseteq E(F, p).$$

We call this form of (1) the *boundedness condition* (for p).

ISBN 0-201-02540-X

Of course the central condition could most easily be met if $\Gamma(\langle H \rangle_i) = \langle \Psi(H) \rangle_i$ for each H and each i. But this is just to require that $\Gamma(H) = \Psi(H)$, that is, that Γ be Herbrand instance preserving on $E(F, \omega)$, and this is just what we cannot require. Naturally, in our constructions we shall try to obtain the identity of $\Gamma(\langle H \rangle_i)$ and $\langle \Psi(H) \rangle_i$ wherever possible, and be content with equivalence under $\mathfrak{A}$ only where necessary. The necessary recourse at *some* juncture to equivalence under $\mathfrak{A}$ rather than identity shows that the construction of Γ and Ψ will depend on the given truth-assignment $\mathfrak{A}$.

Indeed, we shall always be concerned (in this chapter) with constructing, for a given schema F and *each* truth-assignment $\mathfrak{A}$ on $\hat{L}(F, \omega)$, a formula-mapping Γ and an instance-mapping Ψ satisfying the boundedness condition and the central condition. The point of our concern is clear. If the truth-assignment $\mathfrak{A}$ verifies the expansion $E(F, p)$, then the truth-assignment induced by such a Γ verifies the expansion $E(F, \omega)$. We are thus doing more than is required for solvability; for we really need construct Γ and Ψ only for those $\mathfrak{A}$ that do in fact verify $E(F, p)$. However, since our constructions will depend only on the coinstantiations in the expansion of F and not on the truth-functional form of the matrix of F, we need never take into account whether the given truth-assignment verifies $E(F, p)$. If in fact $\mathfrak{A}$ does, then $\mathfrak{A}(\Gamma)$ verifies $E(F, \omega)$; thus all truth-functional considerations are automatically handled by the central condition.

Definition 2.3. An expansion $E(F, p)$ of a schema F is *adequate* iff for each truth-assignment $\mathfrak{A}$ on $\hat{L}(F, \omega)$ there are a formula-mapping Γ and an instance-mapping Ψ for F such that

(1) $\Psi[E(F, \omega)] \subseteq E(F, p)$;
(2) $\mathfrak{A}(\Gamma(\langle H \rangle_i)) = \mathfrak{A}(\langle \Psi(H) \rangle_i)$ for each Herbrand instance H of F and each number i.

A class $\mathfrak{C}$ of schemata is φ-*amenable*, where φ is a function from $\mathfrak{C}$ to the positive integers, iff $E(F, \varphi(F))$ is adequate for each F in $\mathfrak{C}$. A class $\mathfrak{C}$ is *amenable* iff $\mathfrak{C}$ is φ-amenable for some recursive φ.

We have seen that the consistency of an adequate expansion $E(F, p)$ suffices for the consistency of the expansion $E(F, \omega)$. Hence the amenability of a class $\mathfrak{C}$ of schemata implies the solvability of $\mathfrak{C}$.‡ In this chapter we shall prove several classes solvable by showing them to be amenable.

REMARK. In §7 of this chapter we give a theorem on the limits of amenability: any class culled from an unsolvable class by restrictions other

ISBN 0-201-02540-X

---

‡ For each schema F we can *nonconstructively* find an adequate expansion $E(F, p)$; thus every class of schemata is φ-amenable for some function φ. Of course, solvability is established only by φ-amenability for *recursive* φ.

than those on kinds of atomic subformulas—for example, by restrictions on truth-functional form—is not amenable. This reflects the necessity of limiting coinstantiability in order that a class be amenable. The theorem implies the nonamenability of Classes 1.2–1.4. On the other hand, Classes 1.5 and 1.6 are trivially amenable, since for those classes we obtained (Herbrand instance preserving) formula-mappings Γ such that $\Gamma[E(F, \omega)] \subseteq E(F, 1)$. These formula-mappings themselves, when their domains are restricted to Herbrand instances (rather than all of $\tilde{L}(F, \omega)$) are suitable instance-mappings. End of Remark.

Term-mappings γ will continue to play a role in our constructions: both Γ and Ψ will be defined on the basis of term-mappings. Thus our proofs of amenability use three types of mappings: term-mappings on $D(F, \omega)$, formula-mappings on $\tilde{L}(F, \omega)$, and instance-mappings on $E(F, \omega)$.

2. The Monadic, Ackermann, and Minimal Gödel–Kalmár–Schütte Classes

The first class we show amenable is the Monadic Class, a class whose solvability has long been known. Various proofs have been given by Löwenheim (1915), Skolem (1919), Behmann (1922), Hilbert and Ackermann (1928), and Herbrand (1930). An easy proof of its solvability may be obtained thus: for each schema F in the class, we can effectively find, by inverse-prenexing rules and truth-functional manipulations, a schema G equivalent to F such that in G no y-variable governs any x-variable. Since the schema G falls under Class 1.1, the solvability of the Monadic Class is immediate. The inverse-prenexing rules necessary are the following: within a schema we may replace a subformula $\forall v - G_1$ by $-\exists v G_1$, $\exists v - G_1$ by $-\forall v G_1$, $\forall v(G_1 \wedge G_2)$ by $\forall v G_1 \wedge \forall v G_2$, $\exists v(G_1 \vee G_2)$ by $\exists v G_1 \vee \exists v G_2$, and if G_2 does not contain the variable v, $\forall v(G_1 \vee G_2)$ by $\forall v G_1 \vee G_2$, $\exists v(G_1 \wedge G_2)$ by $\exists v G_1 \wedge G_2$. The truth-functional rules needed are essentially de Morgan's laws applied not just to atomic formulas but also to subformulas beginning with quantifiers, which are treated as truth-functionally primitive. However, in order to point out some general features of our method, we shall present a solvability proof by way of amenability.

Class 2.1 (Monadic Class). Let F be any schema containing only monadic, that is, 1-place, predicate letters, and let p be the number of predicate letters in F. Then $E(F, 2^p)$ is adequate.

Proof. Let F be a monadic schema, p the number of predicate letters of F, and $y_1, \ldots, y_n$ all the y-variables of F. Let $\mathfrak{A}$ be any truth-assignment on $\tilde{L}(F, \omega)$. Our aim is to construct for F a formula-mapping Γ and an instance-mapping Ψ such that

(1) $\quad \Psi[E(F, \omega)] \subseteq E(F, 2^p)$;

ISBN 0-201-02540-X

(2) $\mathfrak{A}(\Gamma(\langle H \rangle_i)) = \mathfrak{A}(\langle \Psi(H) \rangle_i)$ for every Herbrand instance H of F and every number i.

To accomplish our aim we shall first construct a term-mapping γ satisfying, among other conditions,

(3) $\gamma[D(F, \omega)] \subseteq D(F, 2^p)$.

Then we shall let Γ be the γ-canonical formula-mapping, that is, $\Gamma(Pt) = P\gamma t$ for each atom Pt in $L(F, \omega)$. And we define Ψ by

$$\Psi(F^*(y_1/t_1,\ldots,y_n/t_n)) = F^*(y_1/\gamma t_1,\ldots,y_n/\gamma t_n)$$

for each Herbrand instance $F^*(y_1/t_1,\ldots,y_n/t_n)$ of F. By (3) the range of Ψ is included in $E(F, 2^p)$, that is, Ψ fulfills condition (1), the boundedness condition.

Moreover, if $H = F^*(y_1/t_1,\ldots,y_n/t_n)$ is any Herbrand instance of F, then condition (2), the central condition, follows immediately for those components $\langle F \rangle_i$ that contain a y-variable. For suppose $\langle F \rangle_i = Py_j$, $1 \leq j \leq n$; then $\langle H \rangle_i = Pt_j$ and $\Gamma(\langle H \rangle_i) = P\gamma t_j$. And, by the definition of Ψ, the substituent in $\Psi(H)$ for y_j is γt_j; hence $\Gamma(\langle H \rangle_i) = \langle \Psi(H) \rangle_i$.

But what if $\langle F \rangle_i$ contains an x-variable? Say $\langle F \rangle_i = Px$. Then $\langle F^* \rangle_i = Pf_x(y_{j_1}\cdots y_{j_k})$, where $y_{j_1},\ldots,y_{j_k}$ are the y-variables governing x in F, $k \geq 0$, and $\langle H \rangle_i = Pf_x(t_{j_1}\cdots t_{j_k})$; so $\Gamma(\langle H \rangle_i) = P\gamma f_x(t_{j_1}\cdots t_{j_k})$. On the other hand, $\langle \Psi(H) \rangle_i = Pf_x(\gamma t_{j_1}\cdots\gamma t_{j_k})$. We would be done, of course, if γ were homomorphic on $f_x(t_{j_1}\cdots t_{j_k})$. But a term-mapping γ homomorphic on all such terms could not fulfill the boundedness condition. (This is just the familiar point that Γ cannot both be Herbrand instance preserving on $E(F, \omega)$ and have finite range.) The solution to this difficulty for the monadic schema F is easy: we need only require of γ that

$$\mathfrak{A}(P\gamma f_x(t_{j_1}\cdots t_{j_k})) = \mathfrak{A}(Pf_x(\gamma t_{j_1}\cdots\gamma t_{j_k})).$$

In fact, define a set-valued function $\xi_{\mathfrak{A}}$ on $D(F, \omega)$ thus: for each term t, let $\xi_{\mathfrak{A}}(t)$ be the (finite) set of predicate letters P of F such that $\mathfrak{A}$ verifies Pt. If γ fulfills the following condition (4), then the central condition is assured for those components $\langle F \rangle_i$ that contain an x-variable.

(4) $\xi_{\mathfrak{A}}(\gamma f(s_1\cdots s_k)) = \xi_{\mathfrak{A}}(f(\gamma s_1\cdots\gamma s_k))$ for each term $f(s_1\cdots s_k)$ in $D(F, \omega)$, $k \geq 0$.

Note that if $k = 0$, that is, the function sign f is a constant c, then (4) amounts to $\xi_{\mathfrak{A}}(\gamma c) = \xi_{\mathfrak{A}}(c)$.

We define the term-mapping γ inductively. Let $f(s_1\cdots s_k)$ be any term in $D(F, \omega)$, $k \geq 0$, and suppose γ is defined on $s_1,\ldots,s_k$ (if $k = 0$, so that $f(s_1\cdots s_k)$ has height 1, then this supposition is vacuous). Then let $\gamma f(s_1\cdots s_k)$ be the earliest term t in $D(F, \omega)$ such that $\xi_{\mathfrak{A}}(t) = \xi_{\mathfrak{A}}(f(\gamma s_1\cdots\gamma s_k))$.

ISBN 0-201-02540-X

Clearly the term-mapping γ fulfills condition (4). It remains only to verify condition (3), that is, to verify that $\gamma[D(F, \omega)] \subseteq D(F, 2^p)$. Since there are p predicate letters of F, there are at most 2^p different values of $\xi_{\mathfrak{A}}(t)$ as t ranges over $D(F, \omega)$. Hence there is a number r, $1 \le r \le 2^p$, with the following property: for each term s in $D(F, r + 1)$ there is a term t in $D(F, r)$ such that $\xi_{\mathfrak{A}}(s) = \xi_{\mathfrak{A}}(t)$. For were there no such r, then $\xi_{\mathfrak{A}}[D(F, q)]$ would be a proper subset of $\xi_{\mathfrak{A}}[D(F, q +1)]$ for each q, $1 \le q \le 2^p$; whence $\xi_{\mathfrak{A}}[D(F, 2^p +1)]$ would have cardinality $2^p + 1$, which is impossible. (The argument just given is an example of the combinatorial *pigeonhole argument,* which we shall discuss in a general setting immediately below.) A simple inductive argument on height now shows that γ takes $D(F, \omega)$ to $D(F, r)$. For let $f(s_1 \cdots s_k)$ be a term, $k \ge 0$, and suppose $\gamma s_1, \ldots, \gamma s_k$ all lie in $D(F, r)$ (again, if $k = 0$, this supposition is vacuous). Then $f(\gamma s_1 \cdots \gamma s_k)$ lies in $D(F, r + 1)$; hence by the property of r there is a term t in $D(F, r)$ with $\xi_{\mathfrak{A}}(t) = \xi_{\mathfrak{A}}(f(\gamma s_1 \cdots \gamma s_k))$. But then $\gamma f(s_1 \cdots s_k)$, which is the earliest such term t, must lie in $D(F, r)$. Thus the proof is concluded and $E(F, 2^p)$ is shown adequate. $\square$

The pigeonhole argument, which will play a key role in all of our amenability proofs, may be stated in abstract form thus: if $S_1 \subseteq S_2 \subseteq \cdots \subseteq S_{n+1}$ is a tower of $n + 1$ nonempty sets and if ξ is any function defined on the members of these sets and whose range contains at most n elements, then there is an $r \le n$ such that $\xi[S_r] = \xi[S_{r+1}]$.

In our applications we generally take the sets S_q to be the domains $D(F, q)$. Then, if r is a number such that $\xi[D(F, r + 1)] = \xi[D(F, r)]$, we can so define the *pigeonhole-mapping* π that, for each term t in $D(F, r + 1)$, the value πt is a term in $D(F, r)$ such that $\xi(\pi t) = \xi(t)$. (We are indifferent to the values of π on terms of height $> r + 1$.)

REMARK. A pigeonhole-mapping could have been used in the proof for the Monadic Class in the following manner. The choice of the number r in that proof yields a pigeonhole-mapping π that takes each term s in $D(F, r + 1)$ to a term πs in $D(F, r)$ with $\xi_{\mathfrak{A}}(\pi s) = \xi_{\mathfrak{A}}(s)$. Define the term-mapping γ thus: $\gamma f(s_1 \cdots s_k) = \pi f(\gamma s_1 \cdots \gamma s_k)$ for each term $f(s_1 \cdots s_k)$ in $D(F, \omega)$. It quickly follows that $\gamma[D(F, \omega)] \subseteq D(F, r)$ and that γ fulfills condition (4).   End of Remark.

The set-valued function $\xi_{\mathfrak{A}}$ used in the proof for the Monadic Class is a simple example of the type of coding function we shall employ in pigeonhole arguments. We now define a more general function on terms in $D(F, \omega)$ that encodes the values a truth-assignment $\mathfrak{A}$ takes on atoms containing those terms. We first set aside an infinite sequence $\mathfrak{z}_1, \ldots, \mathfrak{z}_n, \ldots$ of variables in our formal language for coding purposes only; they are not to appear in any schema.

ISBN 0-201-02540-X

Definition 2.4. Let F be any schema. For each truth-assignment $\mathfrak{A}$ on $\bar{L}(F, \omega)$, each number $n > 0$, and each n-tuple $\langle t_1,\ldots,t_n \rangle$ of (not necessarily distinct) terms in $D(F, \omega)$, let $\mathfrak{p}(F, n, \mathfrak{A}, \langle t_1,\ldots,t_n \rangle)$ be the (finite) set of *all* signed atomic formulas A containing a predicate letter of F and variables among $\mathfrak{z}_1,\ldots,\mathfrak{z}_n$ such that $\mathfrak{A}$ verifies $A(\mathfrak{z}_1/t_1,\ldots,\mathfrak{z}_n/t_n)$; and call $\mathfrak{p}(F, n, \mathfrak{A}, \langle t_1,\ldots,t_n \rangle)$ the *profile* of $\langle t_1,\ldots,t_n \rangle$ under $\mathfrak{A}$ (with respect to F). For a fixed F we shall ordinarily write $\mathfrak{p}_{\mathfrak{A}}(t_1,\ldots,t_n)$ for $\mathfrak{p}(F, n, \mathfrak{A}, \langle t_1,\ldots,t_n \rangle)$.

EXAMPLE. Let F be a schema containing just the 2-place predicate letter P and the 1-place predicate letter Q, let $\mathfrak{A}$ be a truth-assignment on $\bar{L}(F, \omega)$, and let t and s be terms in $D(F, \omega)$ such that $\mathfrak{A}$ verifies

$$Ptt \wedge -Pss \wedge Pts \wedge -Pst \wedge -Qt \wedge Qs.$$

Then

$$\mathfrak{p}(F, 2, \mathfrak{A}, \langle t, s \rangle) = \mathfrak{p}_{\mathfrak{A}}(t, s)$$
$$= \{P\mathfrak{z}_1\mathfrak{z}_1, -P\mathfrak{z}_2\mathfrak{z}_2, P\mathfrak{z}_1\mathfrak{z}_2, -P\mathfrak{z}_2\mathfrak{z}_1, -Q\mathfrak{z}_1, Q\mathfrak{z}_2\};$$

$$\mathfrak{p}(F, 2, \mathfrak{A}, \langle s, t \rangle) = \mathfrak{p}_{\mathfrak{A}}(s, t)$$
$$= \{P\mathfrak{z}_2\mathfrak{z}_2, -P\mathfrak{z}_1\mathfrak{z}_1, P\mathfrak{z}_2\mathfrak{z}_1, -P\mathfrak{z}_1\mathfrak{z}_2, -Q\mathfrak{z}_2, Q\mathfrak{z}_1\};$$

$$\mathfrak{p}(F, 2, \mathfrak{A}, \langle t, t \rangle) = \mathfrak{p}_{\mathfrak{A}}(t, t)$$
$$= \{P\mathfrak{z}_1\mathfrak{z}_1, P\mathfrak{z}_2\mathfrak{z}_2, P\mathfrak{z}_1\mathfrak{z}_2, P\mathfrak{z}_2\mathfrak{z}_1, -Q\mathfrak{z}_1, -Q\mathfrak{z}_2\};$$

$$\mathfrak{p}(F, 1, \mathfrak{A}, \langle t \rangle) = \mathfrak{p}_{\mathfrak{A}}(t) = \{P\mathfrak{z}_1\mathfrak{z}_1, -Q\mathfrak{z}_1\};$$

$$\mathfrak{p}(F, 3, \mathfrak{A}, \langle t, s, t \rangle) = \mathfrak{p}_{\mathfrak{A}}(t, s, t)$$
$$= \{P\mathfrak{z}_1\mathfrak{z}_1, -P\mathfrak{z}_2\mathfrak{z}_2, P\mathfrak{z}_3\mathfrak{z}_3, P\mathfrak{z}_1\mathfrak{z}_2, -P\mathfrak{z}_2\mathfrak{z}_1, P\mathfrak{z}_1\mathfrak{z}_3,$$
$$P\mathfrak{z}_3\mathfrak{z}_1, -P\mathfrak{z}_2\mathfrak{z}_3, P\mathfrak{z}_3\mathfrak{z}_2, -Q\mathfrak{z}_1, Q\mathfrak{z}_2, -Q\mathfrak{z}_3\}.$$

Profiles possess two properties crucial to their role in proofs of amenability.

Basic Property I. For each fixed schema F and each fixed number $n > 0$, $\mathfrak{p}(F, n, \mathfrak{A}, \langle t_1,\ldots,t_n \rangle)$ takes on one of a finite number of values as $\mathfrak{A}$ and $\langle t_1,\ldots,t_n \rangle$ vary.

Definition 2.5. For each schema F and each $n > 0$, let $d(F, n)$ be the number of values of $\mathfrak{p}(F, n, \mathfrak{A}, \langle t_1,\ldots,t_n \rangle)$ as $\mathfrak{A}$ and $\langle t_1,\ldots,t_n \rangle$ vary.

Clearly, for each schema F and each $n > 0$, $d(F, n) \leq 2^{p \cdot n^m}$, where p is the number of predicate letters in F and m is the maximum number of argument places of these letters. If each predicate letter is m-place, then $d(F, n) = 2^{p \cdot n^m}$

Basic Property II. Let F be a schema, let $\mathfrak{A}$ be a truth-assignment on

ISBN 0-201-02540-X

$\tilde{L}(F, \omega)$, and let $\langle t_1,\ldots,t_n \rangle$ and $\langle s_1,\ldots,s_n \rangle$ be n-tuples of terms from $D(F, \omega)$. Then $\mathfrak{p}_{\mathfrak{A}}(t_1,\ldots,t_n) = \mathfrak{p}_{\mathfrak{A}}(s_1,\ldots,s_n)$ iff

$$\mathfrak{A}(Pt_{i_1}\cdots t_{i_k}) = \mathfrak{A}(Ps_{i_1}\cdots s_{i_k})$$

for each k-place predicate letter P of F and each k-tuple $\langle i_1,\ldots,i_k \rangle$ of (not necessarily distinct) numbers from $\{1,\ldots,n\}$, $k > 0$. Moreover, if $\mathfrak{p}_{\mathfrak{A}}(t_1,\ldots,t_n) = \mathfrak{p}_{\mathfrak{A}}(s_1,\ldots,s_n)$, then $\mathfrak{p}_{\mathfrak{A}}(t_1,\ldots,t_n, t_i) = \mathfrak{p}_{\mathfrak{A}}(s_1,\ldots,s_n, s_i)$, $1 \leq i \leq n$.

By using profiles we may recast in more general form the application of the pigeonhole argument made in the proof for the Monadic Class. Given *any* schema F and any truth-assignment $\mathfrak{A}$ on $\tilde{L}(F, \omega)$, there is an $r \leq d(F, 1)$ such that $\{\mathfrak{p}_{\mathfrak{A}}(t) \mid t \in D(F, r+1)\} = \{\mathfrak{p}_{\mathfrak{A}}(t) \mid t \in D(F, r)\}$. Consequently, there is a term-mapping γ fulfilling the following two conditions:

(1) $\gamma[D(F, \omega)] \subseteq D(F, d(F, 1))$;
(2) $\mathfrak{p}_{\mathfrak{A}}(\gamma f(t_1\cdots t_k)) = \mathfrak{p}_{\mathfrak{A}}(f(\gamma t_1 \cdots \gamma t_k))$ for each term $f(t_1\cdots t_k)$ in $D(F, \omega)$.

Note that since $d(F, 1) = 2^p$, where p is the number of predicate letters in F, the bound given in (1) agrees with that we derived for the Monadic Class.

It should be clear that this term-mapping γ, put to work in a proof just like that for the Monadic Class, will yield the adequacy of $E(F, d(F, 1))$ whenever F meets the following specification: Every atomic subformula of F has the form $Pv\cdots v$, that is, all argument places of the predicate letter are occupied by the same variable. (We call atomic formulas of this form *essentially monadic*.)

Moreover, the same term-mapping γ and instance-mapping Ψ yield the amenability of another class: the class of prenex schemata with prefixes $\forall y \exists x_1 \cdots \exists x_n$, whose solvability was first shown by Ackermann (1928). The proof, however, will not be quite as quick as that for the Monadic and Essentially Monadic Classes. In particular, the formula-mapping Γ, although defined from the term-mapping γ, will not be γ-canonical. This feature of the proof is intrinsic: *no* term-mapping γ can support the use of a γ-canonical formula-mapping to establish the amenability of the Ackermann Class. A brief example will show why. Suppose F has prefix $\forall y \exists x$ and that $\langle F \rangle_i = Pyx$. Let $\mathfrak{A}$ be a truth-assignment on $\tilde{L}(F, \omega)$ that verifies an atom Pst iff $t = f_x(s)$. Then $\mathfrak{A}$ verifies $\langle H \rangle_i$ for each Herbrand instance H of F. Hence $\mathfrak{A}$ verifies $\langle \Psi(H) \rangle_i$ for each H and each instance-mapping Ψ, since $\Psi(H)$ is itself always an Herbrand instance. Now if γ is a term-mapping with finite range, then there is a term s such that $\gamma f_x(s) \neq f_x(\gamma s)$. Hence, if Γ is the γ-canonical formula-mapping and if $H = F^*(y/s)$ for this s, then

ISBN 0-201-02540-X

$\Gamma(\langle H\rangle_i) = P\gamma s\ \gamma f_x(s)$. But then $\mathfrak{A}$ falsifies $\Gamma(\langle H\rangle_i)$ and the central condition fails, no matter what instance-mapping we take Ψ to be.

The same reasoning holds for any schema F containing an atomic subformula in which occur both an x-variable and a y-variable governing that x-variable. And all of our amenable classes that follow allow such subformulas.

For expository convenience, in our amenability proof for the Ackermann Class we restrict ourselves to schemata containing dyadic predicate letters only. This restriction is in no way essential to the argument, as we shall soon see.

Class 2.2 (Ackermann Class). Let F be a prenex schema with prefix $\forall y \exists x_1 \cdots \exists x_n$ and matrix containing dyadic predicate letters only. Then $E(F, d(F, 1))$ is adequate.

Proof. The functional form F^* of an Ackermann schema F is $F^M(x_1/f_{x_1}(y),\ldots,x_n/f_{x_n}(y))$; the domain $D(F, \omega)$ is generated from $\dagger$ by the 1-place indicial function signs $f_{x_1},\ldots,f_{x_n}$. Given any truth-assignment $\mathfrak{A}$, let γ be a term-mapping fulfilling conditions (1) and (2) of page 40; these conditions here can be written

(1) $\gamma[D(F, \omega)] \subseteq D(F, d(F, 1))$;
(2) $\mathfrak{p}_{\mathfrak{A}}(\gamma f_{x_j}(t)) = \mathfrak{p}_{\mathfrak{A}}(f_{x_j}(\gamma t))$ for each term t and each j, $1 \le j \le n$.

We define the formula-mapping Γ *noncanonically* from γ thus: for each predicate letter P of F and all terms s and t in $D(F, \omega)$,

$$\Gamma(Pst) = \begin{cases} P\gamma s\ f_{x_j}(\gamma s) & \text{if}\quad t = f_{x_j}(s), 1 \le j \le n; \\ Pf_{x_j}(\gamma t)\ \gamma t & \text{if}\quad s = f_{x_j}(t), 1 \le j \le n; \\ Pf_{x_j}(\gamma u)\ f_{x_k}(\gamma u) & \text{if}\quad s = f_{x_j}(u)\ \text{and}\ t = f_{x_k}(u) \\ & \text{for some term } u \text{ and some} \\ & \text{distinct } j \text{ and } k,\ 1 \le j, k \le n; \\ P\gamma s\ \gamma t & \text{otherwise.} \end{cases}$$

This definition unambiguously determines Γ on $L(F, \omega)$. (Note that Γ is defined to be γ-canonical only on atoms falling under the fourth clause, among which is each essentially monadic atom Pss.)

We let the instance-mapping Ψ take each Herbrand instance $F^*(y/t)$ to $F^*(y/\gamma t)$. The boundedness condition $\Psi[E(F, \omega)] \subseteq E(F, d(F, 1))$ follows immediately from (1). To verify the central condition we proceed by cases. Let $H = F^*(y/t)$.

(a) $\langle F\rangle_i$ is Pyy, Pyx_j, $Px_j y$, or $Px_j x_k$, where j and k are distinct integers $\le n$. In each of these cases $\Gamma(\langle H\rangle_i) = \langle\Psi(H)\rangle_i$. For example, if $\langle F\rangle_i = Pyx_j$, then $\langle H\rangle_i = Pt\ f_{x_j}(t)$ and $\Gamma(\langle H\rangle_i) = P\gamma t\ f_{x_j}(\gamma t)$, which is just $\langle\Psi(H)\rangle_i$.

ISBN 0-201-02540-X

(b) $\langle F \rangle_i$ is $Px_j x_j$, $1 \le j \le n$. Then $\langle H \rangle_i = P f_{x_j}(t) \, f_{x_j}(t)$, and $\Gamma(\langle H \rangle_i) = P \gamma f_{x_j}(t) \, \gamma f_{x_j}(t)$, but $\langle \Psi(H) \rangle_i = P f_{x_j}(\gamma t) \, f_{x_j}(\gamma t)$. The desired identity

$$\mathfrak{A}(\Gamma(\langle H \rangle_i)) = \mathfrak{A}(\langle \Psi(H) \rangle_i)$$

then follows immediately from property (2) of γ. $\square$

Instead of constructing Γ directly from γ, we might first define Ψ as we did and then define Γ thus:

(I)   For each Herbrand instance H and each i, if $\langle H \rangle_i$ contains at least two distinct terms, then let $\Gamma(\langle H_i \rangle) = \langle \Psi(H) \rangle_i$;

(II)   on each atom A for which $\Gamma(A)$ is not determined by (I), let Γ be γ-canonical.

Such a formula-mapping Γ fulfills the central condition: trivially if $\langle H \rangle_i$ falls under (I), and by condition (2) on γ otherwise. We must show, however, that (I) and (II) do in fact define a formula-mapping; in particular, that no ambiguity arises from (I). But no ambiguity can arise from (I) because of the coinstantiability properties of F. Every Ackermann schema obeys a stringent restriction on coinstantiability:

If $\langle F \rangle_i$ and $\langle F \rangle_j$ are coinstantiable, then either $\langle F \rangle_i = \langle F \rangle_j$ or else $\langle F \rangle_i$ and $\langle F \rangle_j$ are essentially monadic.

This restriction follows from the characterization of coinstantiability for Skolem schemata given in the Example on page 26, since F contains only one y-variable.

Indeed, we may easily show the adequacy of $E(F, d(F, 1))$ for *any* schema F obeying the coinstantiability restriction. Given a truth-assignment $\mathfrak{A}$ on $\tilde{L}(F, \omega)$, let the term-mapping γ fulfill conditions (1) and (2) on page 40. Let Ψ take each Herbrand instance $F^*(y_1/t_1, \dots y_m/t_m)$ to $F^*(y_1/\gamma t_1, \dots, y_m/\gamma t_m)$, where $y_1, \dots, y_m$ are all the y-variables of F. Then $\Psi[E(F, \omega)] \subseteq E(F, d(F, 1))$, and it suffices to show the existence of a formula-mapping defined by (I) and (II) above, since such a formula-mapping would fulfill the central condition. That is, we must show that $\langle \Psi(H) \rangle_i = \langle \Psi(H') \rangle_j$ whenever $\langle H \rangle_i = \langle H' \rangle_j$ and $\langle H \rangle_i$ contains at least two distinct terms. But if $\langle H \rangle_i = \langle H' \rangle_j$ and $\langle H \rangle_i$ contains at least two distinct terms, then $\langle F \rangle_i$ and $\langle F \rangle_j$ are coinstantiable though not essentially monadic; so $\langle F \rangle_i = \langle F \rangle_j$. Hence for each variable v in $\langle F \rangle_i$, the substituent for v in H is identical with that in H'. And then the substituent for v in $\Psi(H)$ is identical with that in $\Psi(H')$. Hence $\langle \Psi(H) \rangle_i = \langle \Psi(H') \rangle_j$.

By this more abstract reformulation of the proof for Class 2.2, we have shown the amenability of a class of schemata that extends not only Class

ISBN 0-201-02540-X

2.2 (by allowing arbitrary predicate letters, not just dyadic) but also the Essentially Monadic Class and Class 1.5. Moreover, we may carry the argument further if F is Skolem: we may also allow distinct coinstantiable subformulas $\langle F \rangle_i$ and $\langle F \rangle_j$ provided they are simultaneously coinstantiable. The class of Skolem schemata obeying this weaker restriction is an extension of Class 1.6.

No coinstantiability restriction of such stringency is obeyed by (arbitrary) schemata with prefixes $\forall y_1 \forall y_2 \exists x$. An atomic subformula $P y_1 y_2$ or $P y_2 y_1$ is coinstantiable, but not simultaneously coinstantiable, with subformulas $P y_j x$ and $P x y_j$. Now by the argument on page 40, the presence of atomic subformulas $P y_j x$ and $P x y_j$ precludes the use of a γ-canonical formula-mapping Γ if γ is to have finite range. Moreover, the presence of atomic subformulas $P y_1 y_2$ or $P y_2 y_1$ precludes the use of an instance-mapping Ψ that takes each Herbrand instance $F^*(y_1/s, y_2/t)$ to $F^*(y_1/\gamma s, y_2/\gamma t)$. For were Ψ so defined and were $\langle F \rangle_i$ of the form $P y_1 y_2$, then the central condition would require that $\mathfrak{A}(\Gamma(Pst)) = \mathfrak{A}(P\gamma s\ \gamma t)$. But for the same reasons that Γ cannot be γ-canonical, this requirement cannot be met if γ is to have finite range. In short, $\Psi(H)$ cannot be determined solely by the values of γ on the substituents in H for the y-variables y_1 and y_2. Hence we shall define Ψ by considering the value of γ on the substituent $f_x(s\ t)$ for the x-variable x; and the value γ does take on such a term $f_x(s\ t)$ will not depend just on the values γs and γt.

Schemata with prefixes $\forall y_1 \forall y_2 \exists x$ fall under the Gödel–Kalmár–Schütte Class, that is, the class of all prenex schemata with prefixes $\forall y_1 \forall y_2 \exists x_1 \cdots \exists x_n$, shown solvable by Gödel (1932), Kalmár (1933), and Schütte (1934) and finitely controllable by Gödel (1933) and Schütte (1934a).‡ Again for expository convenience we limit ourselves here to those schemata with the simplest prefix of the Gödel–Kalmár–Schütte type and with matrices containing dyadic predicate letters only. The amenability of the whole Gödel–Kalmár–Schütte Class, Class 2.5, will be shown in §4.

Class 2.3 (**Minimal Gödel–Kalmár–Schütte Class**). Let F be a prenex schema with prefix $\forall y_1 \forall y_2 \exists x$ and matrix containing dyadic predicate letters only. Then $E(F, 2d(F, 2))$ is adequate.

Proof. The functional form F^* of F is $F^M(x/f_x(y_1 y_2))$ and the domain $D(F, \omega)$ is generated from 1 by the 2-place indicial function sign f_x. Recall from Chapter 0 the definitions of the mappings $\hat{y}_1$, $\hat{y}_2$, and $\hat{x}$: if a term t is the substituent in an Herbrand instance H for x, then for $v = y_1$, y_2, or x, $\hat{v}t$ is the substituent in H for the variable v. That is, if $t = f_x(t_1 t_2)$,

ISBN 0-201-02540-X

---

‡ In Kalmár (1933) and Schütte (1934) the class of prenex schemata with prefixes $\exists z_1 \cdots \exists z_i \forall y_1 \forall y_2 \exists x_1 \cdots \exists x_n$ is reduced to the Gödel–Kalmár–Schütte Class. Hence this class is also solvable and finitely controllable. See §5 of this chapter.

then $\hat{y}_1 t = t_1$, $\hat{y}_2 t = t_2$, and $\hat{x}t = t$. (The mappings $\hat{y}_1$, $\hat{y}_2$, and $\hat{x}$ are not defined on the term $\bot$.) Each Herbrand instance H of F has the form $F^*(y_1/\hat{y}_1 t, y_2/\hat{y}_2 t)$ for some unique term t of height >1, and this term t is the substituent in H for x.

Let $\mathfrak{A}$ be any truth-assignment on $\hat{L}(F, \omega)$. For the moment we assume we have a suitable term-mapping γ; we shall construct γ after ascertaining what suitability comes to here. Let $H = F^*(y_1/\hat{y}_1 t, y_2/\hat{y}_2 t)$ be an arbitrary Herbrand instance of F. We define the instance-mapping Ψ so that γt is the substituent for x in $\Psi(H)$; that is, let

$$\Psi(H) = F^*(y_1/\hat{y}_1 \gamma t, y_2/\hat{y}_2 \gamma t).$$

Hence, for each variable v of F, $\hat{v}\gamma t$ is the substituent in $\Psi(H)$ for v. (Thus γt must have height >1 for each term t of height >1.) The boundedness condition requires that $\Psi(H)$ be in $E(F, 2d(F, 2))$, or equivalently that γt have height at most $2d(F, 2) + 1$. Hence we require

(1) $\gamma[D(F, \omega)] \subseteq D(F, 2d(F, 2) + 1)$.

The formula-mapping Γ is to fulfill the central condition

$$\mathfrak{A}(\Gamma(\langle H \rangle_i)) = \mathfrak{A}(\langle \Psi(H) \rangle_i)$$

for each i. Since $\langle F \rangle_i$ may be any atomic formula of the form Pvw, where v and w are among y_1, y_2, and x, we see that the central condition is assured by the following requirement:

(2)   For each predicate letter P of F, each term t of height >1, and all (not necessarily distinct) variables v and w among y_1, y_2, x,

$$\mathfrak{A}(\Gamma(P\hat{v}t\ \hat{w}t)) = \mathfrak{A}(P\hat{v}\gamma t\ \hat{w}\gamma t).$$

Our basic strategy is to satisfy (2) by so defining Γ that $\Gamma(P\hat{v}t\ \hat{w}t) = P\hat{v}\gamma t\ \hat{w}\gamma t$ whenever at least one of v and w is the variable x. Since $\hat{x}\gamma t = \gamma t = \gamma \hat{x}t$ for each term t of height >1, this property of Γ is forthcoming if we let

$$\Gamma(Ps\ t) = \begin{cases} P\hat{y}_j\gamma t\ \gamma t & \text{if } \quad s = \hat{y}_j t,\ j = 1,2; \\ P\gamma s\ \hat{y}_j\gamma s & \text{if } \quad t = \hat{y}_j s,\ j = 1,2; \\ P\gamma s\ \gamma t & \text{otherwise.} \end{cases}$$

The first two clauses of this definition embody the departure from γ-canonicality we should expect from the argument on page 40. Note, however, that all essentially monadic atoms fall under the third clause, and hence Γ is γ-canonical on them.

As it stands, the definition of Γ is subject to ambiguity. Although we cannot have both $s = \hat{y}_j t$ and $t = \hat{y}_k s$, we can have $s = \hat{y}_1 t = \hat{y}_2 t$ or $t = \hat{y}_1 s = \hat{y}_2 s$. But ambiguity is precluded if we require that γ preserve identity

ISBN 0-201-02540-X

of arguments:

(2a)   For each term t of height >1, if $\hat{y}_1 t = \hat{y}_2 t$, then $\hat{y}_1 \gamma t = \hat{y}_2 \gamma t$.‡

We have yet to consider those cases of condition (2) in which neither v nor w is the variable x. The following condition on γ will clearly yield condition (2) when $v = y_1$ and $w = y_2$:

(2b)   For each term t of height >1, if $\Gamma(P\hat{y}_1 t \ \hat{y}_2 t) = Ps_1 s_2$, then

$$\mathfrak{p}_{\mathfrak{A}}(s_1, s_2) = \mathfrak{p}_{\mathfrak{A}}(\hat{y}_1 \gamma t, \hat{y}_2 \gamma t).$$

Moreover, condition (2b) also yields condition (2) when $v = y_2$ and $w = y_1$, since Γ preserves permutations. That is, if $\Gamma(P\hat{y}_1 t \ \hat{y}_2 t) = Ps_1 s_2$, then $\Gamma(P\hat{y}_2 t \ \hat{y}_1 t) = Ps_2 s_1$, and so (2b) implies $\mathfrak{A}(Ps_2 s_1) = \mathfrak{A}(P\hat{y}_2 \gamma t \ \hat{y}_1 \gamma t)$.

Finally, suppose $v = w = y_j$, $j = 1$ or 2. Since Γ is γ-canonical on $P\hat{y}_j t \ \hat{y}_j t$, condition (2) here becomes $\mathfrak{A}(P\gamma\hat{y}_j t \ \gamma\hat{y}_j t) = \mathfrak{A}(P\hat{y}_j \gamma t \ \hat{y}_j \gamma t)$. Hence we require that γ possess the *monadic property*:

(2c)   For each term t of height >1, $\mathfrak{p}_{\mathfrak{A}}(\gamma\hat{y}_j t) = \mathfrak{p}_{\mathfrak{A}}(\hat{y}_j \gamma t)$, $j = 1, 2$.

Thus the adequacy of $E(F, 2d(F, 2))$ will be established if we can construct a term-mapping γ that fulfills conditions (1), (2a), (2b), and (2c). Condition (2b) could most easily be met were γ so defined that $\Gamma(P\hat{y}_1 t \ \hat{y}_2 t) = P\hat{y}_1 \gamma t \ \hat{y}_2 \gamma t$. But, given our definition of Γ, this stipulation would produce a term-mapping γ with infinite range, violating (1). Hence we incorporate into our definition of γ a suitable pigeonhole-mapping π.

For each term t of height >1, let $\xi_{\mathfrak{A}}(t)$ be the pair $\langle \mathfrak{p}_{\mathfrak{A}}(\hat{y}_1 t, \hat{y}_2 t), \eta(t) \rangle$, where $\eta(t) = 0$ if $\hat{y}_1 t = \hat{y}_2 t$ and $\eta(t) = 1$ otherwise. There are at most $2d(F, 2)$ elements in the range of $\xi_{\mathfrak{A}}$. Hence by the pigeonhole argument there is a number r, $2 \le r \le 2d(F, 2) + 1$, and a pigeonhole-mapping π such that for each term t of height >1 in $D(F, r + 1)$,

(π-1) πt is a term in $D(F, r)$ of height >1;
(π-2) $\xi_{\mathfrak{A}}(\pi t) = \xi_{\mathfrak{A}}(t)$.

We define the term-mapping γ inductively.

(γ-1) $\gamma\mathbb{1} = \mathbb{1}$.
(γ-2)   Suppose γ takes $D(F, p)$ into $D(F, r)$ and preserves identity of arguments on $D(F, p)$, $p \ge 1$. Then for each term $f_x(s \ t)$ of height $p + 1$, let

$$\gamma f_x(s \ t) = \begin{cases} \pi f_x(\hat{y}_j \gamma t \ \gamma t) & \text{if } s = \hat{y}_j t, \ j = 1 \text{ or } 2; \\ \pi f_x(\gamma s \ \hat{y}_j \gamma s) & \text{if } t = \hat{y}_j s, \ j = 1 \text{ or } 2; \\ \pi f_x(\gamma s \ \gamma t) & \text{otherwise.} \end{cases}$$

---

‡ We shall reflect further on this requirement in a Remark immediately following this proof.

ISBN 0-201-02540-X

The induction hypotheses are preserved for $p + 1$. Since γ carries $D(F, p)$ to $D(F, r)$, $\gamma f_x(s\ t) = \pi f_x(s_1 s_2)$ for some terms s_1 and s_2 in $D(F, r)$. By (π-1), then, $\gamma f_x(s\ t)$ is in $D(F, r)$ too. Hence γ carries $D(F, p + 1)$ to $D(F, r)$. Second, $\gamma f_x(s\ s) = \pi f_x(\gamma s\ \gamma s)$, so by ($\pi$-2) and the definition of $\xi_{\mathfrak{A}}$, $\hat{y}_1 \gamma f_x(s\ s) = \hat{y}_2 \gamma f_x(s\ s)$. Hence γ preserves identity of arguments on $D(F, p + 1)$.

We have now defined a term-mapping γ on all of $D(F, \omega)$. This term-mapping carries $D(F, \omega)$ to $D(F, r)$, thus fulfilling condition (1), and preserves identity of arguments, thus fulfilling condition (2a). Note too that γ carries terms of height >1 to terms of height >1. Moreover, by the parallel structure of the definitions of γ and Γ, and by property (π-2) of the pigeon-hole-mapping, condition (2b) is also fulfilled. All that remains is to verify the monadic property (2c). We show this by induction on the height of t. Let $p \geq 2$, and assume that for each term u of height $<p$ and >1, $\mathfrak{p}_{\mathfrak{A}}(\gamma \hat{y}_j u) = \mathfrak{p}_{\mathfrak{A}}(\hat{y}_j \gamma u)$, $j = 1, 2$. We show that for each term $t = f_x(t_1 t_2)$ of height p, $\mathfrak{p}_{\mathfrak{A}}(\gamma \hat{y}_j t) = \mathfrak{p}_{\mathfrak{A}}(\hat{y}_j \gamma t)$, that is, $\mathfrak{p}_{\mathfrak{A}}(\gamma t_j) = \mathfrak{p}_{\mathfrak{A}}(\hat{y}_j \gamma t)$, $j = 1, 2$. Now $\gamma t = \pi f_x(s_1 s_2)$ for appropriately selected terms s_1 and s_2 of height $\leq r$. By (π-2), $\mathfrak{p}_{\mathfrak{A}}(\hat{y}_j \gamma t) = \mathfrak{p}_{\mathfrak{A}}(s_j)$; hence it suffices to show that $\mathfrak{p}_{\mathfrak{A}}(\gamma t_j) = \mathfrak{p}_{\mathfrak{A}}(s_j)$, $j = 1, 2$. If $s_j = \gamma t_j$, then we are done. (Note that if $h(t) = 2$, so that $t = f_x(\maltese\ \maltese)$, then both $s_1 = \gamma t_1$ and $s_2 = \gamma t_2$.) But if $s_j \neq \gamma t_j$, then t_j must be $\hat{y}_k t_{3-j}$ for some k, so that $s_j = \hat{y}_k \gamma t_{3-j}$. Since $h(t_{3-j}) < p$, by the induction hypothesis $\mathfrak{p}_{\mathfrak{A}}(\gamma \hat{y}_k t_{3-j}) = \mathfrak{p}_{\mathfrak{A}}(\hat{y}_k \gamma t_{3-j})$, that is, $\mathfrak{p}_{\mathfrak{A}}(\gamma t_j) = \mathfrak{p}_{\mathfrak{A}}(s_j)$.

This concludes the amenability proof for Class 2.3. □

REMARK. We required in our proof that the term-mapping γ preserve identity of arguments: if $\hat{y}_1 t = \hat{y}_2 t$, then $\hat{y}_1 \gamma t = \hat{y}_2 \gamma t$. Although we formulated this requirement in the context of rendering the definition of Γ unambiguous, the requirement flows directly from our basic strategic decision of having $\Gamma(\langle H \rangle_i) = \langle \Psi(H) \rangle_i$ whenever $\langle F \rangle_i$ contains the variable x. For if $\langle F \rangle_i$ and $\langle F \rangle_j$ both contain x and are coinstantiated in $\langle H, H \rangle$, then, since $\Gamma(\langle H \rangle_i) = \Gamma(\langle H \rangle_j)$, our strategy demands that $\langle \Psi(H) \rangle_i = \langle \Psi(H) \rangle_j$. That is, Ψ must preserve simultaneous coinstantiations engendered by atomic subformulas of F that contain x. Now for every Herbrand instance H, atomic subformulas $P y_1 x$ and $P y_2 x$ are coinstantiated in $\langle H, H \rangle$ iff the substituents in H for y_1 and for y_2 are identical; hence the requirement to preserve identity of arguments.

Of course, ambiguity in the definition of Γ could have been precluded by simple fiat. We could stipulate that if $\hat{y}_1 t = \hat{y}_2 t$, then $\Gamma(P \hat{y}_2 t\ t) = P \hat{y}_1 \gamma t\ \gamma t$ and $\Gamma(P t\ \hat{y}_2 t) = P \gamma t\ \hat{y}_1 \gamma t$. But if $\hat{y}_1 \gamma t \neq \hat{y}_2 \gamma t$, then with this stipulation we have abrogated our strategy: for, letting $\langle F \rangle_i = P y_2 x$ and $H = F^*(y_1/\hat{y}_1 t, y_2/\hat{y}_2 t)$, we have $\Gamma(\langle H \rangle_i) = P \hat{y}_1 \gamma t\ \gamma t \neq P \hat{y}_2 \gamma t\ \gamma t = \langle \Psi(H) \rangle_i$. Thus the central condition $\mathfrak{A}(\Gamma(\langle H \rangle_i)) = \mathfrak{A}(\langle \Psi(H) \rangle_i)$ is no longer assured. And

ISBN 0-201-02540-X

the central condition can even fail. For suppose the truth-assignment $\mathfrak{A}$ verifies an atom Ps_1s_2 iff $s_1 = \hat{y}_2s_2$. If $\hat{y}_1\gamma t \neq \hat{y}_2\gamma t$, then $\mathfrak{A}$ falsifies $\Gamma(\langle H\rangle_i)$ but verifies $\langle\Psi(H)\rangle_i$, since $\Gamma(\langle H\rangle_i) = P\hat{y}_1\gamma t\ \gamma t$ and $\langle\Psi(H)\rangle_i = P\hat{y}_2\gamma t\ \gamma t$. Hence simple fiat does not enable us to avoid the identity preservation requirement.

What we have just seen is an extension of our familiar argument that the formula-mapping Γ cannot be γ-canonical. If $\langle F\rangle_i = Py_2x$ then for every Herbrand instance H the atom $\langle H\rangle_i$ has the form $P\hat{y}_2t\ t$. The existence of truth-assignments $\mathfrak{A}$ like that in the previous paragraph shows that if the central condition is to hold, then Γ must carry each atom of the form $P\hat{y}_2t\ t$ to an atom of the same form, that is, to an atom $P\hat{y}_2s\ s$ for some term s. (Similar arguments apply to atoms having the forms $Pt\ \hat{y}_2t$, $P\hat{y}_1t\ t$, and $Pt\ \hat{y}_1t$.) Our basic strategy rests on taking the term s to be γt. End of Remark.

3. The Adequacy Lemma

We begin this brief section by introducing some terminology and notation that facilitate proofs of amenability. Then, using this new terminology, we give conditions that yield the adequacy of expansions. Our first definition is helpful in characterizing atomic formulas and schemata.

Definition 2.6. Let V be a set of variables. An atomic formula A is *V-based* iff V is the set of variables occurring in A. The *basis* $b(F)$ of a schema F is the set of all those sets V of variables such that at least one atomic subformula of F is V-based.

For example, if $A = Py_1y_5y_1x_3y_1$, then A is $\{y_1, y_5, x_3\}$-based. The definition concerns the *set* of variables, and ignores both the order of the variables and the number of occurrences of each variable.

The classes of schemata we treat in §§4–6 are specified by restrictions on both prefix and basis. In our amenability proofs we shall continue to employ formula-mappings Γ that act thus: the image $\Gamma(A)$ of an atom $A = Pt_1\cdots t_p$ is an atom containing the predicate letter P; and the arguments of $\Gamma(A)$ depend only on the terms $t_1,\ldots,t_p$ and not on the predicate letter P. Hence clarity is gained by shifting attention from atoms to sequences of terms.

Definition 2.7. A *sequence-mapping* for a schema F is a function γ^* on the set of all finite sequences of terms in $D(F, \omega)$ such that, for each $p > 0$, γ^* carries $D(F, \omega)^p$ to $D(F, \omega)^p$—that is, γ^* is length preserving.

If γ^* is a sequence-mapping, we usually write $\gamma^*(t_1,\ldots,t_p)$ for $\gamma^*(\langle t_1,\ldots,t_p\rangle)$; if $\gamma^*(t_1,\ldots,t_p) = \langle s_1,\ldots,s_p\rangle$, we write $P\gamma^*(t_1,\ldots,t_p)$ for $Ps_1\cdots s_p$. An *arrangement* of a finite set U of terms or variables is a finite sequence $\langle u_1,\ldots,u_p\rangle$, possibly with repetitions, such that $U = \{u_1,\ldots,u_p\}$.

ISBN 0-201-02540-X

We shall also continue to employ instance-mappings Ψ whose values on $E(F, \omega)$ are determined by the values of term-mappings γ. Consequently, we may easily formulate conditions on term-mappings and sequence-mappings sufficient for the adequacy of an expansion $E(F, q)$ of a schema F. In this formulation we use the mappings $\hat{v}$ defined in Chapter 0, §2. Recall that, for each term $t \neq \dagger$ in $D(F, \omega)$ and each variable v of F, $\chi(t)$ is the x-variable x_k such that t is an x_k-term, and $\hat{v}t$ is defined iff either v governs $\chi(t)$ or else v and $\chi(t)$ are x-variables governed by the same y-variables. If $\hat{v}t$ is defined and if t is the substituent in an Herbrand instance H of F for $\chi(t)$, then $\hat{v}t$ is the substituent in H for v. *Our term-mappings γ will preserve outermost function signs*: $\chi(\gamma t) = \chi(t)$ for each term t. This implies that $\hat{v}\gamma t$ is defined whenever $\hat{v}t$ is defined.

Adequacy Lemma. Let F be a prenex schema whose prefix ends with an existential quantifier $\exists x_n$.‡ An expansion $E(F, q)$ is adequate provided that for each truth-assignment $\mathfrak{A}$ on $\tilde{L}(F, \omega)$ there exist a number $r \leq q + 1$, a term-mapping γ, and a sequence-mapping γ^* fulfilling the following conditions for each term t in $D(F, \omega)$:

(1) $\gamma t \in D(F, r)$;
(2) $\mathfrak{p}_{\mathfrak{A}}(\gamma^*(\hat{v}_1 t,\ldots,\hat{v}_p t)) = \mathfrak{p}_{\mathfrak{A}}(\hat{v}_1 \gamma t,\ldots,\hat{v}_p \gamma t)$ whenever t is an x_n-term and $\langle v_1,\ldots,v_p \rangle$ is an arrangement of some set in $b(F)$.

Proof. Given a truth-assignment $\mathfrak{A}$, let γ and γ^* fulfill (1) and (2). We define Γ and Ψ thus:

(A) $\Gamma(Pt_1\cdots t_p) = P\gamma^*(t_1,\ldots,t_p)$ for each atom $Pt_1\cdots t_p$ in $L(F, \omega)$.
(B) $\Psi(H) = F^*(y_1/\hat{y}_1\gamma t,\ldots,y_m/\hat{y}_m\gamma t)$ for each Herbrand instance H of F, where t is the substituent in H for x_n and $y_1,\ldots,y_m$ are all the y-variables of F.

From (1) we see at once that the instance-mapping Ψ fulfills the boundedness condition $\Psi[E(F, \omega)] \subseteq E(F, q)$. Now suppose $\langle F \rangle_i = Pv_1\cdots v_p$ and let H be the Herbrand instance of F in which t is the substituent for x_n. Then $\Gamma(\langle H \rangle_i) = P\gamma^*(\hat{v}_1 t,\ldots,\hat{v}_p t)$ and $\langle \Psi(H) \rangle_i = P\hat{v}_1\gamma t\cdots\hat{v}_p\gamma t$. Hence, by condition (2), the central condition

$$\mathfrak{A}(\Gamma(\langle H \rangle_i)) = \mathfrak{A}(\langle \Psi(H) \rangle_i)$$

is also fulfilled. □

---

‡ The restriction to such prenex schemata enables us to construct an instance-mapping Ψ from any given term-mapping γ in a straightforward manner. This restriction entails no loss of generality. Given an arbitrary prenex schema F we may simply append an existential quantifier $\exists x_n$ to the prefix of F, where x_n is foreign to F, and then (so that x_n is not vacuously quantified) conjoin the matrix of F with some tautology like $Px_n \vee -Px_n$, where P is a monadic predicate letter. This increases the basis by the addition of $\{x_n\}$: but our amenability proofs always allow arbitrary one-membered sets to occur in bases.

ISBN 0-201-02540-X

Applications of the Adequacy Lemma will be simplified by requiring that the sequence-mapping γ^* preserve permutations and repetitions.

(3)   If $\gamma^*(t_1,\ldots,t_p) = \langle s_1,\ldots,s_p\rangle$, then for each permutation σ of $\{1,\ldots,p\}$ and each i, $1 \le i \le p$, $\gamma^*(t_{\sigma 1},\ldots,t_{\sigma p}) = \langle s_{\sigma 1},\ldots,s_{\sigma p}\rangle$ and $\gamma^*(t_1,\ldots,t_p, t_i) = \langle s_1,\ldots,s_p, s_i\rangle$.

If (3) holds, then condition (2) holds for a term t and *all* arrangements of a given set in $b(F)$ provided that (2) holds for t and any one arrangement of the set. Condition (3) also implies that if two members of a sequence are identical, then so are the homologous members of the image of the sequence under γ^*. (In particular, if $\gamma^*(t) = \langle s\rangle$, then $\gamma^*(t, t,\ldots,t) = \langle s, s,\ldots,s\rangle$.)

Condition (3) reflects our present decision to classify schemata by their bases. Were we to attempt a finer taxonomy, condition (3) would be dropped, and schemata would be classified by an "ordered basis", that is, a set of *sequences* of variables. It does not appear that this more detailed mode of classification would be particularly fruitful in this chapter. In Chapter 5, however, we shall use this mode of classifying schemata to specify a solvable class that is not finitely controllable.

In each of the amenability proofs of the rest of this chapter, a term-mapping γ and a sequence-mapping γ^* will be constructed by simultaneous induction. To carry out such an induction we use two further conditions as induction hypotheses. If $\gamma^*(t_1,\ldots,t_p) = \langle s_1,\ldots,s_p\rangle$, then for each i, $1 \le i \le p$,

(4) $\mathfrak{p}_\mathfrak{A}(s_i) = \mathfrak{p}_\mathfrak{A}(\gamma t_i)$;

(5) $s_i \in D(F, r)$.

(The number r in (5) is to be the same as that in (1). In each proof the appropriate r will be found by the pigeonhole argument.) Thus condition (5) requires that the range of γ^* be bounded. Condition (4) will come into play in showing that one-membered sets satisfy condition (2). It is intimately connected with the *monadic property*: $\mathfrak{p}_\mathfrak{A}(\gamma \hat{y}_i t) = \mathfrak{p}_\mathfrak{A}(\hat{y}_i \gamma t)$ for each y-variable y_i and each x_n-term t (cf. condition (2c) of the amenability proof for Class 2.3).

We shall say, in our inductive constructions, that γ and γ^* fulfill (1)–(5) on $D(F, q)$ iff they fulfill (1)–(5) for each term t in $D(F, q)$ and for each sequence $\langle t_1,\ldots,t_p\rangle$ from $D(F, q)$.

Condition (2) may be viewed as a bundle of subconditions, one for each arrangement of each set $\{v_1,\ldots,v_p\}$ in $b(F)$. The manner in which we contrive to fulfill each subcondition will depend on what kind of set $\{v_1,\ldots,v_p\}$ is. The following definition singles out the two kinds of sets of variables with which we shall principally deal.

Definition 2.8.   Let F be a schema. A set V of variables is *x-focused* iff

ISBN 0-201-02540-X

there is in V an x-variable of F, called the *focus* of V, that is governed in F by every other member of V. An atomic subformula $Pv_1\cdots v_p$ of F or a sequence $\langle v_1,\ldots,v_p\rangle$ of variables of F is *x-focused* iff $\{v_1,\ldots,v_p\}$ is x-focused. Finally, a set of variables of F is a *y-set* iff it contains y-variables only.

If $\{v_1,\ldots,v_p\}$ is a y-set in $b(F)$, then condition (2) for this set acts as a constraint on the construction of the term-mapping γ. For suppose γ and γ^* are defined on a domain $D(F, q)$. If t is an x_n-term of height $q + 1$, then $\langle \hat{v}_1 t,\ldots,\hat{v}_p t\rangle$ is a sequence of terms in $D(F, q)$, and $\gamma^*(\hat{v}_1 t,\ldots,\hat{v}_p t)$ is already determined. To fulfill condition (2), γ must be so extended to $D(F, q + 1)$ that $\mathfrak{p}_{\mathfrak{N}}(\hat{v}_1\gamma t,\ldots,\hat{v}_p\gamma t) = \mathfrak{p}_{\mathfrak{N}}(\gamma^*(\hat{v}_1 t,\ldots,\hat{v}_p t))$. Once γ is extended to $D(F, q + 1)$, we must then extend γ^* to those sequences $\langle t_1,\ldots,t_p\rangle$ from $D(F, q + 1)$ that contain at least one term of height $q + 1$. Among these sequences will be some sequences $\langle \hat{v}_1 t,\ldots,\hat{v}_p t\rangle$, where t is an x_n-term of height $q + 1$ and $\{v_1,\ldots,v_p\}$ is an x-focused set in $b(F)$. Condition (2) for x-focused sets acts primarily as a constraint on the construction of the sequence-mapping γ^*.

4. Skolem Classes

In this section we consider only Skolem schemata, that is, prenex schemata with prefixes $\forall y_1\cdots\forall y_m\exists x_1\cdots\exists x_n$. The basis of a Skolem schema contains just x-focused sets and y-sets. (An x-focused set here is any set of variables that contains at least one x-variable.)

Definition 2.9. If $\langle v_1,\ldots,v_p\rangle$ is an x-focused sequence, then for each x_n-term t we call the sequence $\langle \hat{v}_1 t,\ldots,\hat{v}_p t\rangle$ the *t-instance* of $\langle v_1,\ldots,v_p\rangle$.

Our proofs will exploit the five conditions of the previous section and retain a principal feature of the amenability proof for Class 2.3: if $\langle F\rangle_i$ is x-focused, then $\Gamma(\langle H\rangle_i) = \langle\Psi(H)\rangle_i$. Thus we shall fulfill condition (2) for x-focused sets $\{v_1,\ldots,v_p\}$ by requiring, for each x_n-term t, that γ^* take the t-instance of $\langle v_1,\ldots,v_p\rangle$ to its γt-instance. That is,

$$(2_x) \quad \gamma^*(\hat{v}_1 t,\ldots,\hat{v}_p t) = \langle \hat{v}_1\gamma t,\ldots,\hat{v}_p\gamma t\rangle.$$

Condition (2_x) is by no means trivial, since distinct x-focused sequences may have t-instances in common. For example, if $\hat{y}_1 t = \hat{y}_2 t$, then the t-instance of $\langle y_1, x_n\rangle$ coincides with that of $\langle y_2, x_n\rangle$. In fact, suppose $Pv_1\cdots v_p$ and $Pw_1\cdots w_p$ are x-focused atomic subformulas of F; then the t-instances of $\langle v_1,\ldots,v_p\rangle$ and of $\langle w_1,\ldots,w_p\rangle$ are identical iff $Pv_1\cdots v_p$ and $Pw_1\cdots w_p$ are coinstantiated in $\langle H, H\rangle$, where H is the Herbrand instance in which t is the substituent for x_n.

Now if the t-instances of $\langle v_1,\ldots,v_p\rangle$ and of $\langle w_1,\ldots,w_p\rangle$ are identical, then

ISBN 0-201-02540-X

condition (2_x) requires that $\langle \hat{v}_1 \gamma t, \ldots, \hat{v}_p \gamma t \rangle = \langle \hat{w}_1 \gamma t, \ldots, \hat{w}_p \gamma t \rangle$. This entails the requirement that $Pv_1 \cdots v_p$ and $Pw_1 \cdots w_p$ be coinstantiated in $\langle \Psi(H), \Psi(H) \rangle$. In other words, just as with the Minimal Gödel–Kalmár–Schütte Class, our decision to have $\Gamma(\langle H \rangle_i) = \langle \Psi(H) \rangle_i$ whenever $\langle F \rangle_i$ is x-focused entails that Ψ preserve simultaneous coinstantiations of x-focused atomic subformulas.‡

Note that the t-instance of an x-focused sequence can coincide with the s-instance of another x-focused sequence only if $s = t$. (That is, if two x-focused atomic subformulas of a Skolem schema are coinstantiable, then they engender only simultaneous coinstantiations.) This may be seen thus: Suppose a sequence $\langle t_1, \ldots, t_p \rangle$ of terms is both the t-instance of $\langle v_1, \ldots, v_p \rangle$ and the s-instance of $\langle w_1, \ldots, w_p \rangle$. Let t_j be any term of maximal height among $t_1, \ldots, t_p$. Then both v_j and w_j must be x-variables. Since $\hat{v}_j t = t_j = \hat{w}_j s$, we have $t = \hat{x}_n t_j = s$.

Our first class consists of Skolem schemata no one of whose bases contains distinct y-sets of cardinality >1. Of course, these schemata may have one-membered y-sets in their bases, that is, the schemata may contain arbitrary essentially monadic atomic subformulas.

Class 2.4. Let $F = \forall y_1 \cdots \forall y_m \exists x_1 \cdots \exists x_n F^M$ be a Skolem schema whose basis contains no distinct y-sets of cardinality >1. Then $E(F, m!d(F, m))$ is adequate.

Proof. By permuting y-variables, if necessary, we may assume that the one y-set of cardinality >1 in $b(F)$, if any, is $\{y_1, \ldots, y_l\}$ for some $l \leq m$. Let $\mathfrak{A}$ be any truth-assignment on $\hat{L}(F, \omega)$.

We first construct a pigeonhole-mapping π. Let ι be the mapping that takes each term t of height >1 in $D(F, \omega)$ to the set $\{\langle i, j \rangle \mid \hat{y}_i t = \hat{y}_j t\}$. Then the range of ι contains at most $m!$ members. Hence, by the pigeonhole argument there is a number r, $2 \leq r \leq m!d(F, m) + 1$, and a pigeonhole-mapping π such that for each x_n-term t in $D(F, r + 1)$,

(π-1) πt is an x_n-term in $D(F, r)$;
(π-2) $\iota(\pi t) = \iota(t)$;
(π-3) $\mathfrak{p}_\mathfrak{A}(\hat{y}_1 \pi t, \ldots, \hat{y}_m \pi t) = \mathfrak{p}_\mathfrak{A}(\hat{y}_1 t, \ldots, \hat{y}_m t)$.

We take the number r in the boundedness conditions (1) and (5) to be the one we have just chosen.

We now construct a term-mapping γ and a sequence-mapping γ^* fulfilling conditions (1)–(5). Let γ and γ^* be the identity on terms and sequences from $D(F, 1)$. Clearly these five conditions are fulfilled on this domain.

---

‡ Again, this consequence of our decision and the difficulties it produces might suggest that the decision was ill advised. But, as with Gödel–Kalmár–Schütte schemata, there is little alternative (see the Remark at the end of §2).

ISBN 0-201-02540-X

Assume γ and γ^* fulfill (1)–(5) on $D(F, q)$, $q \geq 1$. We extend γ and γ^* so as to fulfill (1)–(5) on $D(F, q + 1)$.

Suppose t is an x_k-term of height $q + 1$, $1 \leq k \leq n$. Hence each argument $\hat{y}_i t$ of t is of height $\leq q$. Let

$$\gamma t = \hat{x}_k \pi f_{x_n}(s_1 \cdots s_m),$$

where $\langle s_1, \ldots, s_m \rangle$ is determined from $\langle \hat{y}_1 t, \ldots, \hat{y}_m t \rangle$ thus:

(γ-1) $\langle s_1, \ldots, s_l \rangle = \gamma^*(\hat{y}_1 t, \ldots, \hat{y}_l t)$;

(γ-2) for each j, $l < j \leq m$, if there is an $i \leq l$ such that $\hat{y}_j t = \hat{y}_i t$, then let $s_j = s_i$ for the least such i;

(γ-3) for each j, $l < j \leq m$, if s_j is not determined by (γ-2), then let $s_j = \gamma \hat{y}_j t$.

By conditions (1) and (5) for terms and sequences from $D(F, q)$, each s_i has height $\leq r$. Hence, by (π-1), so does γt. Thus our extension of γ fulfills (1) on $D(F, q + 1)$. We shall now see that it also fulfills condition (2) for y-sets, whenever t is an x_n-term. First, for the y-set $\{y_1, \ldots, y_l\}$ we have, by (π-3),

$$\mathfrak{p}_\mathfrak{A}(\hat{y}_1 \gamma t, \ldots, \hat{y}_l \gamma t) = \mathfrak{p}_\mathfrak{A}(s_1, \ldots, s_l).$$

Hence, by (γ-1),

$$\mathfrak{p}_\mathfrak{A}(\hat{y}_1 \gamma t, \ldots, \hat{y}_l \gamma t) = \mathfrak{p}_\mathfrak{A}(\gamma^*(\hat{y}_1 t, \ldots, \hat{y}_l t)),$$

as required. Second, for each y-set $\{y_i\}$ in $b(F)$ we need $\mathfrak{p}_\mathfrak{A}(\gamma^*(\hat{y}_i t)) = \mathfrak{p}_\mathfrak{A}(\hat{y}_i \gamma t)$. By condition (4) for sequences of length 1, $\mathfrak{p}_\mathfrak{A}(\gamma^*(\hat{y}_i t)) = \mathfrak{p}_\mathfrak{A}(\gamma \hat{y}_i t)$. Hence it suffices to show that γ possesses the monadic property: for each i, $1 \leq i \leq m$, $\mathfrak{p}_\mathfrak{A}(\gamma \hat{y}_i t) = \mathfrak{p}_\mathfrak{A}(\hat{y}_i \gamma t)$. By ($\pi$-3), $\mathfrak{p}_\mathfrak{A}(\hat{y}_i \gamma t) = \mathfrak{p}_\mathfrak{A}(s_i)$; hence all we must show is $\mathfrak{p}_\mathfrak{A}(\gamma \hat{y}_i t) = \mathfrak{p}_\mathfrak{A}(s_i)$. But this is immediate from condition (4) if s_i is determined by (γ-1) or by (γ-2); and $s_i = \gamma \hat{y}_i t$ if s_i is determined by (γ-3).

We note two further consequences of our construction. Obviously $\gamma \hat{v} t = \hat{v} \gamma t$ for every x-variable v. Second, γ preserves identity of arguments, that is, if $\hat{y}_i t = \hat{y}_j t$, then $\hat{y}_i \gamma t = \hat{y}_j \gamma t$. For suppose $\hat{y}_i t = \hat{y}_j t$. By ($\pi$-2) it suffices to show that $s_i = s_j$. But if $1 \leq i, j \leq l$, then that $s_i = s_j$ follows from clause (γ-1) and the permutation condition (3). The other two cases are then immediate from this one and clauses (γ-2) and (γ-3).

We have now to extend γ^* to sequences from $D(F, q + 1)$ that contain at least one term of height $q + 1$, and have then to verify condition (2) for x-focused sets and conditions (3), (4), and (5). Let $\langle t_1, \ldots, t_p \rangle$ be such a sequence.

(γ^*-1) · If $\langle t_1, \ldots, t_p \rangle = \langle \hat{v}_1 t, \ldots, \hat{v}_p t \rangle$ for some x_n-term t and some x-

ISBN 0-201-02540-X

focused sequence $\langle v_1,\ldots,v_p\rangle$ (in which case $h(t) = q + 1$), then let $\gamma^*(t_1,\ldots,t_p) = \langle \hat{v}_1\gamma t,\ldots,\hat{v}_p\gamma t\rangle$.

(γ^*-2) Otherwise, let $\gamma^*(t_1,\ldots,t_p) = \langle \gamma t_1,\ldots,\gamma t_p\rangle$.

We must show that γ^* is properly defined, that no ambiguity arises from clause (γ^*-1). But as we saw on page 51, we need be concerned only with the case in which $\langle t_1,\ldots,t_p\rangle$ is the t-instance of distinct x-focused sequences $\langle v_1,\ldots,v_p\rangle$ and $\langle w_1,\ldots,w_p\rangle$, for some x_n-term t (of height $q + 1$). It suffices to show that $\hat{v}_i\gamma t = \hat{w}_i\gamma t$ for each $i \le p$.

(a) Suppose v_i is an x-variable. Then $t_i = \hat{v}_it = \hat{w}_it$ has height $q + 1$; whence w_i is also an x-variable, and so $v_i = \chi(t_i) = w_i$. Hence, obviously, $\hat{v}_i\gamma t = \hat{w}_i\gamma t$.

(b) Suppose v_i is a y-variable. Then $t_i = \hat{v}_it = \hat{w}_it$ has height $\le q$; whence w_i is also a y-variable. Since γ preserves identity of arguments, we have $\hat{v}_i\gamma t = \hat{w}_i\gamma t$.

Thus (γ^*-1) and (γ^*-2) provide an unambiguous extension of γ^* to $D(F, q + 1)$. Clearly, if t is an x_n-term of height $q + 1$, then γ^* takes the t-instance of each x-focused sequence to its γt-instance, and condition (2_x) is fulfilled. Hence condition (2) is fulfilled for all x-focused sets $\{v_1,\ldots,v_p\}$, $p \ge 1$.

The permutation condition (3) also follows: the value of γ^* on $\langle t_1,\ldots,t_p\rangle$ is determined by (γ^*-1) iff it is so determined on any permutation of $\langle t_1,\ldots,t_p\rangle$, and iff it is so determined on any sequence resulting from $\langle t_1,\ldots,t_p\rangle$ by the deletion or addition of multiple occurrences of one or another term.

The boundedness condition (5) on γ^* is immediate from the boundedness condition (1) on γ, since each member of $\gamma^*(t_1,\ldots,t_p)$ is either γt_i or else $\hat{v}\gamma t$ for some variable v and some x_n-term t.

Condition (4) is trivial if $\gamma^*(t_1,\ldots,t_p)$ is determined by clause (γ^*-2). If it is determined by clause (γ^*-1), then for each $i \le p$, its ith member is a term $\hat{v}_i\gamma t$ for some x_n-term t of height $q + 1$ and some variable v_i for which $t_i = \hat{v}_it$. If v_i is an x-variable, then $\hat{v}_i\gamma t = \gamma\hat{v}_it = \gamma t_i$ and condition (4) is again trivial; and if v_i is a y-variable, then $\mathfrak{p}_{\mathfrak{A}}(\hat{v}_i\gamma t) = \mathfrak{p}_{\mathfrak{A}}(\gamma\hat{v}_it)$ by the monadic property of γ.

We have verified that (1)–(5) hold on $D(F, q + 1)$. Thus Class 2.4 is amenable. □

Class 2.4 includes the entire Gödel–Kalmár–Schütte Class, since the y-sets in the basis of a schema in that class are at most $\{y_1, y_2\}$, $\{y_1\}$, and $\{y_2\}$.

Class 2.5 (Gödel–Kalmár–Schütte Class). Let F be a prenex schema with prefix $\forall y_1\forall y_2\exists x_1\cdots\exists x_n$. Then $E(F, 2d(F, 2))$ is adequate. □

ISBN 0-201-02540-X

Two other known solvable classes fall under Class 2.4: the class of those Skolem schemata each of whose atomic subformulas contains either at least one x-variable or else all of the y-variables (Skolem, 1935); and the class extending this one by allowing essentially monadic atomic subformulas (Friedman, 1957; Dreben, 1959).

Class 2.4 is maximal with respect to solvable classes of Skolem schemata in which no restriction is put on x-focused subformulas. Indeed, for each $m \geq 3$ and any distinct sets Y_1, $Y_2 \subseteq \{y_1, \ldots, y_m\}$ of cardinality >1, the class of all schemata $F = \forall y_1 \cdots \forall y_m \exists x F^M$ such that Y_1 and Y_2 are the only y-sets in $b(F)$ is unsolvable (Goldfarb and Lewis, 1975).

It is instructive to examine how the argument for the amenability of Class 2.4 fails if the basis of F contains distinct y-sets of cardinality >1. First let us consider the case of two overlapping y-sets, say $\{y_1, y_2\}$ and $\{y_1, y_3\}$. If condition (2) is to be fulfilled for these y-sets, in the definition

$$\gamma t = \hat{x}_k \pi f_{x_n}(s_1 \cdots s_m)$$

we should, corresponding to clause (γ-1), let $\langle s_1, s_2 \rangle = \gamma^*(\hat{y}_1 t, \hat{y}_2 t)$ and let $\langle s_1, s_3 \rangle = \gamma^*(\hat{y}_1 t, \hat{y}_3 t)$. Hence the first member of $\gamma^*(\hat{y}_1 t, \hat{y}_2 t)$ must be identical with the first member of $\gamma^*(\hat{y}_1 t, \hat{y}_3 t)$. But if condition (2_x) is also to be fulfilled, this cannot be assured: the images under γ^* of two sequences that have some but not all members in common do not themselves always have the homologous members in common. For example, if s and $\hat{y}_1 s$ are both x_n-terms and if (2_x) is fulfilled for the x-focused set $\{y_1, x_n\}$, then the first member of $\gamma^*(\hat{y}_1 s, s)$ is $\hat{y}_1 \gamma s$, whereas the first member of $\gamma^*(\hat{y}_1 s, \hat{y}_1 \hat{y}_1 s)$ is $\gamma \hat{y}_1 s$. And since γ has finite range, there are many terms s for which $\gamma \hat{y}_1 s \neq \hat{y}_1 \gamma s$.

Of course, condition (2) for the y-sets $\{y_1, y_2\}$ and $\{y_1, y_3\}$ requires only that the terms s_1, s_2, and s_3 in the definition of γ be such that $\mathfrak{p}_\mathfrak{A}(s_1, s_2) = \mathfrak{p}_\mathfrak{A}(\gamma^*(\hat{y}_1 t, \hat{y}_2 t))$ and $\mathfrak{p}_\mathfrak{A}(s_1, s_3) = \mathfrak{p}_\mathfrak{A}(\gamma^*(\hat{y}_1 t, \hat{y}_3 t))$. Nevertheless there is still a difficulty: if $\gamma^*(\hat{y}_1 t, \hat{y}_2 t)$ and $\gamma^*(\hat{y}_1 t, \hat{y}_3 t)$ do not share the same first member, there is no assurance that the requisite terms s_1, s_2, and s_3 exist.

Let us now consider the case in which $b(F)$ contains two disjoint y-sets of cardinality >1, say $\{y_1, y_2\}$ and $\{y_3, y_4\}$. In the amenability proof for Class 2.4 the term-mapping γ preserves identity of arguments. But given this property of γ, disjoint y-sets raise the same difficulty as overlapping ones. For if $\hat{y}_1 t = \hat{y}_3 t$, then $\hat{y}_1 \gamma t$ and $\hat{y}_3 \gamma t$ must be identical. Hence to define γt we must find terms s_1, s_2, s_3, and s_4 such that $s_1 = s_3$, $\mathfrak{p}_\mathfrak{A}(s_1, s_2) = \mathfrak{p}_\mathfrak{A}(\gamma^*(\hat{y}_1 t, \hat{y}_2 t))$, and $\mathfrak{p}_\mathfrak{A}(s_3, s_4) = \mathfrak{p}_\mathfrak{A}(\gamma^*(\hat{y}_3 t, \hat{y}_4 t))$. But again, such terms need not exist if $\gamma^*(\hat{y}_1 t, \hat{y}_2 t)$ and $\gamma^*(\hat{y}_3 t, \hat{y}_4 t)$ do not share the same first member.

ISBN 0-201-02540-X

Thus to extend Class 2.4 we explore weakening the requirement that γ preserve identity of arguments. We have already observed that our interest in this requirement stems from condition (2_x). If γ preserves identity of arguments, then for all x-focused sequences $\langle v_1,\ldots,v_p\rangle$ and $\langle w_1,\ldots,w_p\rangle$ and each x_n-term t, $\langle \hat{v}_1\gamma t,\ldots,\hat{v}_p\gamma t\rangle = \langle \hat{w}_1\gamma t,\ldots,\hat{w}_p\gamma t\rangle$ whenever $\langle \hat{v}_1 t,\ldots,\hat{v}_p t\rangle = \langle \hat{w}_1 t,\ldots,\hat{w}_p t\rangle$. Hence the identity preservation requirement assures the existence of a sequence-mapping γ^* that takes t-instances to γt-instances. But such a sequence-mapping γ^* is stronger than is necessary: it fulfills condition (2_x) for *all* x-focused sets $\{v_1,\ldots,v_p\}$, whereas condition (2) of the Adequacy Lemma will be forthcoming provided that γ^* fulfills condition (2_x) merely for all x-focused sets in the basis. In fact, our argument for Class 2.4 proceeded as though every x-focused set of variables of F were in $b(F)$. The unsolvability result cited on page 54 indicates that we cannot continue to proceed in this manner. We must take a closer look at the x-focused sets that are in the basis. Examination of these x-focused sets will allow us to put a more limited identity preservation requirement on the term-mapping γ. To do this we use the following: the t-instances of two x-focused sequences $\langle v_1,\ldots,v_p\rangle$ and $\langle w_1,\ldots,w_p\rangle$ are identical only if the two sequences contain the same x-variables. This is implicitly shown by case (a) of the argument that shows the definition of γ^* in the proof for Class 2.4 to be unambiguous (page 53). (Equivalently: two x-focused atomic subformulas of a Skolem schema are simultaneously coinstantiable only if they contain the same x-variables.)

Definition 2.10. Two y-variables y_i and y_j of a Skolem schema F are *tied* iff there are (not necessarily distinct) x-focused sets V and W in $b(F)$ such that $y_i \in V$, $y_j \in W$, and V and W contain the same x-variables. Two y-variables are *hereditarily tied* iff they are identical or they stand in the ancestral relation (transitive closure) of tied. (See page 58 for an example.)

We shall require of our term-mapping γ that $\hat{y}_i\gamma t = \hat{y}_j\gamma t$ whenever $\hat{y}_i t = \hat{y}_j t$ and y_i and y_j are hereditarily tied. Amenability is then obtainable even if $b(F)$ contains disjoint y-sets of cardinality >1, provided that hereditarily tied y-variables do not compromise this disjointness.

Class 2.6. Let $F = \forall y_1\cdots\forall y_m\exists x_1\cdots\exists x_n F^M$ be a Skolem schema such that, for all distinct y-sets Y and Y' in $b(F)$ of cardinality >1, no member of Y is hereditarily tied to any member of Y'. Then $E(F, m!d(F, m))$ is adequate.

Proof. Note that all distinct y-sets in $b(F)$ of cardinality >1 are disjoint, since each y_i is hereditarily tied to itself.

Let $\mathfrak{A}$ be any truth-assignment on $\tilde{L}(F, \omega)$. For each term t of height > 1, let $\iota(t) = \{\langle i, j\rangle \,|\, \hat{y}_i t = \hat{y}_j t\}$. Then there is a number r,

ISBN 0-201-02540-X

$2 \leq r \leq m! d(F, m) + 1$, and a pigeonhole-mapping π such that for each x_n-term t in $D(F, r + 1)$,

(π-1) πt is an x_n-term in $D(F, r)$;
(π-2) $\iota(\pi t) = \iota(t)$;
(π-3) $\mathfrak{p}_{\mathfrak{A}}(\hat{y}_1 \pi t, \ldots, \hat{y}_m \pi t) = \mathfrak{p}_{\mathfrak{A}}(\hat{y}_1 t, \ldots, \hat{y}_m t)$.

We begin our inductive construction by letting γ and γ^* be the identity on terms and sequences from $D(F, 1)$. Conditions (1)–(5) are then fulfilled on this domain. Assume γ and γ^* fulfill (1)–(5) on $D(F, q)$, $q \geq 1$. We extend γ to $D(F, q + 1)$ as follows. Let t be an x_k-term of height $q + 1$, $1 \leq k \leq n$; then let

$$\gamma t = \hat{x}_k \pi f_{x_n}(s_1 \cdots s_m)$$

where $\langle s_1, \ldots, s_m \rangle$ is determined from $\langle \hat{y}_1 t, \ldots, \hat{y}_m t \rangle$ thus:

(γ-1)   for each y-set $Y = \{y_{i_1}, \ldots, y_{i_p}\}$ in $b(F)$ of cardinality >1, where $i_1 < \cdots < i_p$, let $\langle s_{i_1}, \ldots, s_{i_p} \rangle = \gamma^*(\hat{y}_{i_1} t, \ldots, \hat{y}_{i_p} t)$;
(γ-2)   for each j such that s_j is not determined by (γ-1), if there is an i such that s_i is so determined, $\hat{y}_j t = \hat{y}_i t$, and y_j and y_i are hereditarily tied, then let $s_j = s_i$ for the least such i;
(γ-3)   for each j such that s_j is determined by neither (γ-1) nor (γ-2), let $s_j = \gamma \hat{y}_j t$.

No ambiguity arises from clause (γ-1), since distinct y-sets in $b(F)$ of cardinality >1 are disjoint. From property (π-1) of the pigeonhole-mapping π and the induction hypotheses that γ and γ^* fulfill (1) and (5) on $D(F, q)$, it follows immediately that γt is in $D(F, r)$. Hence condition (1) holds on $D(F, q + 1)$. Property (π-3) and clause (γ-1) immediately yield condition (2) for the y-sets of cardinality >1 in $b(F)$. Condition (2) for the y-sets $\{y_i\}$ in $b(F)$ will be forthcoming provided that γ possesses the monadic property: for each i, $1 \leq i \leq m$, $\mathfrak{p}_{\mathfrak{A}}(\gamma \hat{y}_i t) = \mathfrak{p}_{\mathfrak{A}}(\hat{y}_i \gamma t)$. But just as in the proof for Class 2.4, the monadic property follows from property (π-3) and the induction hypothesis (4). Note that again we have $\gamma \hat{v} t = \hat{v} \gamma t$ for each x-variable v.

We must show that γ fulfills the *weakened identity preservation require-ment*: if y_i and y_j are hereditarily tied and $\hat{y}_i t = \hat{y}_j t$, then $\hat{y}_i \gamma t = \hat{y}_j \gamma t$. By property ($\pi$-2), it is enough to show that $s_i = s_j$. Suppose first that s_i and s_j are both determined by (γ-1). By the specification of Class 2.6, y_i and y_j must be in the same y-set $\{y_{i_1}, \ldots, y_{i_p}\}$. Hence $s_i = s_j$, since γ^* fulfills the permutation condition (3) on $D(F, q)$. Suppose second that one of s_i and s_j, say s_i, is determined by (γ-2). By the transitivity of the relation here-ditarily tied, s_j must be determined by (γ-1) or by (γ-2). If by (γ-1), then

ISBN 0-201-02540-X

the outcome of the first supposition yields $s_i = s_j$. If by (γ-2), then $s_i = s_j$ immediately. Suppose finally that s_i and s_j are both determined by (γ-3). Then $s_i = \gamma\hat{y}_i t = \gamma\hat{y}_j t = s_j$.

We shall now extend the sequence-mapping γ^* to sequences from $D(F, q + 1)$, and verify the remaining conditions. Because γ^* is to fulfill the permutation condition (3), we phrase the definition of γ^* in terms of x-focused sets rather than x-focused sequences.‡ Recall that if V is a set of variables of F and t is a term $\neq \text{⊥}$, then $\hat{V}[t] = \{\hat{v}t \mid v \in V\}$. Note that $\hat{V}[t] = \hat{W}[s]$ iff $\langle \hat{v}_1 t,\ldots,\hat{v}_p t\rangle = \langle \hat{w}_1 s,\ldots,\hat{w}_p s\rangle$ for some arrangements $\langle v_1,\ldots,v_p\rangle$ of V and $\langle w_1,\ldots,w_p\rangle$ of W. Hence if t and s are x_n-terms and V and W are x-focused sets, then $\hat{V}[t] = \hat{W}[s]$ only if $t = s$.

Suppose $\langle t_1,\ldots,t_p\rangle$ is a sequence of terms from $D(F, q + 1)$ containing at least one term of height $q + 1$.

(γ^*-1) If $\{t_1,\ldots,t_p\} = \hat{V}[t]$ for some x_n-term t and some x-focused set V in $b(F)$, let $\gamma^*(t_1,\ldots,t_p) = \langle \hat{v}_1\gamma t,\ldots,\hat{v}_p\gamma t\rangle$, where, for each i, v_i is a member of V and $t_i = \hat{v}_i t$.

(γ^*-2) Otherwise, let $\gamma^*(t_1,\ldots,t_p) = \langle \gamma t_1,\ldots,\gamma t_p\rangle$.

To show that no ambiguity arises from (γ^*-1) it is enough to show the following condition: Suppose $\hat{V}[t] = \hat{W}[t]$ for some (not necessarily distinct) x-focused sets V and W in $b(F)$ and some x_n-term t of height $q + 1$; if $v \in V$, $w \in W$, and $\hat{v}t = \hat{w}t$, then $\hat{v}\gamma t = \hat{w}\gamma t$. Now V and W contain the same x-variables, namely, $\chi(t_i)$ for each i such that $h(t_i) = q + 1$. Hence if $\hat{v}t = \hat{w}t$, then either $v = w = $ some x-variable and $\hat{v}\gamma t = \hat{w}\gamma t$ trivially, or else v and w are both y-variables. But then v and w are tied, so that by weakened preservation of identity, $\hat{v}\gamma t = \hat{w}\gamma t$. (Note that the foregoing condition not only is sufficient but also is necessary to preclude ambiguity.)

Thus there is a sequence-mapping γ^* defined by (γ^*-1) and (γ^*-2); and γ^* fulfills condition (2_x) for x-focused sets in $b(F)$. Moreover, γ^* fulfills the boundedness condition (5), since γ fulfills (1). The permutation condition (3) is also satisfied. For whether $\gamma^*(t_1,\ldots,t_p)$ is determined by (γ^*-1) depends only on the set $\{t_1,\ldots,t_p\}$, and hence is not affected by permuting $\langle t_1,\ldots,t_p\rangle$ or by adding or deleting multiple occurrences of one or another term.

Only condition (4) remains. But just as in the proof for Class 2.4, this condition follows trivially if $\gamma^*(t_1,\ldots,t_p)$ is determined by (γ^*-2); and fol-

---

‡ A sequence $\langle t_1,\ldots,t_p\rangle$ might not be the t-instance of $\langle v_1,\ldots,v_p\rangle$ for any arrangement $\langle v_1,\ldots,v_p\rangle$ of an x-focused set in $b(F)$, yet some arrangement of the set $\{t_1,\ldots,t_p\}$ may be a t-instance of some such $\langle w_1,\ldots,w_q\rangle$. For example, if $b(F)$ contains $\{y_1, y_2, x_1\}$ but not $\{y_1, x_1\}$ or $\{y_2, x_1\}$, and if $\hat{y}_1 t = \hat{y}_2 t$, then $\langle \hat{y}_1 t, \hat{y}_1 t, \hat{x}_1 t\rangle$ is the t-instance of $\langle y_1, y_2, x_1\rangle$ yet $\langle \hat{y}_1 t, \hat{x}_1 t\rangle$ is not the t-instance of any arrangement of an x-focused set in $b(F)$.

ISBN 0-201-02540-X

lows from the monadic property of γ and the fact that $\hat{x}_k\gamma t = \gamma\hat{x}_k t$ for each k if $\gamma^*(t_1,\ldots,t_p)$ is determined by (γ^*-1).

Thus Class 2.6 is shown amenable. $\square$

Class 2.6, unlike Class 2.4, is essentially closed under conjunction and disjunction. If F and G are in Class 2.6, then so are all prenex forms of $F \wedge G$ and $F \vee G$ obtained by prenexing operations that give preference to universal quantifiers over existential quantifiers.

To illustrate the extent of Class 2.6 let us consider Skolem schemata F whose bases contain only one-membered and two-membered sets. (This restriction is similar to, but more liberal than, the restriction to monadic and dyadic predicate letters.) For each such F, two y-variables y_i and y_j are tied iff there is an x-variable x_k such that $\{y_i, x_k\}$ and $\{y_j, x_k\}$ are both in $b(F)$. Hence, for example, if $b(F)$ contains just one-membered sets and $\{y_1, x_1\}$, $\{y_2, x_1\}$, $\{y_3, x_2\}$, $\{y_3, x_3\}$, $\{y_4, x_3\}$, $\{y_5, x_3\}$, $\{y_1, y_2\}$, and $\{y_3, y_4\}$, then F is in Class 2.6.

The two-membered sets in a basis $b(F)$ may be vividly represented by a graph. A small circle represents a variable, with solid black circles for x-variables and unfilled circles for y-variables; an edge connecting two circles represents a two-membered set in $b(F)$. Thus the basis $b(F)$ given in the example of the previous paragraph is represented in Figure 2.1.

Figure 2.1

Two y-variables y_i and y_j are hereditarily tied iff in the graph of $b(F)$ there is a path between y_i and y_j along which y-variables and x-variables strictly alternate. Hence, if the graph of $b(F)$ is as in Figure 2.2, then no y-variables that lie in distinct y-sets of cardinality >1 are hereditarily tied, and F lies in Class 2.6.

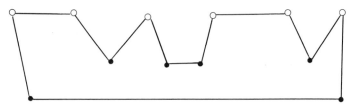

Figure 2.2

ISBN 0-201-02540-X

By way of contrast, we note that the class of Skolem schemata whose bases can be graphed as in Figure 2.3 is unsolvable. The crucial factor is that here, unlike the case in Figure 2.2, y-variables that lie in different y-sets are tied. Tied y-variables can be exploited to prove the unsolvability of this class. We shall elaborate on these points in Chapter 6, §2.

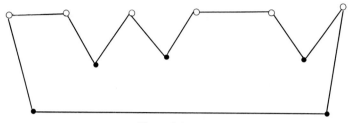

Figure 2.3

The notion of tying was prompted by condition (2_x); more precisely, by the need to avoid ambiguities in (2_x) that could arise when the t-instances of two x-focused sequences coincide. We saw in the amenability proof for Class 2.6 that no ambiguity arises if and only if the term-mapping γ possesses the following property: for all y-variables y_i and y_j and all x_n-terms t,

(#)   if $\hat{y}_i t = \hat{y}_j t$ and there are x-focused sets V and W in $b(F)$ such that $y_i \in V$, $y_j \in W$, and $\hat{V}[t] = \hat{W}[t]$, then $\hat{y}_i \gamma t = \hat{y}_j \gamma t$.

Now the antecedent of (#) holds only if y_i and y_j are tied; hence the weakened identity preservation requirement suffices to preclude ambiguity. Indeed, if $b(F)$ contains only one- and two-membered sets, then the weakened identity preservation requirement is also necessary. For in this case if y_i and y_j are tied, then there is an x-variable x_k such that $\{y_i, x_k\}$ and $\{y_j, x_k\}$ are in $b(F)$. If moreover t is an x_n-term such that $\hat{y}_i t = \hat{y}_j t$, then $\{\hat{y}_i t, \hat{x}_k t\} = \{\hat{y}_j t, \hat{x}_k t\}$. Thus the antecedent of (#) holds, and we must have $\hat{y}_i \gamma t = \hat{y}_j \gamma t$.

But once x-focused sets of cardinality >2 are allowed, the weakened identity preservation requirement is no longer necessary. For example, suppose the x-focused sets in $b(F)$ are just $\{y_1, y_2, x_1\}$ and $\{y_3, y_4, x_1\}$. Then y_i and y_j are tied for all i and j, $1 \le i, j \le 4$. Now suppose t is an x_n-term such that $\hat{y}_1 t = \hat{y}_3 t$. It does not follow that $\{\hat{y}_1 t, \hat{y}_2 t, \hat{x}_1 t\} = \{\hat{y}_3 t, \hat{y}_4 t, \hat{x}_1 t\}$, for we must also have $\hat{y}_2 t = \hat{y}_4 t$. Hence a term-mapping γ suitable for use in an amenability proof need not necessarily preserve the identity of the first and third arguments of t; rather, we need require that $\hat{y}_1 \gamma t = \hat{y}_3 \gamma t$ only when both $\hat{y}_1 t = \hat{y}_3 t$ and $\hat{y}_2 t = \hat{y}_4 t$. Consequently, we shall be able to extend Class 2.6.

Definition 2.11. For each x_n-term t, two y-variables y_i and y_j of a

ISBN 0-201-02540-X

Skolem schema F are *t-united* iff $\hat{y}_i t = \hat{y}_j t$ and there are (not necessarily distinct) *x*-focused sets V and W in $b(F)$ such that $y_i \in V$, $y_j \in W$, and $\hat{V}[t] = \hat{W}[t]$. For each x_n-term t, two *y*-variables are *hereditarily t-united* iff either they are identical or else they stand in the ancestral relation of *t*-united.

We have just noted that, even if two *y*-variables y_i and y_j are tied and $\hat{y}_i t = \hat{y}_j t$, they may not be *t*-united: for them to be *t*-united, other identities among arguments of t may have to hold.

EXAMPLE. Suppose $b(F)$ contains only three-membered sets and t is an x_n-term. Then y_i and y_j are *t*-united just in case $\hat{y}_i t = \hat{y}_j t$ and either

(a)   there are *x*-variables x_p and x_q such that both $\{y_i,\, x_p,\, x_q\}$ and $\{y_j,\, x_p,\, x_q\}$ are in $b(F)$; or

(b)   there are *y*-variables y_k and y_l and an *x*-variable x_p such that $\hat{y}_k t = \hat{y}_l t$ and both $\{y_i,\, y_k,\, x_p\}$ and $\{y_j,\, y_l,\, x_p\}$ are in $b(F)$.

Given a Skolem schema F containing m *y*-variables, we can clearly tell, for each x_n-term t and $1 \le i, j \le m$, whether y_i and y_j are *t*-united. Indeed, whether or not they are depends only on the set $\iota(t) = \{\langle i, j\rangle \mid \hat{y}_i t = \hat{y}_j t\}$. For y_i and y_j are *t*-united just in case there are *x*-focused sets V and W in $b(F)$ with the following properties: $y_i \in V$, $y_j \in W$, and identical sets result from V and from W by replacing each *y*-variable y_p, $1 \le p \le m$, by the *y*-variable $y_{\sigma(p)}$, where $\sigma(p)$ is the least number such that $\langle p, \sigma(p)\rangle \in \iota(t)$.‡ This argument also shows that we can effectively decide, given any dyadic relation on $\{y_1, \ldots, y_m\}$, whether there is an x_n-term t such that the given relation is the relation of *t*-unitedness.

Two *y*-variables y_i and y_j are *t*-united only if $\hat{y}_i t = \hat{y}_j t$ and y_i and y_j are tied; the "only if" becomes "iff" when $b(F)$ contains just one- and two-membered sets. Class 2.7, specified three paragraphs below, is thus a proper extension of Class 2.6, but contains no new schemata whose bases contain just one- and two-membered sets. We give examples of schemata that belong to Class 2.7 but not to Class 2.6 on page 63.

The amenability proof for Class 2.7 will turn on an identity preservation requirement on γ yet weaker than that used for Class 2.6, namely, that $\hat{y}_i \gamma t = \hat{y}_j \gamma t$ whenever y_i and y_j are *t*-united (equivalently, that $\hat{y}_i \gamma t = \hat{y}_j \gamma t$ whenever y_i and y_j are hereditarily *t*-united). As we have seen, condition (2_x) leads to no ambiguities if and only if this weaker requirement is met. (Thus no further weakening of the identity preservation requirement will do.)

---

‡ We note a consequence of this that will be useful later: if s and t are x_n-terms such that $\iota(s) = \{\langle i, j\rangle \mid y_i \text{ and } y_j \text{ are hereditarily } t\text{-united}\}$, then s-unitedness coincides with t-unitedness.

ISBN 0-201-02540-X

Indeed, this weakest identity preservation requirement on γ follows directly from our basic strategy, which demands that the instance-mapping Ψ preserve simultaneous coinstantiations of x-focused atomic subformulas (see page 51). Since in this chapter we characterize schemata just by prefix and basis, we must allow for the possibility that any given schema F contains arbitrary V-based atomic subformulas for each V in $b(F)$. Now assume that y_i and y_j are t-united. Then there may be x-focused atomic subformulas A_1 and A_2 of F such that y_i and y_j occur homologously in A_1 and A_2, respectively, and A_1 and A_2 are coinstantiated in $\langle H, H \rangle$, where H is the Herbrand instance in which t is the substituent for x_n. And if A_1 and A_2 are to be coinstantiated in $\langle \Psi(H), \Psi(H) \rangle$, we must have $\hat{y}_i \gamma t = \hat{y}_j \gamma t$.

Class 2.7. Let $F = \forall y_1 \cdots \forall y_m \exists x_1 \cdots \exists x_n F^M$ be a Skolem schema having the following property: for each x_n-term t in $D(F, \omega)$ and all y-sets Y and Y' in $b(F)$, if $\hat{Y}[t]$ and $\hat{Y}'[t]$ are distinct sets of terms of cardinality >1, then no member of Y is hereditarily t-united to any member of Y'. Then $E(F, m!d(F, m))$ is adequate.

Proof. We first note that no distinct y-sets of cardinality >1 in $b(F)$ overlap. For let t be an x_n-term all of whose arguments are distinct. If Y and Y' were distinct but nondisjoint y-sets of cardinality >1 in $b(F)$, then $\hat{Y}[t]$ and $\hat{Y}'[t]$ would be distinct sets of cardinality >1; and so, since each y-variable is hereditarily t-united to itself, any member of $Y \cap Y'$ would violate the specification of Class 2.7.

Since our adequacy proof closely parallels the previous two we shall merely sketch it. Let $\mathfrak{A}$ be any truth-assignment on $\hat{L}(F, \omega)$. We choose a number r, $2 \le r \le m!d(F, m) + 1$, and a pigeonhole-mapping π that possess properties $(\pi\text{-}1)$, $(\pi\text{-}2)$, and $(\pi\text{-}3)$ of the previous proofs (pages 51 and 56). To begin our inductive construction we let γ and γ^* be the identity on terms and sequences from $D(F, 1)$. Assume γ and γ^* fulfill conditions (1)–(5) on $D(F, q)$, $q \ge 1$. We extend γ to $D(F, q + 1)$ as follows. Let t be an x_k-term of height $q + 1$, $1 \le k \le n$; then let

$$\gamma t = \hat{x}_k \pi f_{x_n}(s_1 \cdots s_m)$$

where $\langle s_1, \ldots, s_m \rangle$ is determined from $\langle \hat{y}_1 t, \ldots, \hat{y}_m t \rangle$ thus:

$(\gamma\text{-}1)$ for each y-set $Y = \{y_{i_1}, \ldots, y_{i_p}\}$ in $b(F)$ such that $|\hat{Y}[t]| > 1$, where $i_1 < \cdots < i_p$, let $\langle s_{i_1}, \ldots, s_{i_p} \rangle = \gamma^*(\hat{y}_{i_1} t, \ldots, \hat{y}_{i_p} t)$;

$(\gamma\text{-}2)$ for each j such that s_j is not determined by $(\gamma\text{-}1)$, if there is an i such that s_i is so determined and y_j and y_i are hereditarily t-united, then let $s_j = s_i$ for the least such i;

$(\gamma\text{-}3)$ for each j such that s_j is determined by neither $(\gamma\text{-}1)$ nor $(\gamma\text{-}2)$, let $s_j = \gamma \hat{y}_j t$.

ISBN 0-201-02540-X

Note that (γ-1) and (γ-3) unambiguously determine $\langle s_1,\ldots,s_m\rangle$. In particular, no ambiguity arises from (γ-1), for as we have seen, no distinct y-sets of cardinality >1 in $b(F)$ overlap.

Just as in the case of Class 2.6, it follows from the induction hypotheses (1) and (5) and property (π-1) that γt is in $D(F, r)$; hence condition (1) is fulfilled on $D(F, q + 1)$. We now show that if y_i and y_j are hereditarily t-united, then $\hat{y}_i\gamma t = \hat{y}_j\gamma t$. By property ($\pi$-2) it is enough to show that if y_i and y_j are hereditarily t-united, then $s_i = s_j$. If one of s_i and s_j is determined by clause (γ-3), so is the other, and the result is immediate. If neither is determined by (γ-3), since hereditary t-unitedness is transitive, the result will follow for all i and j provided it holds for those i and j such that s_i and s_j are both determined by clause (γ-1). In this case there are y-sets Y_1 and Y_2 in $b(F)$ such that $\hat{Y}_1[t]$ and $\hat{Y}_2[t]$ are each of cardinality >1, $y_i \in Y_1$ and $y_j \in Y_2$. But then, by the specification of Class 2.7, $\hat{Y}_1[t] = \hat{Y}_2[t]$. Hence $s_i = s_j$, since γ^* fulfills the permutation condition (3) on $D(F, q)$.

Moreover, γ fulfills condition (2) for each y-set in $b(F)$. If $\hat{Y}[t]$ has cardinality >1, then this condition is, as before, immediate from (π-3) and (γ-1). If $|Y| = 1$, the condition again follows provided that γ possesses the monadic property: $\mathfrak{p}_{\mathfrak{A}}(\gamma\hat{y}_i t) = \mathfrak{p}_{\mathfrak{A}}(\hat{y}_i\gamma t)$, $1 \le i \le m$. And γ does possess the monadic property, by property (π-3) and induction hypothesis (4). What remains to be considered, however, is the case that does not arise in the argument for Class 2.6, namely, the case of y-sets $Y = \{v_1,\ldots,v_p\}$ that contain more than one variable but where $\hat{v}_1 t = \hat{v}_2 t = \cdots \hat{v}_p t$. In this case, by induction hypothesis (3), $\gamma^*(\hat{v}_1 t, \hat{v}_2 t,\ldots,\hat{v}_p t) = \langle u, u,\ldots,u\rangle$, where $\langle u\rangle = \gamma^*(\hat{v}_1 t)$. By induction hypothesis (4) and the monadic property of γ, $\mathfrak{p}_{\mathfrak{A}}(u) = \mathfrak{p}_{\mathfrak{A}}(\gamma\hat{v}_1 t) = \mathfrak{p}_{\mathfrak{A}}(\hat{v}_1\gamma t)$. Hence it suffices to show that $\langle\hat{v}_1\gamma t, \hat{v}_2\gamma t,\ldots,\hat{v}_p\gamma t\rangle = \langle\hat{v}_1\gamma t, \hat{v}_1\gamma t,\ldots,\hat{v}_1\gamma t\rangle$, that is, $\hat{v}_1\gamma t = \hat{v}_2\gamma t = \cdots \hat{v}_p\gamma t$. So assume then that $\hat{v}_i\gamma t \ne \hat{v}_j\gamma t$ for some $i, j \le p$. But then v_i and v_j are not hereditarily t-united; and, given (γ-1), (γ-2), and (γ-3), plus inductive hypothesis (3), at least one of v_i and v_j, say v_i, is hereditarily t-united with a member of a y-set Y' in $b(F)$ such that $|\hat{Y}'[t]| > 1$. We show that this contradicts the specification of Class 2.7. For let s be any x_n-term such that, for all k and $l \le m$, $\hat{y}_k s = \hat{y}_l s$ if *and only if* y_k and y_l are hereditarily t-united. Hence $|\hat{Y}[s]| > 1$ and $|\hat{Y}'[s]| > 1$. Moreover, s-unitedness coincides with t-unitedness (see the footnote on page 60). Hence v_i, a member of Y, is hereditarily s-united with a member of Y', which is impossible.

We now wish to so extend γ^* to sequences from $D(F, q + 1)$ containing at least one term of height $q + 1$ that conditions (3), (4), and (5), and condition (2_x) for x-focused sets in $b(F)$ will be fulfilled on $D(F, q + 1)$. We do this in the same way as in the proof for Class 2.6: the same clauses (γ^*-1) and (γ^*-2) will work (page 57). We need only show that this

ISBN 0-201-02540-X

definition is correct, that no ambiguity arises from (γ*-1): for then the desired conditions follow exactly as before. Thus we must show that if V and W are x-focused sets in $b(F)$, $\hat{V}[t] = \hat{W}[t]$ for some x_n-term t of height $q + 1$, and $\hat{v}t = \hat{w}t$ for $v \in V$ and $w \in W$, then $\hat{v}\gamma t = \hat{w}\gamma t$. But as before, if $\hat{v}t = \hat{w}t$, either v and w are the same x-variable, and so, trivially, $\hat{v}\gamma t = \hat{w}\gamma t$, or else v and w are both y-variables. But then v and w are by definition t-united, and $\hat{v}\gamma t = \hat{w}\gamma t$, since γ preserves the identity of t-united arguments of x_n-terms t of height $q + 1$. $\square$

Two examples, one particular and one more general, will illustrate the kinds of schemata that are in Class 2.7 but not in Class 2.6.

EXAMPLE 1.   Let F be a Skolem schema whose basis contains, aside from one-membered sets, only $\{y_1, y_2, x_1\}$, $\{y_3, y_4, x_1\}$, $\{y_1, y_2\}$, and $\{y_3, y_4\}$. Suppose t is an x_n-term. If $\hat{y}_1 t = \hat{y}_2 t$, then y_1 and y_2 are t-united. If $\hat{y}_3 t = \hat{y}_4 t$, then y_3 and y_4 are t-united. If $\hat{y}_1 t = \hat{y}_3 t$ and $\hat{y}_2 t = \hat{y}_4 t$, then y_1 and y_3 are t-united and y_2 and y_4 are t-united. Only in the last case are there hereditarily t-united y-variables that lie in distinct y-sets of cardinality >1 in $b(F)$, and in this case, $\{\hat{y}_1 t, \hat{y}_2 t\} = \{\hat{y}_3 t, \hat{y}_4 t\}$. Hence F is in Class 2.7. But F is not in Class 2.6, since y_1 and y_3 are tied, as are y_2 and y_4, yet they lie in distinct y-sets of cardinality >1 in $b(F)$.

EXAMPLE 2.   Let F be a Skolem schema whose basis contains, aside from one-membered sets, just

(1)   the disjoint y-sets $Y_1, \ldots, Y_p$ and the (not necessarily disjoint) sets $X_1, \ldots, X_q$ containing x-variables only;

(2)   sets $Y_i \cup X_j$, $1 \le i \le p$ and $1 \le j \le q$;

(3)   for each $j \le q$ at most one set $\{y_k\} \cup X_j$, where y_k is any y-variable.

Suppose t is an x_n-term, and let V and W be x-focused sets in $b(F)$ such that $\hat{V}[t] = \hat{W}[t]$. Then, for some $j \le q$, either

(a) $V = Y_i \cup X_j$ and $W = Y_{i'} \cup X_j$ for some i, $i' \le p$; hence $\hat{Y}_i[t] = \hat{Y}_{i'}[t]$; or

(b) $V = Y_i \cup X_j$ and $W = \{y_k\} \cup X_j$ for some $i \le p$ and some y-variable y_k; hence $\hat{Y}_i[t] = \{\hat{y}_k t\}$, and $|\hat{Y}_i[t]| = 1$; or

(c) $V = \{y_k\} \cup X_j$ and $W = Y_i \cup X_j$ for some $i \le p$ and some y-variable y_k; again $|\hat{Y}_i[t]| = 1$; or

(d) $V = W = \{y_k\} \cup X_j$ for some y-variable y_k.

ISBN 0-201-02540-X

Since the Y_i are all disjoint, it follows that if $\hat{Y}_i[t]$ and $\hat{Y}_k[t]$ are distinct sets of cardinality >1, then no member of Y_i is hereditarily t-united with any member of Y_k. Hence F is in Class 2.7.   End of Examples.

We may extend Example 2 in several ways. We may allow $b(F)$ to contain sets $\{y_i\} \cup X_j$, $1 \le j \le q$, provided that y_i is a y-variable not in $Y_1 \cup \cdots \cup$

Y_p. We may also allow another set X_{q+1} containing just x-variables and distinct from $X_1,\ldots,X_q$, and for some $i \leq p$ all sets $Z \cup X_{q+1}$, where Z is a subset of $Y_i \cup \{y_{l_1},\ldots,y_{l_r}\}$ and $y_{l_1},\ldots,y_{l_r}$ are not in $Y_1 \cup \cdots \cup Y_p$. Of course, these extension procedures may be iterated. Indeed, in the end we could obtain a characterization of the whole of Class 2.7 phrased explicitly in terms of basis restrictions. (It should be clear that whether a schema is in Class 2.7 depends only on the prefix of F and the basis of F, and on nothing else about F.)

We close this section by stating some general facts about the extent of Class 2.7, facts that we shall have occasion to exploit in §5 and in Chapter 6.

If F is a Skolem schema whose basis is the union of two sets S_1 and S_2 that have no variable in common, then no y-variable that occurs in a member of S_1 can be t-united with a y-variable that occurs in a member of S_2. But then Class 2.7 is closed under conjunction and disjunction: if F and G are in Class 2.7, so are any prenex forms of $F \wedge G$ and $F \vee G$ obtained by the usual prenexing rules, giving preference to y-variables over x-variables.

One-membered sets in the basis do not affect t-unitedness. Hence if F is in Class 2.7 and G is a Skolem schema whose basis differs from $b(F)$ only in containing more one-membered sets, then G is also in Class 2.7.

It should also be clear that if F is in Class 2.7 and if G is a Skolem schema whose basis is a subset of the basis of F, then G is in Class 2.7.

If S is a set of sets of variables and $v_1,\ldots,v_k$ are variables, then the set that results from S by *deleting* $v_1,\ldots,v_k$ is simply the set S' of nonempty sets W such that, for some V in S, $W = V - \{v_1,\ldots,v_k\}$. Let F be in Class 2.7, and let G be a Skolem schema whose basis results from $b(F)$ by deleting some y-variables. Then G is in Class 2.7. We leave the verification of this to the reader.

There is, moreover, a solvable class of Skolem schemata that properly includes Class 2.7. In §1 of Chapter 6 we show the solvability of this broader class (Class 6.1), by reducing it to Class 2.7. We do not know whether Class 6.1 is amenable. In §2 of Chapter 6 we go into further detail about the geography of the demarcation line between solvable and unsolvable Skolem classes specified by basis restrictions.

5. Initial x-Variables

The results of the preceding three sections may be extended to schemata containing *initial x-variables,* that is, x-variables not governed by any y-variables. The substituent for an initial x-variable z_i of F is the same in each Herbrand instance of F: z_i is replaced by the 0-place indicial function sign

ISBN 0-201-02540-X

(or, as we shall say, the indicial constant) c_{z_i}. Suppose a schema F contains the initial x-variables $z_1,\ldots,z_l$. The *pruned basis* $b^p(F)$ of F is just the set of sets of variables that results from the basis $b(F)$ of F by deleting $z_1,\ldots,z_l$. "Initially-extended" versions of each of the classes shown amenable in previous sections can be specified by stipulating that the pruned basis of a schema in the extended class must be the basis of some schema in the unextended class.

For example, we have the *Initially-extended Essentially Monadic Class*: the class of schemata F such that each atomic subformula of F contains at most one variable aside from initial x-variables (or, equivalently, such that $b^p(F)$ contains one-membered sets only). If F is such a schema, and F contains the initial x-variables $z_1,\ldots,z_l$, then $E(F, d(F, l + 1))$ is easily shown adequate. All we need require is the existence, for each truth-assignment $\mathfrak{A}$, of a term-mapping γ such that

(i) $\gamma[D(F, \omega)] \subseteq D(F, d(F, l + 1))$;

(ii) $\gamma c_{z_i} = c_{z_i}$ for each i, $1 \le i \le l$;

(iii) $\mathfrak{p}_{\mathfrak{A}}(c_{z_1},\ldots,c_{z_l}, \gamma f(t_1\cdots t_m)) = \mathfrak{p}_{\mathfrak{A}}(c_{z_1},\ldots,c_{z_l}, f(\gamma t_1 \cdots \gamma t_m))$ for each term $f(t_1\cdots t_m)$ in $D(F, \omega)$.

But such a term-mapping can be obtained in a familiar way by means of the pigeonhole argument. We then let Γ be the γ-canonical formula-mapping and Ψ be the instance-mapping that takes $F^*(y_1/t_1,\ldots,y_m/t_m)$ to $F^*(y_1/\gamma t_1,\ldots,y_m/\gamma t_m)$, where $y_1,\ldots,y_m$ are all the y-variables of F. It quickly follows that Γ and Ψ fulfill the boundedness and central conditions.

The initially-extended versions of Classes 2.2–2.7 contain just *initially-extended Skolem schemata*, that is, prenex schemata with prefixes $\exists z_1\cdots\exists z_l\forall y_1\cdots\forall y_m\exists x_1\cdots\exists x_n$. Amenability proofs for each of these extended classes can be obtained by systematically modifying our proofs for the unextended classes. We would require that γ carry each indicial constant to itself, and that γ^* preserve occurrences of indicial constants in sequences. Each use of a profile $\mathfrak{p}_{\mathfrak{A}}(t_1,\ldots,t_k)$ would be replaced by the use of the profile $\mathfrak{p}_{\mathfrak{A}}(c_{z_1},\ldots,c_{z_l}, t_1,\ldots,t_k)$, where $z_1,\ldots,z_l$ are all the initial x-variables of F. Finally, for Classes 2.3–2.7, we would add the requirement on γ that if $\hat{y}_i t$ is an indicial constant, then $\hat{y}_i \gamma t = \hat{y}_i t$. This necessitates a similar requirement on the pigeonhole-mapping π. The stronger properties that π must possess force up the bound needed for adequacy. Thus, for example, in this fashion we would obtain

Initially-extended Class 2.5. Let F be a prenex schema $\exists z_1\cdots\exists z_l\forall y_1\forall y_2\exists x_1\cdots\exists x_n F^M$. Then $E(F, (l + 1)^2 d(F, l + 2))$ is adequate.

REMARK. Initially-extended Class 2.5 and the Bernays–Schönfinkel Class (the prenex subclass of Class 1.1) jointly exhaust the solvable classes

ISBN 0-201-02540-X

specified by quantifier prefix. That is, a class of prenex formulas specified by prefix restrictions alone is solvable if and only if every member of the class is in either Class 1.1 or Initially-extended Class 2.5 (see the Introduction).   End of Remark.

We shall not, however, furnish amenability proofs for our initially-extended classes; the direct method for doing this just outlined involves numerous messy details, particularly in treating Class 2.7. Rather, we prove solvability by use of a reduction method that we give in Chapter 6, §4. This method comes from Kalmár: in 1933 he showed how to reduce any initially-extended Skolem schema F to an unextended Skolem schema G that contains the same number of y-variables as does F. Kalmár then used this reduction to infer the solvability of Initially-extended Class 2.5 from that of unextended Class 2.5.

A slight emendation of Kalmár's method is required for Initially-extended Class 2.7 (which subsumes the extended versions of each of Classes 2.2–2.6). In Chapter 6 we shall reduce any given initially-extended Skolem schema F to a conjunction $G_1 \wedge \cdots \wedge G_k$ of unextended Skolem schemata with the following property: for each i there is a set Y_i of y-variables of F such that the basis $b(G_i)$ results from the pruned basis $b^p(F)$ by deleting all members of Y_i.

Now suppose F is in Initially-extended Class 2.7. That is, $F = \exists z_1 \cdots \exists z_l \forall y_1 \cdots \forall y_m \exists x_1 \cdots \exists x_n F^M$, and (unextended) Class 2.7 contains at least one—and hence every—Skolem schema whose prefix is $\forall y_1 \cdots \forall y_m \exists x_1 \cdots \exists x_n$ and whose basis is $b^p(F)$, the pruned basis of F. Our reduction yields a conjunction $G_1 \wedge \cdots \wedge G_k$ of Skolem schemata; each basis $b(G_i)$ results from $b^p(F)$ by deleting some y-variables. Hence, as we said at the end of §4, each schema G_i is in Class 2.7. Since, as we also said in §4, Class 2.7 is essentially closed under conjunction, our method reduces Initially-extended Class 2.7 to unextended Class 2.7. It follows that Initially-extended Class 2.7 is solvable.

Moreover, the reduction is conservative: F has a finite model iff $G_1 \wedge \cdots \wedge G_k$ has a finite model. Hence the finite controllability of Initially-extended Class 2.7 can be inferred from that of unextended Class 2.7. (We prove the finite controllability of unextended Class 2.7 in §4 of Chapter 3.)

In conclusion we note that some classes do pass from solvable to unsolvable when initial x-variables are allowed. These solvable classes are all nonamenable (see Chapter 5, §2).

6. Arbitrary Prefixes

The relation t-united is defined just for Skolem schemata, but an analogue can be found for prenex schemata with arbitrary prefixes. We shall give the

ISBN 0-201-02540-X

analogue, specify an extension of Class 2.7 in terms of it, and then show the amenability of this Class 2.8. Unfortunately, the specification of the class is far from perspicuous. But by looking at four unsolvable classes we shall try to make clear what leads to the specification.

The class of prenex schemata with prefixes $\forall y_1 \exists x \forall y_2$ and bases $\{\{y_1, y_2\}, \{x, y_2\}, \{y_1\}\}$ is unsolvable (see page 124). Hence we shall require of each schema F in Class 2.8 that

> (I) every member of $b(F)$ be either an x-focused set or a y-set.

The class of prenex schemata with prefixes $\forall y_1 \forall y_2 \exists x_1 \forall y_3 \exists x_2$ and bases $\{\{y_1, y_2, y_3\}, \{x_1\}, \{y_3, x_2\}\}$ is unsolvable (this is Class A of Chapter 6, §3). Yet the basis of each member F of this class falls under the description of the basis of each member of amenable Class 2.4: $b(F)$ contains only x-focused sets and one y-set of cardinality >1 (it does not even contain one-membered y-sets). But there is a difference. The y-set $\{y_1, y_2, y_3\}$ straddles the x-variable x_1. Obviously, no y-set in the basis of any member of Class 2.4 straddles an x-variable, since x-variables in each member of Class 2.4—indeed, in every Skolem schema—are *final*, that is, govern no y-variable. Hence we shall require of each schema F in Class 2.8 that

> (II) no y-set in $b(F)$ contain both a member that governs an x-variable and a member governed by that x-variable.

(It is striking that all that prevents Class A from being solvable is the presence of one nonfinal x-variable occurring only in essentially monadic atomic formulas. See the discussion in Chapter 6, §3.)

To continue this array of unsolvable classes and the resulting restrictions on members of Class 2.8, we must find a relation to play the role of t-united. Our task is eased if we assume that every schema under discussion is *normal*, that is, a prenex schema whose prefix begins with a universal quantifier and ends with an existential one. (No generality is thereby lost: cf. the footnote on page 48.) In our syntactic description of a normal schema we take its y-variables to be $y_1,\ldots,y_m$ and its x-variables to be $x_1,\ldots,x_n$, where both lists are in left-to-right order of occurrence in the prefix. We shall also say that a variable v *dominates* a variable w iff v governs w or $v = w$.

Consider first the straightforward generalization of Definition 2.11, the definition of t-united.

Definition 2.12a. Let F be a normal schema, let v and w be any variables of F, and let t be any x_n-term in $D(F, \omega)$. Then v and w are t-x_n-*united* iff $\hat{v}t = \hat{w}t$ and there are x-focused sets V and W in $b(F)$ such that $\hat{V}[t] = \hat{W}[t]$, $v \in V$, and $w \in W$; and v and w are *hereditarily* t-x_n-*united* iff either $v = w$ or v and w stand in the ancestral relation of t-x_n-united.

ISBN 0-201-02540-X

Clearly, if F is Skolem and v and w are y-variables, then the relation t-x_n-united just given is the relation t-united given in Definition 2.11. If, however, F is not Skolem, then v, say, may be a nonfinal x-variable x_l and w a y-variable y_i governed by x_l, since there are x_n-terms t such that $\hat{x}_l t = \hat{y}_i t$. Of course, no x-variable can be t-x_n-united to any variable governing it; in particular, distinct x-variables cannot be t-x_n-united.

Now consider the class of schemata with prefixes $\forall y_1 \forall y_2 \exists x_1 \forall y_3 \forall y_4 \exists x_2$ and bases $\{\{y_3, y_4\}, \{x_1\}, \{y_1, x_2\}, \{y_2, x_2\}, \{y_3, x_2\}, \{y_4, x_2\}\}$ consisting of only x-focused sets and the one y-set $\{y_3, y_4\}$. This class is unsolvable (Class B, Chapter 6, §3) although the y-set $\{y_3, y_4\}$ does not straddle an x-variable. But a counterpart to straddling is present. For $i = 1, 2$ and $j = 3, 4$, the y-variables y_i and y_j are t-x_2-united whenever t is an x_2-term such that $\hat{y}_i t = \hat{y}_j t$. Hence we shall require of each schema F in Class 2.8 that

(III)   for each y-set Y in $b(F)$ and each x_n-term t in $D(F, \omega)$, if $\hat{Y}[t]$ has cardinality > 1, then no member y of Y is hereditarily t-x_n-united to any variable dominating an x-variable that governs y.

Also, just as in the case of Class 2.7—and for the same reasons—we shall require of each schema in Class 2.8 that

(IV)   if Y and Y' are y-sets in $b(F)$, t is an x_n-term in $D(F, \omega)$, and $\hat{Y}[t]$ and $\hat{Y}'[t]$ are distinct sets of terms of cardinality > 1, then no member of Y is hereditarily t-x_n-united to any member of Y'. (Thus no distinct y-sets in $b(F)$ of cardinality > 1 overlap.)

We are still not done with the need to generalize the relation t-united nor with the need for further restrictions. Here is an unsolvable class (Class C, Chapter 6, §3) that fulfills all restrictions hitherto stipulated: the class of schemata with prefixes $\forall y_1 \forall y_2 \cdots \forall y_7 \exists x_1 \forall y_8 \exists x_2$ and bases $\{\{y_1, y_2, y_3\}, \{y_4, y_5\}, \{x_1\}, \{y_1, y_2, x_2\}, \{y_4, y_5, x_1, x_2\}, \{y_6, y_7, y_8, x_2\}\}$. The y-sets $\{y_1, y_2, y_3\}$ and $\{y_4, y_5\}$ are disjoint, straddle no x-variable, and for no x_2-term t is their disjointness or absence of straddling compromised by t-x_2-unitedness. Nevertheless the disjointness of $\{y_1, y_2, y_3\}$ and $\{y_4, y_5\}$ is fatally compromised by certain x_1-terms. Let s be any x_1-term in $D(F, \omega)$ such that $\hat{y}_1 s = \hat{y}_4 s = \hat{y}_6 s$ and $\hat{y}_2 s = \hat{y}_5 s = \hat{y}_7 s$; let t_1 be the x_2-term such that $\hat{x}_1 t_1 = s$ and $\hat{y}_8 t_1 = \hat{y}_1 s$; and let t_2 be the x_2-term such that $\hat{x}_1 t_2 = s$ and $\hat{y}_8 t_2 = s$. Now y_1 and y_6 are t_1-x_2-united, and y_6 and y_4 are t_2-x_2-united: the terms $\hat{y}_1 t_1$, $\hat{y}_6 t_1$, $y_8 t_1$, $\hat{y}_6 t_2$, and $y_4 t_2$ are identical; whence $\{\hat{y}_1 t_1, \hat{y}_2 t_1, \hat{x}_2 t_1\} = \{\hat{y}_6 t_1, \hat{y}_7 t_1, \hat{y}_8 t_1, \hat{x}_2 t_1\}$ and $\{\hat{y}_6 t_2, \hat{y}_7 t_2, \hat{y}_8 t_2, \hat{x}_2 t_2\} = \{\hat{y}_4 t_2, \hat{y}_5 t_2, \hat{x}_1 t_2, \hat{x}_2 t_2\}$. Hence, y_1 and y_4 (and y_2 and y_5) stand in a relation that is a more subtle form of "unitedness" than t-x_2-united for any x_2-term t. And since $t_1 \neq t_2$, this relation is, of course, neither hereditarily t_1-x_2-united nor hereditarily t_2-x_2-united. Rather, y_1 and y_4 stand in the relation

ISBN 0-201-02540-X

hereditarily s-x_1-united, provided we so define s-x_1-unitedness that, for all variables v and w governing x_1, if there is an x_2-term t such that $s = \hat{x}_1 t$ and v and w are hereditarily t-x_2-united, then v and w are s-x_1-united. For under this definition y_1 and y_6 are s-x_1-united, as are y_6 and y_4. Consequently, y_1 and y_4 are hereditarily s-x_1-united.

Similarly, in general, if F is any normal schema and $1 \leq l < n$, we shall define s-x_l-unitedness in terms of hereditary t-x_{l+1}-unitedness. That is, our definition will be inductive.

Definition 2.12b. Let F be a normal schema, let $1 \leq l < n$, and suppose that hereditary t-x_{l+1}-unitedness is defined for each x_{l+1}-term t. Then, for each x_l-term s and all variables v and w of F that govern x_l, v and w are s-x_l-*united* iff there is an x_{l+1}-term t such that $s = \hat{x}_l t$ and v and w are hereditarily s-x_{l+1}-united; and v and w are *hereditarily s-x_l-united* iff $v = w$ or v and w stand in the ancestral relation of s-x_l-united.

An immediate consequence of the definition is this: if t is an x_k-term, $1 < k \leq n$, and if the variables v and w govern x_{k-1} and are hereditarily t-x_k-united, then v and w are $\hat{x}_{k-1}t$-x_{k-1}-united.

Thus restrictions (III) and (IV) must be reformulated to cover the relation hereditarily t-x_k-united for each k, $1 \leq k \leq n$. But then we are finally in a position to specify Class 2.8 and to show its amenability.

Class 2.8. Let F be a normal schema whose basis contains only x-focused sets and y-sets, and that possesses the following property: for each x_k-term t in $D(F, \omega)$, $1 \leq k \leq n$, and for each y-set Y in $b(F)$ all of whose members govern x_k, if $|\hat{Y}[t]| > 1$ and if a member of Y is hereditarily t-x_k-united with a variable v, then

(1) v dominates no x-variable that governs any member of Y;
(2) v belongs to no y-set Y' in $b(F)$ such that $|\hat{Y}'[t]| > 1$ and $\hat{Y}'[t] \neq \hat{Y}[t]$.

Then $E(F, (m + n)!d(F, m + n))$ is adequate.

Proof. We begin by observing why the argument for Class 2.8 need differ critically from that for Class 2.7. If F is a normal schema, t an x_n-term, and x_j a nonfinal x-variable of F, then there are infinitely many Herbrand instances of F in which the x_j-term $\hat{x}_j t$ is the substituent for x_j: these Herbrand instances share the same substituents for the y-variables governing x_j, but differ in the substituents for the y-variables governed by x_j. In short, there are infinitely many x_n-terms $t' \neq t$ such that $\hat{x}_j t' = \hat{x}_j t$. Hence we shall not be able to demand of the term-mapping γ, constructed to satisfy the Adequacy Lemma of §3, that $\hat{x}_j \gamma t' = \hat{x}_j \gamma t$ whenever $\hat{x}_j t' = \hat{x}_j t$. (Only if x_j is final in F shall we have $\hat{x}_j \gamma t' = \hat{x}_j \gamma t$ whenever $\hat{x}_j t' = \hat{x}_j t$.) Consequently, we shall not have the "commutativity" of γ and $\hat{x}_j$,

ISBN 0-201-02540-X

that is, we shall not have $\gamma \hat{x}_j t = \hat{x}_j \gamma t$ for each x_n-term t and each nonfinal x_j; and we shall not have condition (2_x) on γ and γ^* (page 50), that is, we shall not have the sequence-mapping γ^* carrying $\langle \hat{v}_1 t, ..., \hat{v}_p t \rangle$ to $\langle \hat{v}_1 \gamma t, ..., \hat{v}_p \gamma t \rangle$ for each x_n-term t and each arrangement $\langle v_1, ..., v_p \rangle$ of each x-focused set V in $b(F)$ with a nonfinal focus. For if γ did commute with $\hat{x}_j$, then whenever $\hat{x}_j t' = \hat{x}_j t$ we would have $\hat{x}_j \gamma t' = \gamma \hat{x}_j t' = \gamma \hat{x}_j t = \hat{x}_j \gamma t$. And if γ^* did carry $\langle \hat{v}_1 t, ..., \hat{v}_p t \rangle$ to $\langle \hat{v}_1 \gamma t, ..., \hat{v}_p \gamma t \rangle$ and $\langle \hat{v}_1 t', ..., \hat{v}_p t' \rangle$ to $\langle \hat{v}_1 \gamma t', ..., \hat{v}_p \gamma t' \rangle$, where x_j is the focus of $\{v_1, ..., v_p\}$, and if $\hat{x}_j t' = \hat{x}_j t$, then in this case $\langle \hat{v}_1 t', ..., \hat{v}_p t' \rangle = \langle \hat{v}_1 t, ..., \hat{v}_p t \rangle$; whence we would need to have $\hat{v}_i \gamma t' = \hat{v}_i \gamma t$ for each i, and in particular $\hat{x}_j \gamma t' = \hat{x}_j \gamma t$. (Recall that x_j is the focus of V only if x_j is dominated by every variable in V, and that $\hat{v}t = \hat{v}\hat{x}_j t$ whenever v dominates x_j.)

Once again a bit of notation. For each normal schema F, each truth-assignment $\mathfrak{A}$ on $\check{L}(F, \omega)$, and each x_k-term t in $D(F, \omega)$, $1 \le k \le n$, write $\mathfrak{P}_{\mathfrak{A}}(t)$ for $\mathfrak{p}_{\mathfrak{A}}(\hat{v}_1 t, ... \hat{v}_p t)$, where $v_1, ..., v_p$ are all the variables (in prefix order) that dominate x_k. Note that $\mathfrak{P}_{\mathfrak{A}}(t_1) = \mathfrak{P}_{\mathfrak{A}}(t_2)$ only if $\chi(t_1) = \chi(t_2)$. Note further that if $\mathfrak{P}_{\mathfrak{A}}(t_1) = \mathfrak{P}_{\mathfrak{A}}(t_2)$ and if x_k governs $\chi(t_1)$, then $\mathfrak{P}_{\mathfrak{A}}(\hat{x}_k t_1) = \mathfrak{P}_{\mathfrak{A}}(\hat{x}_k t_2)$, since the variables dominating x_k are a proper subset of those dominating $\chi(t_1)$.

Now instead of the commutativity of γ and $\hat{x}_k$, the term-mapping γ will possess the property

(†)   for each term t and each x-variable x_k dominating $\chi(t)$, $\mathfrak{P}_{\mathfrak{A}}(\hat{x}_k \gamma t) = \mathfrak{P}_{\mathfrak{A}}(\gamma \hat{x}_k t)$.

Thus $\mathfrak{P}_{\mathfrak{A}}(\hat{x}_k \gamma t) = \mathfrak{P}_{\mathfrak{A}}(\hat{x}_k \gamma t')$ whenever t and t' are any terms such that $\hat{x}_k t = \hat{x}_k t'$. Hence γ will possess the property

(††)   if $\hat{x}_k t = \hat{x}_k t'$ and $v_1, ..., v_p$ all dominate x_k, then $\mathfrak{p}_{\mathfrak{A}}(\hat{v}_1 \gamma t, ..., \hat{v}_p \gamma t) = \mathfrak{p}_{\mathfrak{A}}(\hat{v}_1 \gamma t', ..., \hat{v}_p \gamma t')$.

Given property (††), it follows that condition (2) of the Adequacy Lemma will be satisfied for each arrangement $\langle v_1, ..., v_p \rangle$ of each x-focused set in $b(F)$—despite the fact that we shall not have condition (2_x)—provided that we do have

(‡)   for each x_n-term t and each arrangement $\langle v_1, ..., v_p \rangle$ of each x-focused set in $b(F)$ with focus x_k, $1 \le k \le n$, $\gamma^*(\hat{v}_1 t, ..., \hat{v}_p t) = \langle \hat{v}_1 \gamma t', ..., \hat{v}_p \gamma t' \rangle$ for some x_n-term t' such that $\hat{x}_k t = \hat{x}_k t'$.

For (‡) yields $\mathfrak{p}_{\mathfrak{A}}(\gamma^*(\hat{v}_1 t, ..., \hat{v}_p t)) = \mathfrak{p}_{\mathfrak{A}}(\hat{v}_1 \gamma t', ..., \hat{v}_p \gamma t')$, and (††) yields $\mathfrak{p}_{\mathfrak{A}}(\hat{v}_1 \gamma t', ..., \hat{v}_p \gamma t') = \mathfrak{p}_{\mathfrak{A}}(\hat{v}_1 \gamma t, ..., \hat{v}_p \gamma t)$.

A further contrast with the proof for Class 2.7 is that the construction of γ will be by an induction within an induction: the move from terms of height q to x_k-terms, $1 \le k \le n$, of height $q + 1$ will be by an induction on the subscript k. (When $k = 1$, the argument is exactly that of Class 2.7.) This

ISBN 0-201-02540-X

double induction—a part of the simultaneous inductive construction of γ and γ^*—will require as inductive hypotheses three conditions in addition to conditions (1)–(5) of §3.

(6) For each x_l-term t, $1 < l \le n$, $\mathfrak{P}_{\mathfrak{A}}(\hat{x}_{l-1}\gamma t) = \mathfrak{P}_{\mathfrak{A}}(\gamma \hat{x}_{l-1}t)$.

(7) For each x_k-term t, $1 \le k \le n$, and all variables v and w, if v and w are hereditarily t-x_k-united, then $\hat{v}\gamma t = \hat{w}\gamma t$.

(8) For each x_k-term t, $1 \le k \le n$, and each variable v governing x_k, $\mathfrak{p}_{\mathfrak{A}}(\hat{v}\gamma t) = \mathfrak{p}_{\mathfrak{A}}(\gamma \hat{v}t)$.

Condition (6) quickly yields property (†). For let s be an x_k-term. (We have trivially $\mathfrak{P}_{\mathfrak{A}}(\hat{x}_k\gamma s) = \mathfrak{P}_{\mathfrak{A}}(\gamma \hat{x}_k s)$, since $\hat{x}_k\gamma s = \gamma s = \gamma \hat{x}_k s$.) Clearly (6) yields $\mathfrak{P}_{\mathfrak{A}}(\hat{x}_{k-1}\gamma s) = \mathfrak{P}_{\mathfrak{A}}(\gamma \hat{x}_{k-1}s)$; whence $\mathfrak{P}_{\mathfrak{A}}(\hat{x}_{k-2}\gamma s) = \mathfrak{P}_{\mathfrak{A}}(\hat{x}_{k-2}\gamma \hat{x}_{k-1}s)$. But by taking $t = \hat{x}_{k-1}s$, (6) yields $\mathfrak{P}_{\mathfrak{A}}(\hat{x}_{k-2}\gamma \hat{x}_{k-1}s) = \mathfrak{P}_{\mathfrak{A}}(\gamma \hat{x}_{k-2}\hat{x}_{k-1}s) = \mathfrak{P}_{\mathfrak{A}}(\gamma \hat{x}_{k-2}s)$. Hence $\mathfrak{P}_{\mathfrak{A}}(\hat{x}_{k-2}\gamma s) = \mathfrak{P}_{\mathfrak{A}}(\gamma \hat{x}_{k-2}s)$; and so on.

Note that the inference from (6) to property (†), and thence to property (††), can be done within the induction. That is, if (6) holds for each term t in a domain $D(F, q)$, then (†) holds for each term t in $D(F, q)$, and (††) holds for all terms t and t' in $D(F, q)$.

Condition (7) is the weak identity preservation property. Just as condition (2_x) demands that $\langle \hat{v}_1\gamma t, \ldots, \hat{v}_p\gamma t \rangle = \langle \hat{w}_1\gamma t, \ldots, \hat{w}_p\gamma t \rangle$ whenever $\langle \hat{v}_1 t, \ldots, \hat{v}_p t \rangle = \langle \hat{w}_1 t, \ldots, \hat{w}_p t \rangle$, even though $\{v_1, \ldots, v_p\}$ and $\{w_1, \ldots, w_p\}$ might be distinct x-focused sets in $b(F)$, so property (‡) demands that $\langle \hat{v}_1\gamma t', \ldots, \hat{v}_p\gamma t' \rangle = \langle \hat{w}_1\gamma t', \ldots, \hat{w}_p\gamma t' \rangle$ whenever $\langle \hat{v}_1 t, \ldots, \hat{v}_p t \rangle = \langle \hat{v}_1 t', \ldots, \hat{v}_p t' \rangle = \langle \hat{w}_1 t', \ldots, \hat{w}_p t' \rangle$. Hence the term-mapping γ must satisfy (7). In the proofs for Skolem Classes 2.4–2.7 we could show that γ had the appropriate identity preservation property on terms of height $q + 1$ without assuming it had the property on terms of lower height. But now we cannot. For suppose that t is an x_k-term of height $q + 1$, where $k > 1$. Since γt will depend on $\gamma \hat{x}_{k-1}t$, we shall need to know that γ has the property on $\hat{x}_{k-1}t$ in order to show that it has the property on t, and (provided that x_k is immediately preceded in the prefix of F by a y-variable) there are, obviously, x_k-terms t of height $q + 1$ such that $\hat{x}_{k-1}t$ has height $\le q$.

Condition (8) is the monadic property for γ stated in full generality. We know from the proofs for Classes 2.4–2.7 that this property in conjunction with condition (4) is crucial to showing that one-membered sets satisfy condition (2). But here again, unlike those proofs and for exactly the same reason as in the case of condition (7), we cannot show that γ has the property on terms of height $q + 1$ without assuming it on terms of lower height.

We come at last to the explicit construction of γ and γ^*, and as always first we pigeonhole. Let $\mathfrak{A}$ be any truth-assignment on $\hat{L}(F, \omega)$. By the

ISBN 0-201-02540-X

pigeonhole argument there is a number r, $2 \leq r \leq (m + n)!d(F, m+n) + 1$, and a pigeonhole-mapping π such that for each x_n-term t in $D(F, r + 1)$,

(π-1) πt is an x_n-term in $D(F, r)$;
(π-2)   (a) $\mathfrak{P}_{\mathfrak{A}}(\pi t) = \mathfrak{P}_{\mathfrak{A}}(t)$;
        (b)   for all variables v and w, $\hat{v}\pi t = \hat{w}\pi t$ iff $\hat{v}t = \hat{w}t$.

We further stipulate that, for each x_l-term s in $D(F, r + 1)$, when $l < n$,

(π-3) $\pi s = \hat{x}_l \pi t'$, where t' is the earliest x_n-term such that $s = \hat{x}_l t'$.

Hence, if $h(s) \leq r + 1$, then $h(\pi s) \leq r$, $\mathfrak{P}_{\mathfrak{A}}(\pi s) = \mathfrak{P}_{\mathfrak{A}}(s)$, and, for all variables v and w that govern x_l, $\hat{v}\pi s = \hat{w}\pi s$ iff $\hat{v}s = \hat{w}s$.

Let γ and γ^* be the identity on $D(F, 1)$. Now assume γ and γ^* fulfill the inductive hypotheses (1)–(8) on $D(F, q)$, $q \geq 1$. We extend γ to $D(F, q + 1)$. Let t be an x_k-term, $1 \leq k \leq n$, of height $q + 1$. Assume $\gamma \hat{x}_{k-1} t$ is defined, an assumption that is vacuous—as are clauses (γ-1) and (γ-2) immediately below—if $k = 1$, and an assumption that is redundant if $\hat{x}_{k-1}t$ has height $\leq q$. If $k > 1$ and no y-variable both is governed by x_{k-1} and governs x_k, then the mapping $\hat{x}_k$ is defined on all x_{k-1}-terms in $D(F, \omega)$, and we let $\gamma t = \hat{x}_k \gamma \hat{x}_{k-1} t$. (Recall that since F is normal it begins with a y-variable.) If $k = 1$, let $y_{a+1},...,y_{a+b}$ be the y-variables governing x_k (hence $a = 0$); if $k > 1$ and there are y-variables both governed by x_{k-1} and governing x_k, let $y_{a+1},...,y_{a+b}$ be these variables. Then let $\gamma t = \pi s$, where s is the x_k-term of height $\leq r + 1$ specified thus:

(γ-1)   let $\hat{x}_{k-1}s = \gamma \hat{x}_{k-1}t$;
(γ-2)   if y_{a+i}, $1 \leq i \leq b$, is hereditarily t-x_k-united with some variable v that dominates x_{k-1}, then let $\hat{y}_{a+i}s = \hat{v}\gamma \hat{x}_{k-1}t$ for the earliest such variable v.
(γ-3)   if $Y = \{y_{i_1},...,y_{i_p}\}$ is a y-set in $b(F)$ such that $|\hat{Y}[t]| > 1$, where $a < i_1 < \cdots < i_p \leq a + b$, then let $\langle \hat{y}_{i_1}s,...,\hat{y}_{i_p}s \rangle = \gamma^*(\hat{y}_{i_1}t,...,\hat{y}_{i_p}t)$;
(γ-4)   if $\hat{y}_{a+i}s$ is not determined by (γ-3), but y_{a+i} is hereditarily t-x_k-united with some y-variable y_{a+j} such that $\hat{y}_{a+j}s$ is so determined, then let $\hat{y}_{a+i}s = \hat{y}_{a+j}s$ for the least such $a + j$;
(γ-5)   if $\hat{y}_{a+i}s$ is determined by none of (γ-2)–(γ-4), then let $\hat{y}_{a+i}s = \gamma \hat{y}_{a+i}t$.

This definition is unambiguous. No conflicts arise from (γ-3) since no distinct y-sets of cardinality >1 overlap. And none arise between (γ-2) on the one hand and (γ-3) and (γ-4) on the other, because of clause (1) of the specification of Class 2.8. (It should be clear that clauses (γ-3)–(γ-5) exactly parallel clauses (γ-1)–(γ-3) in the definition of the term-mapping γ for Class 2.7.)

ISBN 0-201-02540-X

We verify first that conditions (1) and (6)–(8) are fulfilled for each term t in $D(F, q + 1)$. Suppose t is an x_k-term. We assume not only that conditions (1)–(8) hold on $D(F, q)$, but also that (1) and (6)–(8) hold for the term $\hat{x}_{k-1}t$ (again, if $k = 1$ the assumption is vacuous). Let s be the x_k-term such that, by definition, $\gamma t = \pi s$. By conditions (1) and (5) on $D(F, q)$ and condition (1) for $\hat{x}_{k-1}t$, the term s has height at most $r + 1$. Hence, by the choice of π, $h(\gamma t) \leq r$, which proves condition (1) for s; moreover, if $k > 1$, $\mathfrak{P}_\mathfrak{A}(\hat{x}_{k-1}\gamma t) = \mathfrak{P}_\mathfrak{A}(\hat{x}_{k-1}\pi s) = \mathfrak{P}_\mathfrak{A}(\hat{x}_{k-1}s) = \mathfrak{P}_\mathfrak{A}(\gamma \hat{x}_{k-1}t)$, which proves condition (6) for t. If v is a variable governing x_k, then $\mathfrak{p}_\mathfrak{A}(\hat{v}\gamma t) = \mathfrak{p}_\mathfrak{A}(\hat{v}\pi s) = \mathfrak{p}_\mathfrak{A}(\hat{v}s)$; whence to prove condition (8) for the term t it suffices to show $\mathfrak{p}_\mathfrak{A}(\hat{v}s) = \mathfrak{p}_\mathfrak{A}(\gamma \hat{v}t)$. If v dominates x_{k-1}, then $\hat{v}s = \hat{v}\hat{x}_{k-1}s = \hat{v}\gamma \hat{x}_{k-1}t$ and, since (8) holds for $\hat{x}_{k-1}t$, $\mathfrak{p}_\mathfrak{A}(\hat{v}\gamma \hat{x}_{k-1}t) = \mathfrak{p}_\mathfrak{A}(\gamma \hat{v}\hat{x}_{k-1}t) = \mathfrak{p}_\mathfrak{A}(\gamma \hat{v}t)$. If x_{k-1} governs v, there are three subcases. If $\hat{v}s$ is determined by (γ-2), then that $\mathfrak{p}_\mathfrak{A}(\hat{v}s) = \mathfrak{p}_\mathfrak{A}(\gamma \hat{v}t)$ follows from what we have just seen. If $\hat{v}s$ is determined by (γ-3) or (γ-4), then that $\mathfrak{p}_\mathfrak{A}(\hat{v}s) = \mathfrak{p}_\mathfrak{A}(\gamma \hat{v}t)$ follows from condition (4) on $D(F, q)$. Finally, if $\hat{v}s$ is determined by (γ-5), then $\hat{v}s = \gamma \hat{v}t$, so the result is immediate. We have thus shown condition (8) for t, and we turn to condition (7), the weak identity preservation property. Suppose v and w are hereditarily t-x_k-united. Then $\hat{v}t = \hat{w}t$. We wish to show that $\hat{v}\gamma t = \hat{w}\gamma t$. Again by the choice of π, it suffices to show that $\hat{v}s = \hat{w}s$. We may assume that v and w are distinct: for if $v = w$, the result is trivial. If both v and w govern x_{k-1}, then v and w are hereditarily $\hat{x}_{k-1}t$-x_{k-1}-united; by condition (7) for $\hat{x}_{k-1}t$ we have $\hat{v}s = \hat{v}\hat{x}_{k-1}s = \hat{v}\gamma \hat{x}_{k-1}t = \hat{w}\gamma \hat{x}_{k-1}t = \hat{w}\hat{x}_{k-1}s = \hat{w}s$. Otherwise at least one of v and w, say v, is governed by x_{k-1}. If $\hat{v}s$ is determined by (γ-2), then either $\hat{w}s$ is also determined by (γ-2) or w dominates x_{k-1}. In either case $\hat{v}s = \hat{w}s$. If $\hat{v}s$ is determined by (γ-3) or (γ-4), then $\hat{w}s$ is also determined by one of these two clauses; by reasoning just as in the amenability proof for Class 2.7, from the transitivity of hereditary t-x_k-unitedness, the permutation condition (3) on γ^*, and clause (2) of the specification of Class 2.8, we have $\hat{v}s = \hat{w}s$. Finally, if $\hat{v}s$ is determined by (γ-5), then so is $\hat{w}s$; hence $\hat{v}s = \gamma \hat{v}t = \gamma \hat{w}t = \hat{w}s$. This proves condition (7) for t.

We now verify condition (2) on $D(F, q + 1)$ for y-sets in $b(F)$. Let t be an x_n-term of height $q + 1$, and let $\langle v_1,\ldots,v_p \rangle$ be an arrangement of a y-set in $b(F)$; we must show that $\mathfrak{p}_\mathfrak{A}(\gamma^*(\hat{v}_1 t,\ldots,\hat{v}_p t)) = \mathfrak{p}_\mathfrak{A}(\hat{v}_1\gamma t,\ldots,\hat{v}_p\gamma t)$. By the specification of Class 2.8, all of $v_1,\ldots,v_p$ are governed by x_{k-1} and govern x_k for some k, $1 \leq k \leq n$; and, by property (††) on $D(F, q + 1)$ it suffices to find a term t' of height $\leq q + 1$ such that $\hat{x}_k t = \hat{x}_k t'$ and $\mathfrak{p}_\mathfrak{A}(\gamma^*(\hat{v}_1 t',\ldots,\hat{v}_p t')) = \mathfrak{p}_\mathfrak{A}(\hat{v}_1\gamma t',\ldots,\hat{v}_p\gamma t')$. Suppose first that $\hat{x}_k t$ has height $\leq q$. Then there is an x_n-term t' of height $\leq q$ such that $\hat{x}_k t' = \hat{x}_k t$. (For example, we may take t' to be the unique x_n-term such that $\hat{y}_i t' =$

$\hat{y}_i \hat{x}_k t = \hat{y}_i t$ whenever y_i governs x_k and $\hat{y}_i t' = 4$ whenever y_i is governed by x_k.) By condition (2) on $D(F, q)$ we have $\mathfrak{p}_{\mathfrak{A}}(\gamma^*(\hat{v}_1 t', \ldots, \hat{v}_p t'))$ $= \mathfrak{p}_{\mathfrak{A}}(\hat{v}_1 \gamma t', \ldots, \hat{v}_p \gamma t')$. Suppose second that $\hat{x}_k t$ has height $q + 1$. Then let $t' = \hat{x}_k t$ and let s be the x_k-term such that $\gamma t' = \pi s$. If $\{\hat{v}_1 t', \ldots, \hat{v}_p t'\}$ has cardinality >1, then $\gamma^*(\hat{v}_1 t', \ldots, \hat{v}_p t') = \langle \hat{v}_1 s, \ldots, \hat{v}_p s \rangle$ by clause (γ-3) and the permutation condition (3); by the choice of π, $\mathfrak{p}_{\mathfrak{A}}(\hat{v}_1 s, \ldots, \hat{v}_p s) = \mathfrak{p}_{\mathfrak{A}}(\hat{v}_1 \gamma t', \ldots, \hat{v}_p \gamma t')$, which yields the desired result. If $\{v_1, \ldots, v_p\}$ has cardinality 1, that is, $v_1 = \cdots = v_p$, then we must show $\mathfrak{p}_{\mathfrak{A}}(\gamma^*(\hat{v}_1 t')) = \mathfrak{p}_{\mathfrak{A}}(\hat{v}_1 \gamma t')$; by condition (4) on $D(F, q)$ this is equivalent to $\mathfrak{p}_{\mathfrak{A}}(\gamma \hat{v}_1 t') = \mathfrak{p}_{\mathfrak{A}}(\hat{v}_1 \gamma t')$, which is just the monadic property (8). Finally, if $\{v_1, \ldots, v_p\}$ has cardinality >1 but $\hat{v}_1 t' = \cdots = \hat{v}_p t'$, then, by reasoning just as in the amenability proof for Class 2.7,‡ we have $\hat{v}_1 \gamma t' = \cdots = \hat{v}_p \gamma t'$, so that the desired result again follows from the monadic property.

Thus our extension of γ fulfills conditions (1) and (6)–(8), and condition (2) for y-sets, on the domain $D(F, q + 1)$. Conditions (3), (4), and (5), as well as condition (2) for x-focused sets, await our extension of γ^*.

Let $\langle t_1, \ldots, t_p \rangle$ be a sequence from $D(F, q + 1)$ that contains at least one member of height $q + 1$.

(γ^*-1)   Suppose that $\{t_1, \ldots, t_p\} = \hat{V}[t]$ for some x-focused set V in $b(F)$ and some x_n-term t. Let t' be the earliest such x_n-term, and let V' be any x-focused set in $b(F)$ such that $\{t_1, \ldots, t_p\} = \hat{V}'[t']$. Then let $\gamma^*(t_1, \ldots, t_p)$ $= \langle \hat{v}_1 \gamma t', \ldots, \hat{v}_p \gamma t' \rangle$, where, for each i, v_i is some variable in V' such that $t_i = \hat{v}_i t'$.

(γ^*-2)   Otherwise let $\gamma^*(t_1, \ldots, t_p) = \langle \gamma t_1, \ldots, \gamma t_p \rangle$.

The weak identity preservation property (7) assures that (γ^*-1) engenders no ambiguity. For suppose t' is the earliest x_n-term such that $\{t_1, \ldots, t_p\} = \hat{V}[t']$ for some x-focused set V in $b(F)$, and suppose W is an x-focused set in $b(F)$, not necessarily distinct from V, such that $\hat{V}[t'] = \hat{W}[t']$. If $t_i = \hat{v}t' = \hat{w}t'$ for $v \in V$ and $w \in W$, then by definition v and w are t'-x_n-united. Moreover, since $\{t_1, \ldots, t_p\} \subseteq D(F, q + 1)$, $t' \in D(F, q + 1)$. Hence, by (7), $\hat{v}\gamma t' = \hat{w}\gamma t'$.

The specification of the extension of γ^* to $D(F, q + 1)$ immediately yields the permutation condition (3) and the boundedness condition (5). Condition (4) also follows easily. For suppose $\gamma^*(t_1, \ldots, t_p) = \langle s_1, \ldots, s_p \rangle$; we must show $\mathfrak{p}_{\mathfrak{A}}(\gamma t_i) = \mathfrak{p}_{\mathfrak{A}}(s_i)$, $1 \le i \le p$. But either $s_i = \gamma t_i$, or else

---

‡ That is, we would show that if $\hat{v}_i \gamma t' \ne \hat{v}_j \gamma t'$ for some i and $j \le p$, then there is a term s' such that s'-x_k-unitedness coincides with t'-x_k-unitedness, $\{\hat{v}_1 s', \ldots, \hat{v}_p s'\}$ has cardinality >1, and some v_i is s'-x_k-united either with a variable dominating x_{k-1} or with a variable that lies in some y-set Y' such that $|Y'[s']| > 1$ and $Y'[s'] \ne \{\hat{v}_1 s', \ldots, \hat{v}_p s'\}$. Hence the specification of Class 2.8 would be violated.

ISBN 0-201-02540-X

$s_i = \hat{v}\gamma t'$ for some x_n-term t' of height $\leq q + 1$ and some variable v such that $t_i = \hat{v}t'$. In the second case, by the monadic property (8), $\mathfrak{p}_\mathfrak{A}(\gamma t_i) = \mathfrak{p}_\mathfrak{A}(\gamma \hat{v}t') = \mathfrak{p}_\mathfrak{A}(\hat{v}\gamma t') = \mathfrak{p}_\mathfrak{A}(s_i)$, as required.

We need finally to verify condition (2) for x-focused sets. Let t be an x_n-term of height $q + 1$, let $\langle v_1,\ldots,v_p \rangle$ be an arrangement of an x-focused set in $b(F)$, and let x_k be the focus of $\{v_1,\ldots,v_p\}$. Suppose first that $\hat{x}_k t$ has height $\leq q$ (so that $\langle \hat{v}_1 t,\ldots,\hat{v}_p t \rangle$ is a sequence from $D(F, q)$). Then there is an x_n-term t' of height $\leq q$ such that $\hat{x}_k t' = \hat{x}_k t$. By induction hypothesis (2) on $D(F, q)$, $\mathfrak{p}_\mathfrak{A}(\gamma^*(\hat{v}_1 t',\ldots,\hat{v}_p t')) = \mathfrak{p}_\mathfrak{A}(\hat{v}_1\gamma t',\ldots,\hat{v}_p\gamma t')$. Thus, given property (††) on $D(F, q + 1)$, condition (2) holds for t. Suppose second that $\hat{x}_k t$ has height $q + 1$. Let t' be the earliest x_n-term such that $\{\hat{v}_1 t,\ldots,\hat{v}_p t\} = \hat{V}[t']$ for some x-focused set V in $b(F)$. Thus $h(t') \leq q + 1$, and there is a variable v in V such that $\hat{x}_k t = \hat{v}t'$. But then $v = x_k$: for if v were a y-variable, then $h(\hat{v}t') \leq q$; and if v were an x-variable other than x_k, then $\chi(\hat{v}t') = v \neq x_k$. Hence $\hat{x}_k t = \hat{x}_k t'$, and so $\langle \hat{v}_1 t,\ldots,\hat{v}_p t \rangle = \langle \hat{v}_1 t',\ldots,\hat{v}_p t' \rangle$. (We have just implicitly shown that γ^* possesses property (‡).) By (γ^*-1), $\gamma^*(\hat{v}_1 t,\ldots,\hat{v}_p t) = \langle \hat{v}_1\gamma t',\ldots,\hat{v}_p\gamma t' \rangle$. Hence property (††) on $D(F, q + 1)$ again yields $\mathfrak{p}_\mathfrak{A}(\gamma^*(\hat{v}_1 t,\ldots,\hat{v}_p t)) = \mathfrak{p}_\mathfrak{A}(\hat{v}_1\gamma t,\ldots,\hat{v}_p\gamma t)$, which is condition (2).

All the conditions on the extensions of γ and γ^* have been verified, and the proof of amenability is concluded. $\square$

In Chapter 6, §4 we shall show that a modification of Kalmár's method for eliminating initial x-variables (cf. Chapter 2, §5) suffices to reduce the initially-extended version of Class 2.8 to Class 2.8.

7. Failure of Amenability

We conclude this chapter with a theorem on the limits of the amenability method. We assume that the classes of schemata under discussion are effectively specified, that is, are recursive sets of schemata.

Definition 2.13.   A class $\mathfrak{C}_2$ of schemata is *full* in a class $\mathfrak{C}_1$ of schemata iff for each schema F_1 in $\mathfrak{C}_1$ there is a schema F_2 in $\mathfrak{C}_2$ such that

(a) F_1 and F_2 share the same quantifier structure, that is, F_1 and F_2 contain the same y-variables and the same x-variables, and each x-variable is governed by the same y-variables in F_1 and in F_2;
(b)   every atomic subformula of F_1 is an atomic subformula of F_2.

Theorem.   If an amenable class $\mathfrak{C}_2$ is full in a class $\mathfrak{C}_1$, then $\mathfrak{C}_1$ is amenable.

Proof.   Suppose $\mathfrak{C}_2$ is φ-amenable, where φ is a recursive function. Let F_1 be any schema in $\mathfrak{C}_1$, with y-variables $y_1,\ldots,y_n$. Since $\mathfrak{C}_1$ and $\mathfrak{C}_2$ are effectively specified classes, and (a) and (b) of Definition 2.13 are clearly effective conditions, there is an effective method for finding, given any F_1

ISBN 0-201-02540-X

in $\mathfrak{C}_1$, a schema F_2 in $\mathfrak{C}_2$ fulfilling (a) and (b). It is then enough to show that the expansion $E(F_1, \varphi(F_2))$ is adequate.

Since F_1 and F_2 share the same quantifier structure, $D(F_1, p) = D(F_2, p)$ for each p. Hence there is a bijection $\theta\colon E(F_1, \omega) \to E(F_2, \omega)$ defined thus:

$$\theta(F_1{}^*(y_1/t_1, \ldots, y_n/t_n)) = F_2{}^*(y_1/t_n, \ldots, y_n/t_n)$$

for all terms $t_1, \ldots, t_n$ in $D(F_1, \omega)$. Clearly $\theta[E(F_1, p)] = E(F_2, p)$ for every p, and if $\langle F_1 \rangle_i = \langle F_2 \rangle_j$, then $\langle H \rangle_i = \langle \theta(H) \rangle_j$ for every Herbrand instance H of F_1.

Let $\mathfrak{A}$ be any truth-assignment on $\hat{L}(F_1, \omega)$. We extend $\mathfrak{A}$ arbitrarily to $\hat{L}(F_2, \omega)$ (every predicate letter of F_1 is a predicate letter of F_2, but not necessarily conversely). Since $\mathfrak{C}_2$ is φ-amenable, there are for F_2 a formula-mapping Γ and an instance-mapping Ψ such that

(1) $\Psi[E(F_2, \omega)] \subseteq E(F_2, \varphi(F_2))$; and
(2) $\mathfrak{A}(\Gamma(\langle H \rangle_i)) = \mathfrak{A}(\langle \Psi(H) \rangle_i)$ for every Herbrand instance H of F_2 and every i.

Define an instance-mapping Ψ' for F_1 thus: for every Herbrand instance H of F_1, $\Psi'(H) = \theta^{-1}(\Psi(\theta(H)))$, where θ^{-1} is the inverse of θ. By the boundedness condition (1) on Ψ, the boundedness condition

$$\Psi'[E(F_1, \omega)] \subseteq E(F_1, \varphi(F_2))$$

on Ψ' is fulfilled. It remains to show the central condition on Γ and Ψ', that is,

$$\mathfrak{A}(\Gamma(\langle H \rangle_i)) = \mathfrak{A}(\langle \Psi'(H) \rangle_i)$$

for each Herbrand instance H of F_1 and each i. Let H be any Herbrand instance of F_1 and i any number. Since every atomic subformula of F_1 is an atomic subformula of F_2, there is a j such that $\langle F_1 \rangle_i = \langle F_2 \rangle_j$, and hence $\langle H \rangle_i = \langle \theta(H) \rangle_j$. Thus

$$\mathfrak{A}(\Gamma(\langle H \rangle_i)) = \mathfrak{A}(\Gamma(\langle \theta(H) \rangle_j)).$$

By the central condition (2) on Γ and Ψ,

$$\mathfrak{A}(\Gamma(\langle \theta(H) \rangle_j)) = \mathfrak{A}(\langle \Psi(\theta(H)) \rangle_j).$$

By the choice of i and j,

$$\langle \Psi(\theta(H)) \rangle_j = \langle \theta^{-1}(\Psi(\theta(H))) \rangle_i.$$

Hence, by the definition of Ψ',

$$\mathfrak{A}(\Gamma(\langle H \rangle_i)) = \mathfrak{A}(\langle \Psi'(H) \rangle_i),$$

as desired. $\square$

ISBN 0-201-02540-X

Corollary. A class $\mathfrak{C}_2$ full in an unsolvable class is not amenable.

Proof. No unsolvable class is amenable. $\square$

The Corollary yields the nonamenability of solvable Classes 1.2, 1.3, and 1.4, the condensable classes of Chapter 1. Indeed, Classes 1.2 and 1.3 are full in the class of all schemata. Class 1.4, the class of schemata with matrices in conjunctive normal form containing no more than two occurrences of any predicate letter, is full in the class of all schemata containing no more than two occurrences of any predicate letter; the latter class, as we noted in Chapter 1, is unsolvable.

There are more interesting solvable classes whose amenability is ruled out by the Corollary. Chapter 4 will be entirely given over to the *Maslov Class*: the class of Skolem schemata with Krom matrices (a Krom matrix is a conjunction of binary disjunctions of signed atomic formulas). The Maslov Class is full in the class of all Skolem schemata. And in §2 of Chapter 5 we treat the class of prenex schemata with prefixes $\forall y_1 \exists x \forall y_2$ and Krom matrices, and the class of prenex schemata with prefixes $\forall y_1 \exists x \forall y_2$ and Horn matrices (a Horn matrix is a conjunction of disjunctions of signed atomic formulas no conjunct of which contains more than one positively signed atomic formula). Each of these classes is full in the (unsolvable) class of all prenex schemata with prefixes $\forall y_1 \exists x \forall y_2$.

ISBN 0-201-02540-X

_____

CHAPTER 3

Finite Controllability

A class of schemata is said to be *finitely controllable* iff each schema in the class that has a model at all has a finite model. The class of schemata having finite models is recursively enumerable (but not, as shown by Trakhtenbrot (1950) recursive); so too is the class of unsatisfiable schemata. Hence any effectively specifiable class that is finitely controllable is also solvable. The converse fails. Indeed, *ad hoc* counterexamples abound—for instance, any class $\{F\}$, where F is a schema having just infinite models. Moreover, in Chapter 5 we present three "naturally specified" solvable classes that are not finitely controllable; one of those classes is also amenable. In this chapter we show that all the solvable classes of Chapters 1 and 2 are finitely controllable.

1. The Finite Model Lemma

It is easy to see why the consistency of the Herbrand expansion of a schema F does not in general yield the existence of a finite model for F. Let F be $\forall y_1 \forall y_2 \exists x F^M$, and let S be a subset of $D(F, \omega)$. A truth-assignment $\mathfrak{A}$ provides a model for F over S just in case for all t_1 and t_2 in S there is an s in S such that $\mathfrak{A}$ verifies $F^M(y_1/t_1, y_2/t_2, x/s)$. Now if $\mathfrak{A}$ verifies $E(F, \omega)$, then for all t_1 and t_2 in $D(F, \omega)$ there is such an s in $D(F, \omega)$, namely, $s = f_x(t_1 t_2)$. Hence $\mathfrak{A}$ provides a model for F over $D(F, \omega)$; indeed, $\mathfrak{A}$ provides a model for F over each subset S of $D(F, \omega)$ that contains $f_x(t_1 t_2)$ whenever it contains t_1 and t_2. But clearly each S with this property is infinite. However, a finite model is forthcoming if $D(F, \omega)$ can be shrunk to finite size without sacrifice of consistency: if there is a term-mapping δ with finite range such that the set of instances

$$\{F^M(y_1/\delta t_1, y_2/\delta t_2, x/\delta f_x(t_1 t_2)) | t_1, t_2 \in D(F, \omega)\}$$

is consistent. The analogous fact holds for arbitrary schemata.

Burton Dreben and Warren D. Goldfarb, The Decision Problem: Solvable Classes of Quantificational Formulas

ISBN 0-201-02540-X

Finite Model Lemma.   A schema F has a finite model if there is a term-mapping δ such that $\delta[D(F, \omega)]$ is finite and $\Delta[E(F, \omega)]$ is consistent, where Δ is the δ-canonical formula-mapping.

Proof. F has a model over a universe U if there are interpretations over U of the predicate letters and indicial function signs of the functional form F^* under which the universal closure of F^* comes out true in U. Suppose that $\delta[D(F, \omega)]$ is finite and $\Delta[E(F, \omega)]$ is consistent. We shall construct a model for F over the finite universe $U = \delta[D(F, \omega)]$.

For each $t \in U$, let $\delta^{-1}t$ be the earliest term s in $D(F, \omega)$ such that $\delta s = t$. Let $\mathfrak{A}$ be any truth-assignment verifying $\Delta[E(F, \omega)]$. If P is an n-place predicate letter, assign to P the set $\{\langle t_1,\ldots,t_n\rangle \,|\, t_1,\ldots,t_n \in U$ and $\mathfrak{A}$ verifies $Pt_1\cdots t_n\}$. If f is an n-place indicial function sign, assign to f the function taking each n-tuple $\langle t_1,\ldots,t_n\rangle$ from U to $\delta f(\delta^{-1}t_1\cdots\delta^{-1}t_n)$. Now suppose $y_1,\ldots,y_m$ are all the y-variables of F. Then for the values $t_1,\ldots,t_m$ in U of the variables $y_1,\ldots,y_m$, the functional form F^* is true under the interpretations just given whenever $\mathfrak{A}$ verifies $\Delta(F^*y_1/\delta^{-1}t_1,\ldots,y_m/\delta^{-1}t_m))$. Since $\mathfrak{A}$ verifies $\Delta(H)$ for every Herbrand instance H of F, it follows that the universal closure of F^* is true in U under the interpretations.   □

The converse of the Finite Model Lemma also holds: if F has a finite model then there is a term-mapping δ fulfilling the requirements of the lemma. We shall never make use of the general form of the converse, although a special case is used in Chapter 5, §3.

Throughout the rest of this chapter we restrict the use of the letter "Δ" to δ-canonical formula-mappings; the term-mapping δ will always be clear from the context.

2. Classes 1.1–1.6

From the Finite Model Lemma we infer at once the finite controllability of Class 1.1.

Class 1.1.   Let F be a schema in which no y-variable governs any x-variable. If F has a model, then F has a model whose cardinality is that of $D(F, 1)$.

Proof.   If F is in Class 1.1, then $D(F, \omega) = D(F, 1)$. Hence the identity term-mapping meets the requirements of the Finite Model Lemma, provided that F has a model.   □

When $D(F, \omega)$ is infinite, nontrivial term-mappings δ are required. In this section we treat Classes 1.2–1.6 by constructing equivalence relations between terms that can, in certain contexts, replace the relation of identity between terms. To obtain term-mappings δ meeting the requirements of the Finite Model Lemma it is essential that each of these equivalence relations

ISBN 0-201-02540-X

be of *finite order*, that is, each relation partitions $D(F, \omega)$ into a finite number of equivalence classes.

Equivalence Lemma. For each schema F and each positive integer n, there is an equivalence relation $\simeq$ of finite order on $D(F, \omega)$ with the following property: for each n-tuple $\langle H_1,\ldots,H_n \rangle$ of Herbrand instances of F there is a term-mapping γ such that

(1) for all s and t if $s \simeq t$ then $\gamma s = \gamma t$;

(2) the γ-canonical formula-mapping is Herbrand instance preserving on $\{H_1,\ldots,H_n\}$.

Proof. Suppose F contains m x-variables. We shall construct the desired equivalence relation by means of a number-valued function ξ on $D(F, \omega)$. The value of ξ on a term t is to be distinct from its values on the arguments of t, from its values on the arguments of the arguments of t, and so on, until "argument of" is repeated $m \cdot n$ times.

For the purposes of this proof, and this proof only, we let $\hat{v}t$ be $\dagger$ when v is a y-variable that does not govern $\chi(t)$. (Normally $\hat{v}t$ is undefined in this case.) Let I be the set of sequences of y-variables of F of length $\leq m \cdot n$, including the empty sequence. If $\bar{v} = \langle v_1,\ldots,v_k \rangle$ is any member of I and t is any term, then we write $\bar{v}t$ for the term $\hat{v}_1\hat{v}_2\cdots\hat{v}_k t$; if $k = 0$, that is, $\bar{v}$ is the empty sequence, then $\bar{v}t = t$. We define the function ξ thus:

(1) $\xi(\dagger) = 1$;

(2) If $t \neq \dagger$, then $\xi(t)$ is the least positive integer not in $\{\xi(\bar{v}t) \mid \bar{v} \in I$ and $\bar{v}$ not the empty sequence$\}$.

The equivalence relation $\simeq$ is determined by the values of both ξ and χ on a term, on its arguments, on the arguments of its arguments, and so on, again to a "depth" of $m \cdot n$.

For all terms s and t in $D(F, \omega)$, let $s \simeq t$ iff $\xi(\bar{v}s) = \xi(\bar{v}t)$ and $\chi(\bar{v}s) = \chi(\bar{v}t)$ for each $\bar{v}$ in I.

Clearly $\simeq$ is of finite order; indeed, its order is at most $(|I| \cdot (m + 1))^{|I|}$. We show that $\simeq$ has the required property.

Let $\langle H_1,\ldots,H_n \rangle$ be any n-tuple of Herbrand instances of F. Let Ξ be the set $\{\xi(s) \mid s$ is the substituent in at least one of $H_1,\ldots,H_n$ for some x-variable of $F\}$. Thus Ξ is a set of at most $m \cdot n$ numbers. We define a term-mapping γ thus:

(γ-1) if $\xi(t) \notin \Xi$, then $\gamma t = \dagger$;

(γ-2) if $\xi(t) \in \Xi$, then γ is homomorphic on t.

By (γ-2), the γ-canonical formula-mapping is Herbrand instance preserving on $\{H_1,\ldots,H_n\}$. It remains to show that, for all s and t in $D(F, \omega)$, if

ISBN 0-201-02540-X

$s \simeq t$, then $\gamma s = \gamma t$. Here we use the following obvious consequence of the definition of γ:

(#)   If $\xi(t_1) = \xi(t_2)$ and $\chi(t_1) = \chi(t_2)$ but yet $\gamma t_1 \neq \gamma t_2$, then $\xi(t_1) \in \Xi$ and $\gamma \hat{v} t_1 \neq \gamma \hat{v} t_2$ for some y-variable v.

Now suppose $s \simeq t$; then $\xi(\hat{v}s) = \xi(\hat{v}t)$ and $\chi(\hat{v}s) = \chi(\hat{v}t)$ for each $\hat{v}$ in I. Suppose moreover $\gamma s \neq \gamma t$. By (#) then, $\xi(s) \in \Xi$ and there is a y-variable v_1 such that $\gamma \hat{v}_1 s \neq \gamma \hat{v}_1 t$. Another application of (#) shows that $\xi(\hat{v}_1 s) \in \Xi$ and there is a y-variable v_2 such that $\gamma \hat{v}_2 \hat{v}_1 s \neq \gamma \hat{v}_2 \hat{v}_1 t$. Iterating (#) $m \cdot n - 2$ more times, we obtain a sequence $\langle v_{m \cdot n}, \ldots, v_1 \rangle$ in I such that for each q, $1 \leq q \leq m \cdot n$, $\gamma \hat{v}_q \cdots \hat{v}_1 s \neq \gamma \hat{v}_q \cdots \hat{v}_1 t$. Also, $\xi(s)$, $\xi(\hat{v}_1 s)$, $\xi(\hat{v}_2 \hat{v}_1 s), \ldots, \xi(\hat{v}_{m \cdot n} \cdots \hat{v}_1 s)$ are all members of Ξ. By the definition of ξ they are all distinct numbers. But then $|\Xi| > m \cdot n$, a contradiction. We conclude that $\gamma s = \gamma t$. □

The finite controllability of each class containing only condensable schemata—and hence the finite controllability of Classes 1.2–1.4—follows easily from the Equivalence Lemma (condensability was defined on page 20). For each schema F and each $n > 0$, let $\simeq_{F,n}$ be the equivalence relation on $D(F, \omega)$ constructed in the proof of that lemma, and let $e(F, n)$ be the order of $\simeq_{F,n}$. Note that e is a primitive recursive function of F and n.

Corollary 1.   Every satisfiable n-condensable schema F, $n > 0$, has a finite model of cardinality $e(F, n)$.

Proof.   Suppose F is satisfiable and n-condensable. Let δ be the term-mapping that takes each t in $D(F, \omega)$ to the earliest s in $D(F, \omega)$ such that $s \simeq_{F,n} t$. Thus $|\delta[D(F, \omega)]| = e(F, n)$, and it suffices to show the consistency of $\Delta[E(F, \omega)]$.

Since each member of $\Delta[E(F, \omega)]$ is an instance of F, from n-condensability we know that $\Delta[E(F, \omega)]$ is inconsistent only if there is an inconsistent conjunction of n or fewer of its members. Hence it is enough to show the consistency of every conjunction of n or fewer members of $\Delta[E(F, \omega)]$. (Not every member of $\Delta[E(F, \omega)]$ is an Herbrand instance of F. Thus it is crucial that the definition of n-condensability talks of arbitrary instances of F and not just Herbrand instances.)

Any such conjunction has the form $\Delta(H_1) \wedge \cdots \wedge \Delta(H_n)$ for some (not necessarily distinct) Herbrand instances $H_1, \ldots, H_n$ of F. Let γ be the term-mapping furnished by the Equivalence Lemma: γ takes $\simeq_{F,n}$-equivalent terms to the same term and the γ-canonical formula-mapping Γ is Herbrand instance preserving on $\{H_1, \ldots, H_n\}$. By the definition of δ, for all terms t in $D(F, \omega)$, $\gamma \delta t = \gamma t$. Hence $\Gamma(\Delta(K)) = \Gamma(K)$ for all K in $\tilde{L}(F, \omega)$. But then $\Gamma(\Delta(H_1)) \wedge \cdots \wedge \Gamma(\Delta(H_n))$ is a conjunction of Herbrand instances of F, and so is consistent, since F is satisfiable. By the Consistency Lemma of Chapter 1, §3, $\Delta(H_1) \wedge \cdots \wedge \Delta(H_n)$ is consistent. □

ISBN 0-201-02540-X

A similar proof will yield the finite controllability of Classes 1.5 and 1.6. We showed in §3 of Chapter 1 that the schemata in these classes meet the antecedent of the next corollary.

Corollary 2. Suppose the schema F possesses an Herbrand instance H_0 such that $\langle H_0 \rangle_i = \langle H_0 \rangle_j$ whenever $\langle F \rangle_i$ and $\langle F \rangle_j$ are coinstantiable. Then F has a model only if F has a model of cardinality $e(F, 2)$.

Proof. Let δ be the term-mapping that takes each t in $D(F, \omega)$ to the earliest s in $D(F, \omega)$ such that $s \simeq_{F,2} t$. Then $|\delta[D(F, \omega)]| = e(F, 2)$, and it suffices to show that $\Delta[E(F, \omega)]$ is consistent.

If F has a model, then the Herbrand instance H_0 is consistent. By the Consistency Lemma it suffices to show the existence of a formula-mapping that carries every member of $\Delta[E(F, \omega)]$ to H_0. Such a formula-mapping will exist provided that, for all H and H' in $E(F, \omega)$ and all numbers i and j, if $\Delta(\langle H \rangle_i) = \Delta(\langle H' \rangle_j)$ then $\langle H_0 \rangle_i = \langle H_0 \rangle_j$. Suppose then that $\Delta(\langle H \rangle_i) = \Delta(\langle H' \rangle_j)$. By the Equivalence Lemma there is a term-mapping γ such that γ takes $\simeq_{F,2}$-equivalent terms to the same term and the γ-canonical formula-mapping Γ is Herbrand instance preserving on $\{H, H'\}$. Obviously $\Gamma(\Delta(\langle H \rangle_i)) = \Gamma(\Delta(\langle H' \rangle_j))$; and, just as in the proof of Corollary 1, we have $\Gamma(H) = \Gamma(\Delta(H))$ and $\Gamma(H') = \Gamma(\Delta(H'))$. Hence $\Gamma(\langle H \rangle_i) = \Gamma(\langle H' \rangle_j)$. But $\Gamma(H)$ and $\Gamma(H')$ are Herbrand instances. Thus $\langle F \rangle_i$ and $\langle F \rangle_j$ are coinstantiable, and so $\langle H_0 \rangle_i = \langle H_0 \rangle_j$. □

3. Amiability; the Monadic, Ackermann, and Minimal Gödel–Kalmár–Schütte Classes

Our method for showing the finite controllability of the classes of Chapter 2 differs from what we have just seen in much the same way that the amenability method for solvability differs from that of Chapter 1, §3. In particular, our term-mappings δ will be defined relative to given truth-assignments $\mathfrak{A}$. A simple example is furnished by the Monadic Class.

Class 2.1. Let F be a schema containing only monadic predicate letters, and let p be the number of predicate letters in F. If F has a model, then F has a model of cardinality at most 2^p.

Proof. If F has a model, then there is a truth-assignment $\mathfrak{A}$ verifying $E(F, \omega)$. Let δ be the term-mapping that takes each term t to the earliest term s such that $\mathfrak{p}_{\mathfrak{A}}(s) = \mathfrak{p}_{\mathfrak{A}}(t)$. Thus $|\delta[D(F, \omega)]| \leq d(F, 1) = 2^p$. It suffices to show that $\mathfrak{A}$ verifies $\Delta[E(F, \omega)]$. Since $\mathfrak{A}$ verifies $E(F, \omega)$, we need only show that $\mathfrak{A}(\Delta(\langle H \rangle_i)) = \mathfrak{A}(\langle H \rangle_i)$ for each Herbrand instance H of F and each i. Suppose $\langle H \rangle_i = Pt$; then $\Delta(\langle H \rangle_i) = P\delta t$, and since $\mathfrak{p}_{\mathfrak{A}}(t) = \mathfrak{p}_{\mathfrak{A}}(\delta t)$, the desired equivalence under $\mathfrak{A}$ is immediate. □

The proof just given is in a sense unnecessary. In the amenability proof

ISBN 0-201-02540-X

for Class 2.1 we constructed, for each truth-assignment $\mathfrak{A}$, a term-mapping γ such that $\gamma[D(F, \omega)] \subseteq D(F, 2^p)$ and such that if $\mathfrak{A}$ verifies $E(F, 2^p)$ then $\mathfrak{A}$ verifies $\Gamma[E(F, \omega)]$, where Γ is the γ-canonical formula-mapping. Thus the amenability proof itself establishes that every satisfiable monadic schema F has a finite model of cardinality at most $|D(F, 2^p)|$. Of course, the proof in the preceding paragraph yields the considerably smaller—and well-known—bound of 2^p on the size of the model; moreover, it is somewhat simpler than the amenability proof. This is because in the amenability proof the bound 2^p is on the *height* of terms in the image $\gamma[D(F, \omega)]$, while for finite controllability we are indifferent to a bound on the height of terms in $\delta[D(F, \omega)]$, but are concerned only with the size of this set.

We saw in Chapter 2 that canonical formula-mappings are of no use in amenability proofs beyond the Essentially Monadic Class. For the same reason, our remaining finite controllability proofs cannot be as direct as that for the Monadic Class. Given a schema F in one of our more extensive classes, there will ordinarily be truth-assignments verifying $E(F, \omega)$ but falsifying $\Delta[E(F, \omega)]$ whenever δ has finite range. For example, let $F = \forall y \exists x Pyx$, and let $\mathfrak{A}$ verify Pst iff $s = \hat{y}t$ (that is, iff s is the argument of t). Then $\mathfrak{A}$ verifies $E(F, \omega)$, but if δ has finite range then $\mathfrak{A}$ must falsify $P\delta\hat{y}t\,\delta t$ for some term t.

The point of the example is general. Given an arbitrary schema F, a truth-assignment will verify $\Delta[E(F, \omega)]$, for some δ with finite range, only if that truth-assignment treats some atoms not of the form $P\hat{y}_j t\,t$ as though they were of the form $P\hat{y}_j t\,t$, and treats some atoms not of the form $Pt\,\hat{y}_j t$ as though they were of the form $Pt\,\hat{y}_j t$ (similar requirements, naturally, hold for atoms containing k-place predicate letters, $k > 2$). Hence, given an arbitrary truth-assignment $\mathfrak{A}$ on $\bar{L}(F, \omega)$, we shall seek a formula-mapping Γ such that *many* atoms $Ps\,t$ receive the same truth-value under the induced truth-assignment $\mathfrak{A}(\Gamma)$ that the atom $P\hat{y}_j t\,t$ receives under the given truth-assignment $\mathfrak{A}$, and many atoms $Pt\,s$ receive the same truth-value under $\mathfrak{A}(\Gamma)$ that $Pt\,\hat{y}_j t$ receives under $\mathfrak{A}$. To specify which atoms $Ps\,t$ and $Pt\,s$ are to be so treated, we shall use equivalence relations of finite order on $D(F, \omega)$.

The Ackermann Class provides a particularly easy illustration of this line of argument. Since an atom $Ps\,t$ occurs in the expansion of an Ackermann schema only if the terms s and t differ in height by at most one, the formula-mapping Γ may be chosen to leave invariant each atom occurring in $E(F, \omega)$ (but not, of course, each atom in $L(F, \omega)$).

Class 2.2. Let F be a prenex· schema with a prefix $\forall y \exists x_1 \cdots \exists x_n$ and matrix containing dyadic predicate letters only. If F has a model, then F has a model whose cardinality is at most $3n \cdot d(F, 1) + 1$.

Proof. For all terms s and t in $D(F, \omega)$, let $s \simeq t$ iff $h(s) \equiv h(t) \pmod 3$.

Define the formula-mapping Γ thus:

$$\Gamma(Ps\ t) = \begin{cases} P\hat{y}t\ t & \text{if } s \simeq \hat{y}t; \\ Ps\ \hat{y}s & \text{if } t \simeq \hat{y}s; \\ Ps\ t & \text{otherwise.} \end{cases}$$

The use of congruence mod 3 assures us that Γ is well defined: we cannot have both $s \simeq \hat{y}t$ and $t \simeq \hat{y}s$.

Let $\mathfrak{A}$ be any truth-assignment verifying $E(F, \omega)$, and let δ be the term-mapping on $D(F, \omega)$ defined inductively thus: $\delta 4 = 4$; for each k, $1 \leq k \leq n$, $\delta f_{x_k}(t) = f_{x_k}(s)$, where s is the earliest term in $D(F, \omega)$ such that $s \simeq \delta t$ and $\mathfrak{p}_{\mathfrak{A}}(s) = \mathfrak{p}_{\mathfrak{A}}(\delta t)$.

There are at most $3d(F, 1)$ terms s such that, for no term u earlier than s, are both $s \simeq u$ and $\mathfrak{p}_{\mathfrak{A}}(s) = \mathfrak{p}_{\mathfrak{A}}(u)$. Hence $|\delta[D(F, \omega)]| \leq 3n \cdot d(F, 1) + 1$. We must show that $\mathfrak{A}(\Gamma)$ verifies $\Delta[E(F, \omega)]$ or, equivalently, that $\mathfrak{A}$ verifies $\Gamma(\Delta(H))$ for each Herbrand instance H. It suffices to define an instance-mapping Ψ such that the following modified form of the central condition holds for each H and each i:

$$\mathfrak{A}(\Gamma(\Delta(\langle H \rangle_i))) = \mathfrak{A}(\langle \Psi(H) \rangle_i).$$

Let $H = F^*(y/t)$; then we let $\Psi(H) = F^*(y/s)$, where s is the earliest term in $D(F, \omega)$ such that $s \simeq \delta t$ and $\mathfrak{p}_{\mathfrak{A}}(s) = \mathfrak{p}_{\mathfrak{A}}(\delta t)$.

(a)   Suppose $\langle F \rangle_i$ contains x-variables only. Then $\Delta(\langle H \rangle_i) = \langle \Psi(H) \rangle_i$. Since the terms in $\Delta(\langle H \rangle_i)$ are all of the same height, $\Gamma(\Delta(\langle H \rangle_i)) = \Delta(\langle H \rangle_i)$.

(b)   Suppose $\langle F \rangle_i = Pyx_j$. Then $\Delta(\langle H \rangle_i) = P\delta t\ \delta f_{x_j}(t) = P\delta t\ f_{x_j}(s)$. Since $\delta t \simeq s$, $\Gamma(\Delta(\langle H \rangle_i)) = Ps\ f_{x_j}(s) = \langle \Psi(H) \rangle_i$.

(c)   Suppose $\langle F \rangle_i = Px_jy$. As in (b), $\Gamma(\Delta(\langle H \rangle_i)) = \langle \Psi(H) \rangle_i$.

(d)   Suppose $\langle F \rangle_i = Pyy$. Then $\Gamma(\Delta(\langle H \rangle_i)) = \Delta(\langle H \rangle_i) = P\delta t\ \delta t$, but $\langle \Psi(H) \rangle_i = Ps\ s$. Since $\mathfrak{p}_{\mathfrak{A}}(\delta t) = \mathfrak{p}_{\mathfrak{A}}(s)$, the desired equivalence under $\mathfrak{A}$ follows. $\square$

The modification of the central condition introduced in the preceding proof will command our attention in this chapter.

Definition 3.1.   A schema F is *r-amiable*, $r > 0$, iff for each truth-assignment $\mathfrak{A}$ on $\bar{L}(F, \omega)$ there are a term-mapping δ, a formula-mapping Γ, and an instance-mapping Ψ such that

(1) $|\delta[D(F, \omega)]| \leq r$;

(2) $\mathfrak{A}(\Gamma(\Delta(\langle H \rangle_i))) = \mathfrak{A}(\langle \Psi(H) \rangle_i)$ for each Herbrand instance H of F and each i, where Δ is the δ-canonical formula-mapping.

If F is r-amiable, then F either is unsatisfiable or else has a finite model of cardinality $\leq r$. For suppose F satisfiable. Then there is a truth-assignment $\mathfrak{A}$ verifying $E(F, \omega)$. Let δ, Γ, and Ψ be as in Definition 3.1. Then $\mathfrak{A}$

ISBN 0-201-02540-X

verifies $\Psi[E(F, \omega)]$; hence, by the modified central condition (2), $\mathfrak{A}$ verifies $\Gamma(\Delta(H))$ for each Herbrand instance H of F. That is, the induced truth-assignment $\mathfrak{A}(\Gamma)$ verifies $\Delta[E(F, \omega)]$. Therefore $\Delta[E(F, \omega)]$ is consistent, and by the Finite Model Lemma the schema F has a model with universe $\delta[D(F, \omega)]$.

Definition 3.2. A class of schemata is *amiable* iff every schema in the class is *r*-amiable for some *r*.

If a class of schemata is amiable, then the class is finitely controllable. Now in Definition 3.2 we do not require a recursive function φ such that, for each F in the class, F is $\varphi(F)$-amiable. Finite controllability will follow even without such a φ. However, in all our amiability proofs we shall in fact provide such a function φ that is primitive recursive. (Thus the relation between the amiability of a class and the *r*-amiability of its members differs from that between the amenability of a class and the adequacy of expansions $E(F, p)$ of its members. In order to yield solvability, amenability must require recursive uniformity.) The amiability of a class of schemata does not appear to imply its amenability, even if we require recursively uniform amiability in the sense just mentioned. In Definition 3.1 we place no restriction on the range of Ψ, whereas in the analogous definition of adequacy it is essential that $\Psi[E(F, \omega)]$ be bounded.‡

Just as in our adequacy proofs, the use of instance-mappings avoids the issue of whether the given truth-assignment $\mathfrak{A}$ does or does not verify the expansion of F. Our constructions do not depend on any particular attributes of $\mathfrak{A}$. Of course, a finite model for F is assured only if $\mathfrak{A}$ does verify the expansion.

The modified central condition may be rewritten thus:

$$\mathfrak{A}(\Gamma)(\Delta(\langle H \rangle_i)) = \mathfrak{A}(\langle \Psi(H) \rangle_i).$$

This form of the condition indicates more clearly that our aim in constructing the formula-mapping Γ is to induce a suitable truth-assignment $\mathfrak{A}(\Gamma)$. But note that the modified central condition does *not* demand that $\mathfrak{A}(\Gamma)$ verify $E(F, \omega)$, even when $\mathfrak{A}$ does verify $E(F, \omega)$. We shall return to this possible divergence in the actions of $\mathfrak{A}$ and $\mathfrak{A}(\Gamma)$ on $E(F, \omega)$ during the course of our amiability proof for the Minimal Gödel–Kalmár–Schütte Class.

Class 2.3. Let F be a prenex schema with prefix $\forall y_1 \forall y_2 \exists x$ and matrix containing dyadic predicate letters only. Then F is $(49d(F, 2) + 1)$-amiable.

Proof.   Let $\mathfrak{A}$ be any truth-assignment on $\tilde{L}(F, \omega)$. The role that congruence mod 3 had in the amiability proof for the Ackermann Class is now

ISBN 0-201-02540-X

---

‡ Nonetheless, an argument just like that in §7 of Chapter 2 establishes that if a class $\mathfrak{C}_1$ is full in a class $\mathfrak{C}_2$ and if $\mathfrak{C}_2$ is not finitely controllable, then $\mathfrak{C}_1$ is not amiable.

taken by a more powerful combinatorial device, first used by Gödel in his finite controllability proof for the Gödel–Kalmár–Schütte Class (1933).

Combinatorial Lemma.   There is a function α: $D(F, \omega) \rightarrow \{1,\dots,7\}$ such that, for all terms t and s, if $\alpha(t) = \alpha(\hat{y}_j s)$, then $\alpha(s) \neq \alpha(\hat{y}_k t)$, $1 \leq j, k \leq 2$.

Proof.   Let $\alpha(t) = 1$ and, if $h(t) > 1$, let $\alpha(t)$ be the least i such that both $\alpha(\hat{y}_1 t)$ and $\alpha(\hat{y}_2 t)$ occur on the ith line of Table 3.1. The required property of α is easily verified.   □

Table 3.1

2	3	5
3	4	6
4	5	7
5	6	1
6	7	2
7	1	3
1	2	4

We define the equivalence relation $\simeq$:

$$s \simeq t \quad \text{iff} \quad \alpha(s) = \alpha(t) \quad \text{and} \quad \mathfrak{p}_{\mathfrak{A}}(s) = \mathfrak{p}_{\mathfrak{A}}(t).$$

(Thus $\simeq$ depends on the given truth-assignment $\mathfrak{A}$.) From the Combinatorial Lemma it follows that for no terms s and t are both $s \simeq \hat{y}_j t$ and $t \simeq \hat{y}_k s$, $1 \leq j, k \leq 2$, and hence also that for no term t is $t \simeq \hat{y}_j t$, $j = 1, 2$. Call a term t *tight* iff $\hat{y}_1 t \simeq \hat{y}_2 t$ only if $\hat{y}_1 t = \hat{y}_2 t$, and let Γ be the formula-mapping defined thus: for each atom $Ps\, t$ in $L(F, \omega)$,

$$\Gamma(Ps\ t) = \begin{cases} P\hat{y}_j t\ t & \text{if } s \simeq \hat{y}_j t \text{ and } t \text{ is tight, } j = 1, 2; \\ Ps\ \hat{y}_j s & \text{if } t \simeq \hat{y}_j s \text{ and } s \text{ is tight, } j = 1, 2; \\ Pu\ u & \text{if } s \simeq t \text{ but } s \neq t, \text{ where } u \text{ is the} \\ & \text{earliest term such that } s \simeq u \simeq t; \\ Ps\ t & \text{otherwise.} \end{cases}$$

No ambiguity arises from this definition of Γ. If $Ps\ t$ falls under the first clause for both values of j or under the second clause for both values of j then the requirement that the terms t and s be tight assures us that $\hat{y}_1 t = \hat{y}_2 t$ or $\hat{y}_1 s = \hat{y}_2 s$; and the Combinatorial Lemma implies that no atom $Ps\ t$ falls under more than one of the four clauses. (Each essentially monadic atom $Pt\ t$ falls under the fourth clause.)

The induced truth-assignment $\mathfrak{A}(\Gamma)$ may well falsify $E(F, \omega)$. For there are terms t such that Γ carries neither $P\hat{y}_1 t\ \hat{y}_2 t$ nor $P\hat{y}_2 t\ \hat{y}_1 t$ to itself; yet, for all terms t—whether tight or not—the formula-mapping Γ does carry $P\hat{y}_j t\ t$ and $Pt\ \hat{y}_j t$ to themselves, $j = 1, 2$. Hence $\mathfrak{A}(\Gamma)$ agrees with $\mathfrak{A}$ on those atomic subformulas of the Herbrand instance $H = F^*(y_1/\hat{y}_1 t, y_2/\hat{y}_2 t)$ that contain t, but need not agree with $\mathfrak{A}$ on those atomic subformulas

ISBN 0-201-02540-X

that contain both $\hat{y}_1 t$ and $\hat{y}_2 t$. But then, even if $\mathfrak{A}$ verifies every Herbrand instance of F, there is no guarantee that $\mathfrak{A}(\Gamma)$ verifies H.‡

However, the action of $\mathfrak{A}(\Gamma)$ on $E(F, \omega)$ is not our concern. Rather, we seek a term-mapping δ with finite range such that $\mathfrak{A}(\Gamma)$ acts appropriately on each instance $F^M(y_1/\delta\hat{y}_1 t, y_2/\delta\hat{y}_2 t, x/\delta t)$ of F. Since we shall not, in general, pick $\delta t = f_x(\delta\hat{y}_1 t\ \delta\hat{y}_2 t)$, such instances of F will not, in general, be Herbrand instances. To obtain the term-mapping δ we first establish the following

Existence Condition. There is a term-mapping μ such that, for each term $t \neq \dagger$,

(a) μt is tight;

(b) $\hat{y}_i t \simeq \hat{y}_i \mu t$, $i = 1, 2$;

(c) $\mathfrak{A}(\Gamma(P\hat{y}_i t\ \hat{y}_j t)) = \mathfrak{A}(P\hat{y}_i \mu t\ \hat{y}_j \mu t)$ for each predicate letter P of F and $1 \leq i, j \leq 2$.

Proof.   Let $\mu t = f_x(s_1 s_2)$, where $\Gamma(P\hat{y}_1 t\ \hat{y}_2 t) = Ps_1 s_2$. (In the amenability proof for this class we let $\gamma t = \pi f_x(s_1 s_2)$, where $\Gamma(P\hat{y}_1 t\ \hat{y}_2 t) = Ps_1 s_2$; of course the Γ there is not the Γ here.) From the definition of Γ we see that $\hat{y}_1 t \simeq s_1$ and $\hat{y}_2 t \simeq s_2$, that $s_1 \simeq s_2$ only if $s_1 = s_2$, and that $\Gamma(P\hat{y}_2 t\ \hat{y}_1 t) = Ps_2 s_1$. Hence the term μt fulfills conditions (a) and (b), and fulfills condition (c) when $\langle i, j \rangle = \langle 1, 2 \rangle$ and when $\langle i, j \rangle = \langle 2, 1 \rangle$. But μt also fulfills condition (c) when $\langle i, j \rangle = \langle 1, 1 \rangle$ or $\langle 2, 2 \rangle$; for then $\Gamma(P\hat{y}_i t\ \hat{y}_j t) = P\hat{y}_i t\ \hat{y}_j t$, and since $\hat{y}_i t \simeq \hat{y}_i \mu t$, $\mathfrak{p}_{\mathfrak{A}}(\hat{y}_i t) = \mathfrak{p}_{\mathfrak{A}}(\hat{y}_i \mu t)$. $\square$

Properties (a), (b), and (c) of μ in conjunction with the definition of Γ quickly yield

(I) $\mathfrak{A}(\Gamma(\langle F^M(y_1/\hat{y}_1 t, y_2/\hat{y}_2 t, x/\mu t)\rangle_i)) = \mathfrak{A}(\langle F^*(y_1/\hat{y}_1 \mu t, y_2/\hat{y}_2 \mu t)\rangle_i)$

for each $t \neq \dagger$ and each i.

Hence $\mathfrak{A}(\Gamma)$ verifies the instance $F^M(y_1/\hat{y}_1 t, y_2/\hat{y}_2 t, x/\mu t)$ whenever $\mathfrak{A}$ verifies the Herbrand instance $F^*(y_1/\hat{y}_1 \mu t, y_2/\hat{y}_2 \mu t)$. Thus if $\mathfrak{A}$ verifies $E(F, \omega)$, then—even though, as we have emphasized, $\mathfrak{A}(\Gamma)$ need not verify $E(F, \omega)$—$\mathfrak{A}(\Gamma)$ provides a model for F over $D(F, \omega)$: for all terms t_1 and t_2 there is a term s, namely, $s = \mu f_x(t_1 t_2)$, such that $\mathfrak{A}(\Gamma)$ verifies $F^M(y_1/t_1, y_2/t_2, x/s)$. But our task is to show that if $\mathfrak{A}$ verifies $E(F, \omega)$ then $\mathfrak{A}(\Gamma)$ provides a model for F over a finite subset of $D(F, \omega)$. Hence we again define an equivalence relation on $D(F, \omega)$ of finite order. For all terms s and $t \neq \dagger$, let $s \sim t$ iff $\mathfrak{p}_{\mathfrak{A}}(\hat{y}_1 s, \hat{y}_2 s) = \mathfrak{p}_{\mathfrak{A}}(\hat{y}_1 t, \hat{y}_2 t)$ and $\alpha(\hat{y}_i s) = \alpha(\hat{y}_i t)$,

---

‡ For example, suppose F^M is $(Py_1 y_2 \Leftrightarrow -Py_1 x)$. Clearly there is a truth-assignment $\mathfrak{A}$ that verifies $E(F, \omega)$ and for which there are a tight term t and a term $s \simeq \hat{y}_1 t$ such that $\mathfrak{A}$ verifies $P\hat{y}_1 t\ t$ and falsifies $Ps\ t$. (Since $\mathfrak{A}$ verifies $E(F, \omega)$, $\mathfrak{A}$ then verifies $Ps\ f_x(s\ t)$.) Now $\Gamma(Ps\ t) = P\hat{y}_1 t\ t$; hence $\mathfrak{A}(\Gamma)$ verifies $Ps\ t$. And $\Gamma(Ps\ f_x(s\ t)) = Ps\ f_x(s\ t)$; hence $\mathfrak{A}(\Gamma)$ verifies $Ps\ f_x(s\ t)$. But then $\mathfrak{A}(\Gamma)$ falsifies the Herbrand instance $F^*(y_1/s, y_2/t)$.

ISBN 0-201-02540-X

$i = 1, 2$. Clearly there are at most $49d(F, 2)$ $\sim$-equivalence classes of terms $\neq \mathbf{1}$, and if $s \sim t$, then $\hat{y}_i s \simeq \hat{y}_i t$, $i = 1, 2$. Equally clearly, for all terms s and t, if $\mu t \sim s$, then s fulfills (b) and (c) of the Existence Condition, that is, if $\mu t \sim s$ then

(b) $\hat{y}_i t \simeq \hat{y}_i s$, $\mathbf{1} = 1, 2$;

(c) $\mathfrak{A}(\Gamma(P\hat{y}_i t \ \hat{y}_j t)) = \mathfrak{A}(P\hat{y}_i s \ \hat{y}_j s)$ for each predicate letter P of F and $1 \le i, j \le 2$.

And then, if $\mu t \sim s$ and s is tight, s fulfills

(II) $\mathfrak{A}(\Gamma(\langle F^M(y_1/\hat{y}_1 t, \ y_2/\hat{y}_2 t, \ x/s)\rangle_i)) = \mathfrak{A}(\langle F^*(y_1/\hat{y}_1 s, \ y_2/\hat{y}_2 s)\rangle_i)$ for each i.

(For if $\langle F \rangle_i$ contains the variable x, then by condition (b) and the definition of Γ, $\Gamma(\langle F^M(y_1/\hat{y}_1 t, \ y_2/\hat{y}_2 t, \ x/s)\rangle_i) = \langle F^*(y_1/\hat{y}_1 s, \ y_2/\hat{y}_2 s)\rangle_i$; and if $\langle F \rangle_i$ contains only y-variables, then (II) follows immediately from condition (c).) Consequently there is a finite subset S of $D(F, \omega)$, a subset that contains at most one tight representative of each $\sim$-equivalence class, with the following property: if $\mathfrak{A}$ verifies $E(F, \omega)$, then for all terms t_1 and t_2 there is a term s in S such that $\mathfrak{A}(\Gamma)$ verifies $F^M(y_1/t_1, \ y_2/t_2, x/s)$. To construct S explicitly (and to complete the amiability proof for Class 2.3), we define inductively the term-mapping δ: let $\delta \mathbf{1} = \mathbf{1}$; for each $t \neq \mathbf{1}$ let δt be the earliest tight term s such that $s \sim \mu f_x(\delta \hat{y}_1 t \ \delta \hat{y}_2 t)$. From (II) we have at once

(III) $\mathfrak{A}(\Gamma(\langle F^M(y_1/\delta\hat{y}_1 t, \ y_2/\delta\hat{y}_2 t, \ x/\delta t)\rangle_i)) = \mathfrak{A}(\langle F^*(y_1/\hat{y}_1\delta t, \ y_2/\hat{y}_2\delta t)\rangle_i)$ for each term $t \neq \mathbf{1}$ and each i.

But, if $H = F^*(y_1/\hat{y}_1 t, \ y_2/\hat{y}_2 t)$ and we set $\Psi(H) = F^*(y_1/\hat{y}_1\delta t, \ y_2/\hat{y}_2\delta t)$, then (III) is the modified central condition. Thus we have shown F to be $(49d(F, 2) + 1)$-amiable, since $|\delta[D(F, \omega)]| \le 49d(F, 2) + 1$. $\square$

4. The Amiability Lemma

The treatment of Classes 2.4–2.8 is, as in Chapter 2, simplified by shifting attention from formula-mappings and instance-mappings to sequence-mappings and term-mappings.

Amiability Lemma.   Let F be a prenex schema whose prefix ends with $\exists x_n$. Then F is r-amiable if there exist, for each truth-assignment $\mathfrak{A}$ on $\tilde{L}(F, \omega)$, a term-mapping δ and a sequence-mapping β^* such that

(1) $|\delta[D(F, \omega)]| \le r$;

(2) $\mathfrak{p}_{\mathfrak{A}}(\beta^*(\delta\hat{v}_1 t, \ldots, \delta\hat{v}_p t)) = \mathfrak{p}_{\mathfrak{A}}(\hat{v}_1\delta t, \ldots, \hat{v}_p \delta t)$ whenever t is an x_n-term in $D(F, \omega)$ and $\langle v_1, \ldots, v_p \rangle$ is an arrangement of some set in $b(F)$.

ISBN 0-201-02540-X

Proof.   Given a truth-assignment $\mathfrak{A}$, let δ and β^* fulfill (1) and (2). Let Γ be the formula-mapping that carries each atom $Pt_1\cdots t_p$ in $L(F, \omega)$ to $P\beta^*(t_1,\ldots,t_p)$. Let Ψ be the instance-mapping that carries each Herbrand instance H of F to $F^*(y_1/\hat{y}_1\delta t,\ldots,y_m/\hat{y}_m\delta t)$, where t is the substituent in H for x_n and $y_1,\ldots,y_m$ are all the y-variables of F.

If t is the substituent in H for x_n and if $\langle F\rangle_i = Pv_1\cdots v_p$, then $\Gamma(\Delta(\langle H\rangle_i))$ $= P\beta^*(\delta\hat{v}_1t,\ldots,\delta\hat{v}_pt)$ and $\langle\Psi(H)\rangle_i = P\hat{v}_1\delta t\cdots\hat{v}_p\delta t$. By condition (2) we have $\mathfrak{A}(\Gamma(\Delta(\langle H\rangle_i))) = \mathfrak{A}(\langle\Psi(H)\rangle_i)$. Thus the modified central condition is fulfilled and so, by condition (1), F is r-amiable. $\square$

The sequence-mappings β^* will always preserve permutations and repetitions. That is, β^* will fulfill condition (3) of Chapter 2, §3. Given this property of β^*, condition (2) will hold for all arrangements of a set in $b(F)$ provided that it holds for any one arrangement of this set.

We introduced our discussion of amiability by observing that "an [arbitrary] truth-assignment $\mathfrak{A}$ will verify $\Delta[E(F, \omega)]$ for some δ with finite range" only if there is a "formula-mapping Γ such that *many* atoms Pst receive the same truth-value under the induced truth-assignment $\mathfrak{A}(\Gamma)$ that the atom $P\hat{y}_jt\,t$ receives under the given truth-assignment $\mathfrak{A}$" (page 83). This observation should make clear that our primary concern in the construction of a sequence-mapping β^* is to fulfill condition (2) of the Amiability Lemma for x-focused sets $\{v_1,\ldots,v_p\}$. We cannot preclude from our consideration truth-assignments $\mathfrak{A}$ that verify an atom $Pt_1\cdots t_p$ iff $\langle t_1,\ldots,t_p\rangle$ is $\langle\hat{v}_1s,\ldots,\hat{v}_ps\rangle$ for some term s. But, since δ is to have finite range, there will be x_n-terms t such that $\langle\delta\hat{v}_1t,\ldots,\delta\hat{v}_pt\rangle$ is not $\langle\hat{v}_1s,\ldots,\hat{v}_ps\rangle$ for any term s. Hence to fulfill condition (2), the sequence-mapping β^* must carry some sequences that are not $\langle\hat{v}_1s,\ldots,\hat{v}_ps\rangle$ for any s to sequences that are $\langle\hat{v}_1s,\ldots,\hat{v}_ps\rangle$ for some s. To specify which sequences are to be subject to such special treatment, we shall use equivalence relations $\simeq$ of finite order on $D(F, \omega)$. We must avoid asking too much of β^*: we must not require that β^* carry a sequence of terms both to $\langle\hat{v}_1s,\ldots,\hat{v}_ps\rangle$ and to $\langle\hat{w}_1t,\ldots,\hat{w}_pt\rangle$ unless we are sure that $\langle\hat{v}_1s,\ldots,\hat{v}_ps\rangle = \langle\hat{w}_1t,\ldots,\hat{w}_pt\rangle$. To assure that we do not ask too much, we shall rely on combinatorial properties of the equivalence relations $\simeq$ and various notions of tightness of terms. (The notion of tightness plays a role like that of the identity preservation requirement in the amenability proofs. To obtain the amenability of the progressively more extensive classes of Chapter 2 we progressively weakened that requirement; similarly, to obtain amiability we shall progressively weaken the notion of a tight term.) Given a suitable sequence-mapping β^* we shall be able to obtain condition (2) for x-focused sets by so constructing δ that if $\langle v_1,\ldots,v_p\rangle$ is an arrangement of an x-focused set in $b(F)$, then $\langle\delta\hat{v}_1t,\ldots,\delta\hat{v}_pt\rangle$ is among the sequences treated specially by β^*.

The argument for Class 2.3 remains our model. We shall construct δ in

ISBN 0-201-02540-X

two stages. First we shall find a term-mapping μ such that μt is tight (in some appropriate sense), $\hat{y}_i t \simeq \hat{y}_i \mu t$ for each y-variable y_i, and μt fulfills a condition like condition (2) for y-sets. (For each of Classes 2.4–2.8 the definition of μ will closely parallel that of the term-mapping γ in the amenability proof, save for the absence of a pigeonhole-mapping π.) The properties of μ will imply an analogue to condition (I) of the argument for Class 2.3. Thus, if $\mathfrak{A}$ verifies $E(F, \omega)$ and if Γ is defined from β^* as in the proof of the Amiability Lemma, then $\mathfrak{A}(\Gamma)$ provides a model for F over $D(F, \omega)$. But we shall not stop to prove this analogue to (I): our interest is in obtaining a finite model for F, and since μ will have infinite range, it will not yield such a model. Rather, we immediately define the term-mapping δ from the term-mapping μ, using an equivalence relation of finite order to make the range of δ finite. (For Classes 2.4 and 2.6, just as for Class 2.3, an equivalence relation distinct from that used in the construction of β^* is used for δ. For Classes 2.7 and 2.8, however, the equivalence relation needed for β^* also yields δ.)

5. Skolem Classes

For the two definitions and three lemmas that follow, let $F = \forall y_1 \cdots \forall y_m \exists x_1 \cdots \exists x_n F^M$ be an arbitrary but fixed Skolem schema.

Combinatorial Lemma 1. There is a number κ_m and a function $\alpha \colon D(F, \omega) \to \{1, \ldots, \kappa_m\}$ such that, for all terms s and t, if $\alpha(s) = \alpha(\hat{y}_i t)$ then $\alpha(t) \neq \alpha(\hat{y}_j s)$, $1 \le i, j \le m$, and $\alpha(s) = \alpha(\hat{x}_k s)$, $1 \le k \le n$.

Proof. See §2 of the Appendix. Note that the properties of α also imply that $\alpha(t) \neq \alpha(\hat{y}_i t)$ for each $i \le m$. Incidentally, our proof yields the upper bound of $(m^2 + m + 1)^{m+1}$ on κ_m; and we have seen that we may take $\kappa_1 = 3$ and $\kappa_2 = 7$. $\square$

Definition 3.3. An equivalence relation $\simeq$ on $D(F, \omega)$ is *faithful* iff, for all terms s and t, $s \simeq t$ only if $\alpha(s) = \alpha(t)$ and $\chi(s) = \chi(t)$, where α is the function given by Combinatorial Lemma 1.

Notational Convention. Let $\simeq$ be an equivalence relation on $D(F, \omega)$. Then we automatically extend $\simeq$ in the natural way to an equivalence relation on sequences from $D(F, \omega)$: $\langle s_1, \ldots, s_p \rangle \simeq \langle t_1, \ldots, t_p \rangle$ iff $s_i \simeq t_i$ for every $i \le p$.

Faithfulness Lemma 1. Let $\simeq$ be a faithful equivalence relation on $D(F, \omega)$, let s and t be x_n-terms, and let $\langle v_1, \ldots, v_p \rangle$ and $\langle w_1, \ldots, w_p \rangle$ be x-focused sequences of variables such that $\langle \hat{v}_1 s, \ldots, \hat{v}_p s \rangle \simeq \langle \hat{w}_1 t, \ldots, \hat{w}_p t \rangle$. Then, for each $i \le p$, either $v_i = w_i =$ some x-variable or else v_i and w_i are both y-variables.

Proof. Case 1. Suppose v_i and w_i are both x-variables. Then $v_i = \chi(\hat{v}_i s) = \chi(\hat{w}_i t) = w_i$.

ISBN 0-201-02540-X

Case 2. Suppose v_i is an x-variable and w_i is a y-variable. Then $\alpha(s) = \alpha(\hat{v}_i s) = \alpha(\hat{w}_i t)$. Since $\langle w_1, \ldots, w_p \rangle$ is x-focused, some w_j is an x-variable, and $\alpha(\hat{w}_j t) = \alpha(t)$. If v_j is also an x-variable, then $\alpha(s) = \alpha(\hat{v}_j s) = \alpha(\hat{w}_j t) = \alpha(t)$, which, since $\alpha(s) = \alpha(\hat{w}_i t)$, contradicts Combinatorial Lemma 1. If v_j is a y-variable, then $\alpha(\hat{v}_j s) = \alpha(\hat{w}_j t) = \alpha(t)$, which, since $\alpha(s) = \alpha(\hat{w}_i t)$, also contradicts Combinatorial Lemma 1. Hence Case 2 is impossible.

Case 3. Suppose v_i is a y-variable and w_i is an x-variable. Just as in Case 2, this is impossible.

It follows that if v_i and w_i do not fall under Case 1, then they are both y-variables. $\square$

Definition 3.4. Let $\simeq$ be an equivalence relation on $D(F, \omega)$ and let $\langle v_1, \ldots, v_p \rangle$ be an x-focused sequence of variables of F. A sequence $\langle t_1, \ldots, t_p \rangle$ of terms is a *t-$\simeq$-instance* of $\langle v_1, \ldots, v_p \rangle$ iff t is an x_n-term and, for each $i \leq p$, (1) if v_i is a y-variable then $t_i \simeq \hat{v}_i t$, and (2) if v_i is an x-variable then $t_i = \hat{v}_i t$.

If $\langle v_1, \ldots, v_p \rangle$ contains at least one y-variable, then for each x_n-term t, there may be many t-$\simeq$-instances of $\langle v_1, \ldots, v_p \rangle$. If $\langle t_1, \ldots, t_p \rangle$ is a t-$\simeq$-instance of $\langle v_1, \ldots, v_p \rangle$, then $\langle t_1, \ldots, t_p \rangle \simeq \langle \hat{v}_1 t, \ldots, \hat{v}_p t \rangle$.

In the amiability proofs for Classes 2.4 and 2.6, the sequence-mapping β^* will carry t-$\simeq$-instances of x-focused sequences $\langle v_1, \ldots, v_p \rangle$ to sequences $\langle \hat{v}_1 t, \ldots, \hat{v}_p t \rangle$. That is, β^* will carry t-$\simeq$-instances to t-instances (see page 50). Moreover, we shall so construct the term-mapping δ that if $\langle v_1, \ldots, v_p \rangle$ is an arrangement of an x-focused set in $b(F)$, then $\langle \delta\hat{v}_1 t, \ldots, \delta\hat{v}_p t \rangle$ is a δt-$\simeq$-instance of $\langle v_1, \ldots, v_p \rangle$, and is carried by β^* to the δt-instance of $\langle v_1, \ldots, v_p \rangle$. (Clearly this yields condition (2) of the Amiability Lemma for x-focused sets.) We shall avoid the danger of conflicting requirements on β^* by exploiting suitable notions of tightness, Faithfulness Lemma 1, and

Faithfulness Lemma 2. Let $\simeq$ be a faithful equivalence relation on $D(F, \omega)$, and let s and t be x_n-terms. If $s \neq t$, then no s-$\simeq$-instance of any x-focused sequence is also a t-$\simeq$-instance of an x-focused sequence.

Proof. Suppose $\langle t_1, \ldots, t_p \rangle$ is both an s-$\simeq$-instance of $\langle v_1, \ldots, v_p \rangle$ and a t-$\simeq$-instance of $\langle w_1, \ldots, w_p \rangle$. Hence $\langle t_1, \ldots, t_p \rangle \simeq \langle \hat{v}_1 s, \ldots, \hat{v}_p s \rangle \simeq \langle \hat{w}_1 t, \ldots, \hat{w}_p t \rangle$. Let v_i be an x-variable; by Faithfulness Lemma 1, $v_i = w_i$. Hence, by clause (2) of Definition 3.4, $\hat{v}_i s = t_i = \hat{w}_i t$. Since s and t are x_n-terms, $s = t$. $\square$

Class 2.4. Let $F = \forall y_1 \cdots \forall y_m \exists x_1 \cdots \exists x_n F^M$ be a Skolem schema whose basis contains no distinct y-sets of cardinality >1. Then F is $(n \cdot ((n + 1) \cdot \kappa_m)^m d(F, m) + 1)$-amiable.

Proof. We may assume that the one y-set of cardinality >1 in $b(F)$, if any, is $\{y_1, \ldots, y_l\}$ for some $l \leq m$. Let $\mathfrak{A}$ be any truth-assignment on $\bar{L}(F, \omega)$. We define an equivalence relation $\simeq$ on $D(F, \omega)$ thus: $s \simeq t$ iff $\alpha(s)$

ISBN 0-201-02540-X

$= \alpha(t)$, $\chi(s) = \chi(t)$, and $\mathfrak{p}_{\mathfrak{A}}(s) = \mathfrak{p}_{\mathfrak{A}}(t)$. Hence $\approx$ is faithful. A sequence of terms is *tight* iff it contains no distinct but equivalent members. A term t is *tight* iff $\langle \hat{y}_1 t,\dots,\hat{y}_m t \rangle$ is tight. Let η be the term-mapping that takes each term t to its earliest equivalent.

We construct a sequence-mapping β^* that possesses the following four properties for all sequences $\langle t_1,\dots,t_p \rangle$ from $D(F, \omega)$, $p \geq 1$:

(β^*-1)   if, for some tight x_n-term t, $\langle t_1,\dots,t_p \rangle$ is a t-$\approx$-instance of an x-focused sequence $\langle v_1,\dots,v_p \rangle$, then $\beta^*(t_1,\dots,t_p) = \langle \hat{v}_1 t,\dots,\hat{v}_p t \rangle$;

(β^*-2) β^* preserves permutations and repetitions;

(β^*-3) $\beta^*(t_1,\dots,t_p) \approx \langle t_1,\dots,t_p \rangle$;

(β^*-4) $\beta^*(t_1,\dots,t_p)$ is a tight sequence.

Indeed, if $\langle t_1,\dots,t_p \rangle$ is a t-$\approx$-instance of $\langle v_1,\dots,v_p \rangle$, where t is a tight x_n-term and $\langle v_1,\dots,v_p \rangle$ is an x-focused sequence, then let $\beta^*(t_1,\dots,t_p) = \langle \hat{v}_1 t,\dots,\hat{v}_p t \rangle$; otherwise let $\beta^*(t_1,\dots,t_p) = \langle \eta t_1,\dots,\eta t_p \rangle$. To show that no ambiguity arises in this construction, it suffices by Faithfulness Lemma 2 to show that if $\langle v_1,\dots,v_p \rangle$ and $\langle w_1,\dots,w_p \rangle$ have a common t-$\approx$-instance, where t is a tight x_n-term, then $\langle \hat{v}_1 t,\dots,\hat{v}_p t \rangle = \langle \hat{w}_1 t,\dots,\hat{w}_p t \rangle$. But if they have a common t-$\approx$-instance, then $\langle \hat{v}_1 t,\dots,\hat{v}_p t \rangle \approx \langle \hat{w}_1 t,..,\hat{w}_p t \rangle$. By Faithfulness Lemma 1, for each $i \leq p$, either $v_i = w_i =$ some x-variable or else v_i and w_i are y-variables. In the former case $\hat{v}_i t = \hat{w}_i t$ immediately. In the latter case $\hat{v}_i t = \hat{w}_i t$ since t is tight. Hence $\langle \hat{v}_1 t,\dots,\hat{v}_p t \rangle = \langle \hat{w}_1 t,\dots,\hat{w}_p t \rangle$.

We explicitly constructed the sequence-mapping β^* so as to fulfill (β^*-1). Clearly β^* also fulfills (β^*-2) and (β^*-3). To show (β^*-4), two observations suffice. First, each sequence $\langle \eta t_1,\dots,\eta t_p \rangle$ is tight. Second, if t is tight and $\langle v_1,\dots,v_p \rangle$ is x-focused, then $\langle \hat{v}_1 t,..,\hat{v}_p t \rangle$ is tight. For if $\hat{v}_i t \approx \hat{v}_j t$ then, by Faithfulness Lemma 1 applied, say, to the x-focused sequences $\langle v_i, x_n \rangle$ and $\langle v_j, x_n \rangle$, either $v_i = v_j =$ some x-variable or else v_i and v_j are y-variables. In either case, $\hat{v}_i t = \hat{v}_j t$. Hence $\langle \hat{v}_1 t,\dots,\hat{v}_p t \rangle$ contains no distinct but equivalent members.

Existence Condition.   There is a term-mapping μ such that, for every x_n-term t,

(a) μt is a tight x_n-term;

(b) $\hat{y}_i t \approx \hat{y}_i \mu t$, $1 \leq i \leq m$;

(c) $\mathfrak{p}_{\mathfrak{A}}(\beta^*(\hat{y}_1 t,\dots,\hat{y}_l t)) = \mathfrak{p}_{\mathfrak{A}}(\hat{y}_1 \mu t,\dots,\hat{y}_l \mu t)$.

Proof.   For each x_n-term t let $\mu t = f_{x_n}(s_1 \cdots s_m)$, where $\langle s_1,\dots,s_m \rangle$ is determined from $\langle \hat{y}_1 t,\dots,\hat{y}_m t \rangle$ by mimicking clauses (γ-1)–(γ-3) of the definition of γ in the amenability proof for Class 2.4. That is,

(μ-1) $\langle s_1,\dots,s_l \rangle = \beta^*(\hat{y}_1 t,\dots,\hat{y}_l t)$;

(μ-2)   for each j, $l < j \leq m$, if there is an $i \leq l$ such that $\hat{y}_j t \approx \hat{y}_i t$, then let $s_j = s_i$ for the least such i;

ISBN 0-201-02540-X

(μ-3)   for each j, $l < j \le m$, if s_j is not determined by (μ-2), then let $s_j = \eta \hat{y}_j t$.

From properties (β^*-3) and (β^*-4) it follows that $\hat{y}_i t \simeq s_i$, $1 \le i \le l$, and that $\langle s_1, \ldots, s_l \rangle$ is a tight sequence. Hence $\hat{y}_i t \simeq s_i$, $1 \le i \le m$, whence μt fulfills condition (b); and $\langle s_1, \ldots, s_m \rangle$ is a tight sequence, whence μt fulfills condition (a). Condition (c) is immediate. $\square$

Construction of δ.   First we define an equivalence relation $\sim$ on x_n-terms in $D(F, \omega)$: for all x_n-terms s and t, let $s \sim t$ iff $\mathfrak{p}_\mathfrak{A}(\hat{y}_1 s, \ldots, \hat{y}_m s) = \mathfrak{p}_\mathfrak{A}(\hat{y}_1 t, \ldots, \hat{y}_m t)$, $\alpha(\hat{y}_i s) = \alpha(\hat{y}_i t)$, and $\chi(\hat{y}_i s) = \chi(\hat{y}_i t)$, $1 \le i \le m$. Recall that $\chi[D(F, \omega)]$ has cardinality $n + 1$ (see page 13). Hence there are at most $(\kappa_m \cdot (n + 1))^m d(F, m)$ $\sim$-equivalence classes of x_n-terms. Note that if $s \sim t$, then $\hat{y}_i s \simeq \hat{y}_i t$ for each $i \le m$, since $\alpha(\hat{y}_i s) = \alpha(\hat{y}_i t)$, $\chi(\hat{y}_i s) = \chi(\hat{y}_i t)$, and $\mathfrak{p}_\mathfrak{A}(\hat{y}_i s) = \mathfrak{p}_\mathfrak{A}(\hat{y}_i t)$.

Let $\delta\mathbf{1} = \mathbf{1}$. For each x_n-term t and each $k \le n$ let $\delta \hat{x}_k t = \hat{x}_k t'$, where t' is the earliest tight $\sim$-equivalent of $\mu f_{x_n}(\delta \hat{y}_1 t \cdots \delta \hat{y}_m t)$; such a t' exists, since by (a) of the Existence Condition $\mu f_{x_n}(\delta \hat{y}_1 t \cdots \delta \hat{y}_m t)$ is tight. Clearly the range of δ contains at most $n \cdot (\kappa_m \cdot (n + 1))^m d(F, m) + 1$ members; this yields condition (1) of the Amiability Lemma.

Let t be an x_n-term, and let $s = f_{x_n}(\delta \hat{y}_1 t \cdots \delta \hat{y}_m t)$. Thus $\delta t \sim \mu s$. And then $\hat{y}_i \delta t \simeq \hat{y}_i \mu s$ for each $i \le m$. Since $\hat{y}_i \mu s \simeq \hat{y}_i s = \delta \hat{y}_i t$ by (b) of the Existence Condition, we have $\hat{y}_i \delta t \simeq \delta \hat{y}_i t$, $1 \le i \le m$. Moreover, by the construction of δ, $\delta \hat{x}_k t = \hat{x}_k \delta t$ for each $k \le n$. Hence if $\langle v_1, \ldots, v_p \rangle$ is x-focused, then $\langle \delta \hat{v}_1 t, \ldots, \delta \hat{v}_p t \rangle$ is a δt-$\simeq$-instance of $\langle v_1, \ldots, v_p \rangle$. Consequently, by ($\beta^*$-1) and the tightness of δt, if $\langle v_1, \ldots, v_p \rangle$ is an arrangement of an x-focused set in $b(F)$, then $\beta^*(\delta \hat{v}_1 t, \ldots, \delta \hat{v}_p t) = \langle \hat{v}_1 \delta t, \ldots, \hat{v}_p \delta t \rangle$. Hence $\mathfrak{p}_\mathfrak{A}(\beta^*(\delta \hat{v}_1 t, \ldots, \delta \hat{v}_p t)) = \mathfrak{p}_\mathfrak{A}(\hat{v}_1 \delta t, \ldots, \hat{v}_p \delta t)$, and condition (2) of the Amiability Lemma is verified for the x-focused sets in $b(F)$.

To verify condition (2) for the y-sets in $b(F)$ it suffices to consider just the sequence $\langle y_1, \ldots, y_l \rangle$ and for each $i \le m$, the sequence $\langle y_i \rangle$. By (c) of the Existence Condition $\mathfrak{p}_\mathfrak{A}(\beta^*(\delta \hat{y}_1 t, \ldots, \delta \hat{y}_l t)) = \mathfrak{p}_\mathfrak{A}(\hat{y}_1 \mu s, \ldots, \hat{y}_l \mu s)$; since $\mu s \sim \delta t$, $\mathfrak{p}_\mathfrak{A}(\hat{y}_1 \mu s, \ldots, \hat{y}_l \mu s) = \mathfrak{p}_\mathfrak{A}(\hat{y}_1 \delta t, \ldots, \hat{y}_l \delta t)$. Hence $\mathfrak{p}_\mathfrak{A}(\beta^*(\delta \hat{y}_1 t, \ldots, \delta \hat{y}_l t)) = \mathfrak{p}_\mathfrak{A}(\hat{y}_1 \delta t, \ldots, \hat{y}_l \delta t)$. For each $i \le m$, $\beta^*(\delta \hat{y}_i t) \simeq \langle \delta \hat{y}_i t \rangle$ by (β^*-3); and as we have seen $\delta \hat{y}_i t \simeq \hat{y}_i \delta t$. Hence $\mathfrak{p}_\mathfrak{A}(\beta^*(\delta \hat{y}_i t)) = \mathfrak{p}_\mathfrak{A}(\delta \hat{y}_i t) = \mathfrak{p}_\mathfrak{A}(\hat{y}_i \delta t)$. Thus condition (2) for the y-sets in $b(F)$ is verified, and amiability is established.$\ddagger$ $\square$

An immediate consequence is the amiability of the Gödel–Kalmár–Schütte Class.

$\ddagger$ A slight modification in the definition of δ will yield a smaller bound on the size of its range. Namely, let $\delta\mathbf{1}$ be the earliest x_n-term none of whose arguments is $\mathbf{1}$. Then $\mathbf{1}$ is not in the range of δ, nor is any term of which $\mathbf{1}$ is an argument. Consequently $|\delta[D(F, \omega)]|$ is no greater than $n \cdot (n \cdot \kappa_m)^m d(F, m)$. This yields the bound of $49 n^3 d(F, 2)$ for the Gödel–Kalmár–Schütte Class. The same procedure may be used for Class 2.6 as well.

ISBN 0-201-02540-X

Class 2.5. Let F be a prenex schema with prefix $\forall y_1 \forall y_2 \exists x_1 \cdots \exists x_n$. Then F is $(49n(n + 1)^2 d(F, 2) + 1)$-amiable. $\square$

The generalization of the proof for Class 2.4 to Class 2.6 is straightforward. To obtain condition (2) of the Amiability Lemma for x-focused sets, it is enough that the sequence-mapping β^* take t-$\simeq$-instances of $\langle v_1, \ldots, v_p \rangle$ to the t-instance of $\langle v_1, \ldots, v_p \rangle$ when $\langle v_1, \ldots, v_p \rangle$ is an x-focused set in $b(F)$. We may therefore weaken the notion of tightness of terms, just as in the amenability proof for Class 2.6 we weakened the requirement that γ preserve identity of arguments. However, since β^* must preserve permutations and repetitions, β^* must also take $\langle t_1, \ldots, t_p \rangle$ to a t-instance when some arrangement of the set $\{t_1, \ldots, t_p\}$ is a t-$\simeq$-instance of some arrangement of an x-focused set in $b(F)$, even when $\langle t_1, \ldots, t_p \rangle$ is not itself a t-$\simeq$-instance of any arrangement of any x-focused set in $b(F)$ (cf. the footnote on page 57). For this reason, it is more convenient to talk of sets of terms than of sequences.

If $\simeq$ is an equivalence relation on $D(F, \omega)$, we extend $\simeq$ in the natural way to subsets of $D(F, \omega)$: $S \simeq T$ iff the same equivalence classes are represented in S as in T.

Faithfulness Lemma 1 may then be restated: Let $\simeq$ be a faithful equivalence relation on $D(F, \omega)$, let V and W be x-focused sets, and let s and t be terms such that $\hat{V}[s] \simeq \hat{W}[t]$. If $v \in V$, $w \in W$, and $\hat{v}s \simeq \hat{w}t$, then either $v = w =$ some x-variable or else v and w are y-variables.

For if $\hat{V}[s] \simeq \hat{W}[t]$, then there are arrangements $\langle v_1, \ldots, v_p \rangle$ and $\langle w_1, \ldots, w_p \rangle$ of V and W such that $\langle \hat{v}_1 s, \ldots, \hat{v}_p s \rangle \simeq \langle \hat{w}_1 t, \ldots, \hat{w}_p t \rangle$. If $v \in V$, $w \in W$, and $\hat{v}s \simeq \hat{w}t$, then there are such arrangements with $v = v_1$ and $w = w_1$. Hence the original form of Faithfulness Lemma 1 yields this restated form. It also follows that if $\hat{V}[s] \simeq \hat{W}[t]$, then V and W share the same x-variables.

A set T of terms is a t-$\simeq$-*instance* of an x-focused set V iff some arrangement of T is a t-$\simeq$-instance of some arrangement of V. If T is a t-$\simeq$-instance of V, then $T \simeq \hat{V}[t]$.

Faithfulness Lemma 2 may also be restated: Let $\simeq$ be a faithful equivalence relation on $D(F, \omega)$. If $s \neq t$, then no s-$\simeq$-instance of an x-focused set is also a t-$\simeq$-instance of an x-focused set.

The notions tied and hereditarily tied are defined on page 55.

Class 2.6. Let $F = \forall y_1 \cdots \forall y_m \exists x_1 \cdots \exists x_n F^M$ be a Skolem schema such that, for all distinct y-sets Y and Y' in $b(F)$ of cardinality >1, no member of Y is hereditarily tied to any member of Y'. Then F is $(n \cdot ((n + 1) \cdot \kappa_m)^m d(F, m) + 1)$-amiable.

Proof.   Let $\mathfrak{A}$ be any truth-assignment on $\bar{L}(F, \omega)$, and define a faithful equivalence relation $\simeq$ on $D(F, \omega)$ thus: $s \simeq t$ iff $\alpha(s) = \alpha(t)$, $\chi(s) = \chi(t)$, and $\mathfrak{p}_{\mathfrak{A}}(s) = \mathfrak{p}_{\mathfrak{A}}(t)$. A sequence or set of terms is *tight* iff it contains no

ISBN 0-201-02540-X

distinct but equivalent members. A term t is *tight* iff for all hereditarily tied y-variables y_i and y_j, $\hat{y}_i t \simeq \hat{y}_j t$ only if $\hat{y}_i t = \hat{y}_j t$. Let η be the term-mapping that takes each term t to its earliest equivalent.

We shall construct a sequence-mapping β^* that possesses the following four properties for all sequences $\langle t_1,\ldots,t_p\rangle$ from $D(F, \omega)$:

(β^*-1) if, for some tight x_n-term t and some arrangement $\langle v_1,\ldots,v_p\rangle$ of an x-focused set in $b(F)$, $\langle t_1,\ldots,t_p\rangle$ is a t-$\simeq$-instance of $\langle v_1,\ldots,v_p\rangle$, then $\beta^*(t_1,\ldots,t_p) = \langle \hat{v}_1 t,\ldots,\hat{v}_p t\rangle$;

(β^*-2) β^* preserves permutations and repetitions;

(β^*-3) $\beta^*(t_1,\ldots,t_p) \simeq \langle t_1,\ldots,t_p\rangle$;

(β^*-4) $\beta^*(t_1,\ldots,t_p)$ is a tight sequence.

Indeed, we define β^* thus:

(i) if $\{t_1,\ldots,t_p\}$ is a t-$\simeq$-instance of V for some tight x_n-term t and some x-focused set V in $b(F)$, then let $\beta^*(t_1,\ldots,t_p) = \langle \hat{v}_1 t,\ldots,\hat{v}_p t\rangle$ where, for each $i \le p$, v_i is a member of V such that $t_i \simeq \hat{v}_i t$;

(ii) otherwise let $\beta^*(t_1,\ldots,t_p) = \langle \eta t_1,\ldots,\eta t_p\rangle$.

To show that clause (i) gives rise to no ambiguity, it suffices, by Faithfulness Lemma 2, to show: let t be a tight x_n-term, let V and W be x-focused sets in $b(F)$ such that $\hat{V}[t] \simeq \hat{W}[t]$, let $v \in V$, let $w \in W$, and let $\hat{v}t \simeq \hat{w}t$; then $\hat{v}t = \hat{w}t$. Now by Faithfulness Lemma 1 either $v = w =$ some x-variable or else v and w are y-variables. In the former case $\hat{v}t = \hat{w}t$ immediately. In the latter case v and w are tied y-variables; for $\hat{V}[t] \simeq \hat{W}[t]$, and so V and W share the same x-variables. Since t is tight, $\hat{v}t = \hat{w}t$ in this case too.

The sequence-mapping β^* clearly possesses properties (β^*-1), (β^*-2), and (β^*-3). It also possesses property (β^*-4): the argument of the preceding paragraph shows that $\hat{v}t = \hat{w}t$ whenever t is tight, V is an x-focused set in $b(F)$, $v \in V$, $w \in V$, and $\hat{v}t \simeq \hat{w}t$. Hence the set $\hat{V}[t]$ is tight, and so if $\beta^*(t_1,\ldots,t_p)$ is determined by clause (i), then $\beta^*(t_1,\ldots,t_p)$ is a tight sequence.

Existence Condition. There is a term-mapping μ such that, for every x_n-term t,

(a) μt is a tight x_n-term;

(b) $\hat{y}_i t \simeq \hat{y}_i \mu t$, $1 \le i \le m$;

(c) $\wp_\mathfrak{N}(\beta^*(\hat{y}_{i_1} t,\ldots,\hat{y}_{i_p} t)) = \wp_\mathfrak{N}(\hat{y}_{i_1}\mu t,\ldots,\hat{y}_{i_p}\mu t)$ whenever $Y = \{y_{i_1},\ldots,y_{i_p}\}$ is a y-set of cardinality >1 in $b(F)$ and $i_1 <\cdots <i_p$.

Proof. For each x_n-term t let $\mu t = f_{x_n}(s_1\cdots s_m)$, where $\langle s_1,\ldots,s_m\rangle$ is determined from $\langle \hat{y}_1 t,\ldots,\hat{y}_m t\rangle$ thus:

(μ-1) for each y-set $Y = \{y_{i_1},\ldots,y_{i_p}\}$ of cardinality >1 in $b(F)$, where $i_1 <\cdots < i_p$, let $\langle s_{i_1},\ldots,s_{i_p}\rangle = \beta^*(\hat{y}_{i_1} t,\ldots,\hat{y}_{i_p} t)$;

ISBN 0-201-02540-X

(μ-2)   for each j such that s_j is not determined by (μ-1), if there is an i such that s_i is so determined, $\hat{y}_j t \simeq \hat{y}_i t$, and y_i and y_j are hereditarily tied, then let $s_j = s_i$ for the least such i;

(μ-3)   for each j such that s_j is determined by neither (μ-1) nor (μ-2), let $s_j = \eta \hat{y}_j t$.

The verification that μt fulfills conditions (a), (b), and (c) is routine.   □

Construction of δ.   We proceed just as for Class 2.4. First, for all x_n-terms s and t let $s \sim t$ iff $\mathfrak{p}_{\mathfrak{A}}(\hat{y}_1 s, \ldots, \hat{y}_m s) = \mathfrak{p}_{\mathfrak{A}}(\hat{y}_1 t, \ldots, \hat{y}_m t)$, $\alpha(\hat{y}_i s) = \alpha(\hat{y}_i t)$, and $\chi(\hat{y}_i s) = \chi(\hat{y}_i t)$, $1 \le i \le m$. Second, let $\delta 4 = 4$ and, for each x_n-term t and each k, $1 \le k \le n$, let $\delta \hat{x}_k t = \hat{x}_k t'$, where t' is the earliest tight $\sim$-equivalent of $\mu f_{x_n}(\delta \hat{y}_1 t \cdots \delta \hat{y}_m t)$. Hence $|\delta[D(F, \omega)]| \le n \cdot (\kappa_m \cdot (n + 1))^m d(F, m) + 1$.

Let t be an x_n-term. As before, if $\langle v_1, \ldots, v_p \rangle$ is x-focused, then $\langle \delta \hat{v}_1 t, \ldots, \delta \hat{v}_p t \rangle$ is a δt-$\simeq$-instance of $\langle v_1, \ldots, v_p \rangle$. Hence condition (2) of the Amiability Lemma for the x-focused sets in $b(F)$ follows from (β*-1) and the tightness of δt.

By the permutation property (β*-2), to verify condition (2) of the Amiability Lemma for the y-sets in $b(F)$ it suffices to consider just the sequences $\langle y_{i_1}, \ldots, y_{i_p} \rangle$, where $\{y_{i_1}, \ldots, y_{i_p}\}$ is a y-set of cardinality >1 in $b(F)$ and $i_1 < \cdots < i_p$, and the sequences $\langle y_i \rangle$ for $1 \le i \le m$. Just as in the argument for Class 2.4, condition (2) for the former sequences follows from (c) of the Existence Condition and the fact that $\delta t \sim \mu f_{x_n}(\delta \hat{y}_1 t \ldots \delta \hat{y}_m t)$, and condition (2) for the latter sequences holds since $\beta*(\delta \hat{y}_i t) \simeq \langle \delta \hat{y}_i t \rangle \simeq \langle \hat{y}_i \delta t \rangle$.

Thus both conditions of the Amiability Lemma are fulfilled, and amiability is established.   □

A rather subtle refinement of our method is needed to establish the amiability of Class 2.7. An example suggests why. Let F be a Skolem schema whose basis contains $\{y_1, y_2, x_1\}$, $\{y_3, y_4, x_1\}$, $\{y_1, y_2\}$, and $\{y_3, y_4\}$. We saw in Chapter 2, page 63, that there are such schemata in Class 2.7, but not in Class 2.6. Suppose we define a faithful equivalence relation $\simeq$ as in the argument for Class 2.6, and let t be an x_n-term. Hitherto our strategy to obtain condition (2) of the Amiability Lemma for the x-focused sets $\{y_1, y_2, x_1\}$ and $\{y_3, y_4, x_1\}$ would have been to secure the following: if t is tight, in some appropriate sense of tightness, then $\beta*$ carries each t-$\simeq$-instance of $\langle y_1, y_2, x_1 \rangle$ to $\langle \hat{y}_1 t, \hat{y}_2 t, \hat{x}_1 t \rangle$, and carries each t-$\simeq$-instance of $\langle y_3, y_4, x_1 \rangle$ to $\langle \hat{y}_3 t, \hat{y}_4 t, \hat{x}_1 t \rangle$. Now suppose that $\langle \hat{y}_1 t, \hat{y}_2 t \rangle \simeq \langle \hat{y}_3 t, \hat{y}_4 t \rangle$; then the sequences $\langle y_1, y_2, x_1 \rangle$ and $\langle y_3, y_4, x_1 \rangle$ have t-$\simeq$-instances in common, whence we must have $\langle \hat{y}_1 t, \hat{y}_2 t, \hat{x}_1 t \rangle = \langle \hat{y}_3 t, \hat{y}_4 t, \hat{x}_1 t \rangle$. Thus we would have to require of each tight x_n-term t that if $\langle \hat{y}_1 t, \hat{y}_2 t \rangle \simeq \langle \hat{y}_3 t, \hat{y}_4 t \rangle$ then $\langle \hat{y}_1 t, \hat{y}_2 t \rangle = \langle \hat{y}_3 t, \hat{y}_4 t \rangle$. Further, since $\{y_1, y_2\}$ and $\{y_3, y_4\}$ are in $b(F)$, the Existence Condition would require, for each x_n-term t, the

ISBN 0-201-02540-X

existence of a tight x_n-term μt such that $\hat{y}_i t \simeq \hat{y}_i \mu t$, $1 \le i \le 4$, $\mathfrak{p}_{\mathfrak{A}}(\beta^*(\hat{y}_1 t,$ $\hat{y}_2 t)) = \mathfrak{p}_{\mathfrak{A}}(\hat{y}_1 \mu t, \hat{y}_2 \mu t)$, and $\mathfrak{p}_{\mathfrak{A}}(\beta^*(\hat{y}_3 t, \hat{y}_4 t)) = \mathfrak{p}_{\mathfrak{A}}(\hat{y}_3 \mu t, \hat{y}_4 \mu t)$. Hence, if $\langle \hat{y}_1 t, \hat{y}_2 t \rangle \simeq \langle \hat{y}_3 t, \hat{y}_4 t \rangle$ then $\langle \hat{y}_1 \mu t, \hat{y}_2 \mu t \rangle \simeq \langle \hat{y}_3 \mu t, \hat{y}_4 \mu t \rangle$; and then, as we have seen, for μt to be tight we would have to require that $\langle \hat{y}_1 \mu t, \hat{y}_2 \mu t \rangle = \langle \hat{y}_3 \mu t, \hat{y}_4 \mu t \rangle$. But clearly there will be no such tight x_n-term μt unless $\mathfrak{p}_{\mathfrak{A}}(\beta^*(\hat{y}_1 t, \hat{y}_2 t)) = \mathfrak{p}_{\mathfrak{A}}(\beta^*(\hat{y}_3 t, \hat{y}_4 t))$ whenever $\langle \hat{y}_1 t, \hat{y}_2 t \rangle \simeq \langle \hat{y}_3, \hat{y}_4 t \rangle$.

More generally, to establish the amiability of Class 2.7 we shall have to require of the equivalence relation $\simeq$ and the sequence-mapping β^* that

(i) if $\langle s_1, \ldots, s_p \rangle \simeq \langle t_1, \ldots, t_p \rangle$, then $\mathfrak{p}_{\mathfrak{A}}(\beta^*(s_1, \ldots, s_p)) = \mathfrak{p}_{\mathfrak{A}}(\beta^*(t_1, \ldots, t_p))$.

Given requirement (i), we no longer need be concerned (explicitly) with t-$\simeq$-instances. Instead, we require of β^* that

(ii) for each tight x_n-term t (in the appropriate sense of tightness) and each arrangement $\langle v_1, \ldots, v_p \rangle$ of an x-focused set in $b(F)$, $\mathfrak{p}_{\mathfrak{A}}(\beta^*(\hat{v}_1 t, \ldots, \hat{v}_p t)) = \mathfrak{p}_{\mathfrak{A}}(\hat{v}_1 t, \ldots, \hat{v}_p t)$.

Requirements (i) and (ii), taken together, are quite demanding. In fact, they imply the following condition on the equivalence relation $\simeq$, a condition not met by any equivalence relation we have so far constructed:

(#) if s and t are tight x_n-terms, and $\langle v_1, \ldots, v_p \rangle$ and $\langle w_1, \ldots, w_p \rangle$ are arrangements of x-focused sets in $b(F)$ such that $\langle \hat{v}_1 s, \ldots, \hat{v}_p s \rangle \simeq \langle \hat{w}_1 t, \ldots, \hat{w}_p t \rangle$, then $\mathfrak{p}_{\mathfrak{A}}(\hat{v}_1 s, \ldots, \hat{v}_p s) = \mathfrak{p}_{\mathfrak{A}}(\hat{w}_1 t, \ldots, \hat{w}_p t)$.

Hence if we are to construct a sequence-mapping β^* meeting requirements (i) and (ii), we first have to make certain that $\simeq$ fulfills (#). There is no difficulty in obtaining (#) when $\langle v_1, \ldots, v_p \rangle = \langle w_1, \ldots, w_p \rangle$. We need merely ask that $s \simeq t$ only if $\mathfrak{p}_{\mathfrak{A}}(\hat{y}_1 s, \ldots, \hat{y}_m s, \hat{x}_1 s, \ldots, \hat{x}_n s) = \mathfrak{p}_{\mathfrak{A}}(\hat{y}_1 t, \ldots, \hat{y}_m t, \hat{x}_1 t, \ldots, \hat{x}_n t)$. For then $\mathfrak{p}_{\mathfrak{A}}(\hat{v}_1 s, \ldots, \hat{v}_p s) = \mathfrak{p}_{\mathfrak{A}}(\hat{v}_1 t, \ldots, \hat{v}_p t)$ whenever $\hat{x}_k s \simeq \hat{x}_k t$ for some variable x_k. But difficulty does arise when $\langle v_1, \ldots, v_p \rangle \ne \langle w_1, \ldots, w_p \rangle$. To reduce this case to that in which $\langle v_1, \ldots, v_p \rangle = \langle w_1, \ldots, w_p \rangle$, we shall employ another Combinatorial Lemma, a generalization of Hilfssatz 1 in Gödel's finite controllability proof for the Gödel–Kalmár–Schütte Class (1933, page 435). The Lemma furnishes equivalence relations $\simeq$ such that if the antecedent of (#) holds, then $\langle \hat{v}_1 s, \ldots, \hat{v}_p s \rangle \simeq \langle \hat{w}_1 s, \ldots, \hat{w}_p s \rangle \simeq \langle \hat{v}_1 t, \ldots, \hat{v}_p t \rangle \simeq \langle \hat{w}_1 t, \ldots, \hat{w}_p t \rangle$. And if s and t are tight in the appropriate sense, then we shall be able to infer that $\langle \hat{v}_1 s, \ldots, \hat{v}_p s \rangle = \langle \hat{w}_1 s, \ldots, \hat{w}_p s \rangle$ and $\langle \hat{v}_1 t, \ldots, \hat{v}_p t \rangle = \langle \hat{w}_1 t, \ldots, \hat{w}_p t \rangle$.

Combinatorial Lemma 2. Let E be an equivalence relation of finite order on $D(F, \omega)$. Then there is an equivalence relation $\simeq$ on $D(F, \omega)$, also of finite order, such that, for all terms s and t and all y-variables y_i and y_j,

(1)   if $s \simeq t$ then sEt;

(2)   if $s \simeq t$ and $\hat{y}_i s \simeq \hat{y}_j t$ then $\hat{y}_i s \simeq \hat{y}_j s \simeq \hat{y}_j t \simeq \hat{y}_i t$.

Proof.   See §2 of the Appendix. We note here that if E has order r, then $\simeq$ has order at most $r(2rm^4 + 1)^{m^2}$, where m is the number of y-variables of F.   □

In the amiability proof for Class 2.7 we shall so construct the sequence-mapping β^* that $\beta^*(s_1,...,s_p) = \beta^*(t_1,...,t_p)$ whenever $\langle s_1,...,s_p \rangle \simeq \langle t_1,...,t_p \rangle$. Clearly, then, β^* meets requirement (i). We shall also so construct β^* that if t is a tight x_n-term and $\langle v_1,...,v_p \rangle$ is an arrangement of an x-focused set in $b(F)$, then $\beta^*(\hat{v}_1 t,...,\hat{v}_p t) = \langle \hat{w}_1 s,...,\hat{w}_p s \rangle$ for some x-focused set $\{w_1,...,w_p\}$ and some tight x_n-term s (not necessarily identical with $\{v_1,...,v_p\}$ and with t) such that $\langle \hat{v}_1 t,...,\hat{v}_p t \rangle \simeq \langle \hat{w}_1 s,...,\hat{w}_p s \rangle$. By condition (#), the sequence-mapping β^* then meets requirement (ii).

Class 2.7.   Let $F = \forall y_1 \cdots \forall y_m \exists x_1 \cdots \exists x_n F^M$ be a Skolem schema having the following property: for each x_n-term t in $D(F, \omega)$ and all y-sets Y and Y' in $b(F)$, if $\hat{Y}[t]$ and $\hat{Y}'[t]$ are distinct sets of terms of cardinality >1, then no member of Y is hereditarily t-united to any member of Y'. Then F is $(r(2rm^4 + 1)^{m^2} + 1)$-amiable where $r = n \cdot \kappa_m \cdot d(F, m + n) + 1$.

Proof.   Let $\mathfrak{A}$ be any truth-assignment on $\tilde{L}(F, \omega)$. Let E be the equivalence relation on $D(F, \omega)$ defined thus: sEt iff either $s = t = \dashv$ or else $\alpha(s) = \alpha(t)$, $\chi(s) = \chi(t)$, and $\mathfrak{p}_{\mathfrak{A}}(\hat{y}_1 s,...,\hat{y}_m s,\hat{x}_1 s,...,\hat{x}_n s) = \mathfrak{p}_{\mathfrak{A}}(\hat{y}_1 t, ...,\hat{y}_m t,\hat{x}_1 t,...,\hat{x}_n t)$. Let $\simeq$ be the equivalence relation on $D(F, \omega)$ given by Combinatorial Lemma 2. Thus $\simeq$ is faithful; and if $s \simeq t$, then $\mathfrak{p}_{\mathfrak{A}}(\hat{v}_1 s,...,\hat{v}_p s) = \mathfrak{p}_{\mathfrak{A}}(\hat{v}_1 t,...,\hat{v}_p t)$ for all variables $v_1,...,v_p$ of F. Moreover, if $s \simeq t$ then $\mathfrak{p}_{\mathfrak{A}}(s) = \mathfrak{p}_{\mathfrak{A}}(t)$. Let η be the term-mapping that takes each term t to its earliest $\simeq$-equivalent. The range of η has cardinality at most $r(2rm^4 + 1)^{m^2}$.

For each x_n-term t, two y-variables y_i and y_j are t-$\simeq$-*united* iff $\hat{y}_i t \simeq \hat{y}_j t$ and there are x-focused sets V and W in $b(F)$ such that $y_i \in V$, $y_j \in W$, and $\hat{V}[t] \simeq \hat{W}[t]$. Two y-variables are *hereditarily* t-$\simeq$-*united* iff they are identical or they stand in the ancestral relation of the relation t-$\simeq$-united. An x_n-term t is *tight* iff $\hat{y}_i t = \hat{y}_j t$ for all hereditarily t-$\simeq$-united y-variables y_i and y_j. A set of sequence of terms is *tight* iff it contains no distinct but $\simeq$-equivalent members.

Our amiability proof divides into four principal steps. First, from the faithfulness of $\simeq$ and Faithfulness Lemma 1 we derive three properties of tight terms. Second, we use these properties to construct a sequence-mapping β^* meeting requirements (i) and (ii). Third, we prove an Existence Condition similar to those in the previous amiability proofs. Finally, we construct the term-mapping δ, and verify conditions (1) and (2) of the Amiability Lemma.

Tightness Properties.   Let s be a tight x_n-term.

ISBN 0-201-02540-X

(T-1)   If V and W are x-focused sets in $b(F)$ such that $\hat{V}[s] \simeq \hat{W}[s]$, and if $v \in V$ and $w \in W$ are variables such that $\hat{v}s \simeq \hat{w}s$, then $\hat{v}s = \hat{w}s$, and hence $\hat{V}[s] = \hat{W}[s]$.

(T-2)   Two y-variables are s-$\simeq$-united iff they are s-united.

(T-3)   If t is an x_n-term such that $\hat{y}_i s \simeq \hat{y}_i t$ for each $i \le m$, then t-$\simeq$-unitedness and s-unitedness coincide.

Proof.   (T-1)   By Faithfulness Lemma 1 either $v = w =$ some x-variable or else v and w are y-variables. In the former case $\hat{v}s = \hat{w}s$ immediately. In the latter case v and w are by definition s-$\simeq$-united. Since s is tight, $\hat{v}s = \hat{w}s$.

(T-2)   Suppose y_i and y_j are s-$\simeq$-united. Hence $\hat{y}_i s \simeq \hat{y}_j s$ and there are x-focused sets V and W in $b(F)$ such that $\hat{V}[s] \simeq \hat{W}[s]$, $y_i \in V$, and $y_j \in W$. By (T-1), $\hat{V}[s] = \hat{W}[s]$ and $\hat{y}_i s = \hat{y}_j s$. Hence y_i and y_j are s-united. Conversely, clearly two y-variables are s-united only if they are s-$\simeq$-united.

(T-3)   By (T-2) it is enough to show that t-$\simeq$-unitedness and s-$\simeq$-unitedness coincide. Suppose y_i and y_j are t-$\simeq$-united. Then $\hat{y}_i t \simeq \hat{y}_j t$ and there are x-focused V and W in $b(F)$ such that $\hat{V}[t] \simeq \hat{W}[t]$, $y_i \in V$, and $y_j \in W$. By Faithfulness Lemma 1, V and W contain the same x-variables and $\{\hat{y}_k t \mid y_k \in V\} \simeq \{\hat{y}_k t \mid y_k \in W\}$. Since by hypothesis $\hat{y}_k t \simeq \hat{y}_k s$ for each $k \le m$, $\hat{V}[s] \simeq \hat{W}[s]$. Hence y_i and y_j are s-$\simeq$-united. Symmetrically, if y_i and y_j are s-$\simeq$-united, then they are t-$\simeq$-united.   □

Properties of β^*.   There is a sequence-mapping β^* such that for all sequences $\langle t_1,\ldots,t_p \rangle$ from $D(F, \omega)$, $p \ge 1$,

(β^*-1)   for each tight x_n-term t and each arrangement $\langle v_1,\ldots,v_p \rangle$ of an x-focused set in $b(F)$, $\mathfrak{p}_{\mathfrak{A}}(\beta^*(\hat{v}_1 t,\ldots,\hat{v}_p t)) = \mathfrak{p}_{\mathfrak{A}}(\hat{v}_1 t,\ldots,\hat{v}_p t)$;

(β^*-2) β^* preserves permutations and repetitions;

(β^*-3) $\beta^*(t_1,\ldots,t_p) \simeq \langle t_1,\ldots,t_p \rangle$;

(β^*-4) $\beta^*(s_1,\ldots,s_p) = \beta^*(t_1,\ldots,t_p)$ whenever $\langle s_1,\ldots,s_p \rangle \simeq \langle t_1,\ldots,t_p \rangle$.

Proof.   Define β^* thus:

(i)   Suppose $\{t_1,\ldots,t_p\} \simeq \hat{V}[t]$ for some tight x_n-term t and some x-focused set V in $b(F)$. Let s be the earliest tight x_n-term for which such a set exists; let W be any such x-focused set and, for each $i \le p$, let $w_i \in W$ be a variable such that $\hat{w}_i s \simeq t_i$. Finally, let $\beta^*(t_1,\ldots,t_p) = \langle \hat{w}_1 s,\ldots,\hat{w}_p s \rangle$.

(ii)   Otherwise let $\beta^*(t_1,\ldots,t_p) = \langle \eta t_1,\ldots,\eta t_p \rangle$.

By (T-1), no ambiguity arises from clause (i). Given any sequence $\langle t_1,\ldots,t_p \rangle$, whether or not $\beta^*(t_1,\ldots,t_p)$ is determined by clause (i)—and if so, for which x_n-term s it is so determined—depends only on the equivalence classes represented in $\{t_1,\ldots,t_p\}$. Properties (β^*-2) and (β^*-4) follow easily. Moreover, (β^*-3) is immediate.

It remains to verify (β^*-1); here, and only here, Combinatorial Lemma 2

ISBN 0-201-02540-X

comes into play. Let t be a tight x_n-term and let $\langle v_1,\ldots,v_p \rangle$ be an arrangement of an x-focused set V in $b(F)$. Then $\beta^*(\hat{v}_1 t,\ldots,\hat{v}_p t)$ is determined by clause (i): $\beta^*(\hat{v}_1 t,\ldots,\hat{v}_p t) = \langle \hat{w}_1 s,\ldots,\hat{w}_p s \rangle$ for some tight x_n-term s, some x-focused set W in $b(F)$ such that $\hat{W}[s] \simeq \hat{V}[t]$, and some variables $w_1,\ldots,w_p$ in W such that $\langle \hat{v}_1 t,\ldots,\hat{v}_p t \rangle \simeq \langle \hat{w}_1 s,\ldots,\hat{w}_p s \rangle$. It suffices to show that $\mathfrak{p}_{\mathfrak{A}}(\hat{v}_1 t,\ldots,\hat{v}_p t) = \mathfrak{p}_{\mathfrak{A}}(\hat{w}_1 s,\ldots,\hat{w}_p s)$. Suppose that $v \in V$ and $w \in W$ are variables such that $\hat{v} t \simeq \hat{w} s$. By Faithfulness Lemma 1 either $v = w =$ some x-variable or else v and w are y-variables. Since V and W contain at least one x-variable, $\hat{x}_k t \simeq \hat{x}_k s$ for some $k \le n$. But then, by clause (2) of Combinatorial Lemma 2, if v and w are y-variables such that $\hat{v} t \simeq \hat{w} s$, then $\hat{v} t \simeq \hat{w} t \simeq \hat{v} s \simeq \hat{w} s$. (Recall that $\hat{v}\hat{x}_k t = \hat{v} t$ and $\hat{v}\hat{x}_k s = \hat{v} s$ for all variables v of F.) Hence $\hat{V}[t] \simeq \hat{W}[t] \simeq \hat{W}[s] \simeq \hat{V}[s]$, and in particular $\langle \hat{v}_1 t,\ldots,\hat{v}_p t \rangle \simeq \langle \hat{w}_1 t,\ldots,\hat{w}_p t \rangle$. By (T-1), $\langle \hat{v}_1 t,\ldots,\hat{v}_p t \rangle = \langle \hat{w}_1 t,\ldots,\hat{w}_p t \rangle$. Since $\hat{x}_k t \simeq \hat{x}_k s$, $\mathfrak{p}_{\mathfrak{A}}(\hat{w}_1 t,\ldots,\hat{w}_p t) = \mathfrak{p}_{\mathfrak{A}}(\hat{w}_1 s,\ldots,\hat{w}_p s)$. Hence $\mathfrak{p}_{\mathfrak{A}}(\hat{v}_1 t,\ldots,\hat{v}_p t) = \mathfrak{p}_{\mathfrak{A}}(\hat{w}_1 s,\ldots,\hat{w}_p s)$, and $(\beta^*\text{-1})$ is verified. $\square$

Existence Condition. There is a term-mapping μ such that, for every x_n-term t,

   (a) μt is a tight x_n-term;

   (b) $\hat{y}_i t \simeq \hat{y}_i \mu t$, $1 \le i \le m$;

   (c) $\mathfrak{p}_{\mathfrak{A}}(\beta^*(\hat{y}_{i_1} t,\ldots,\hat{y}_{i_p} t)) = \mathfrak{p}_{\mathfrak{A}}(\hat{y}_{i_1}\mu t,\ldots,\hat{y}_{i_p}\mu t)$ whenever $\{y_{i_1},\ldots,y_{i_p}\}$ is a y-set in $b(F)$, $i_1 < \cdots < i_p$, $p \ge 1$.

Proof. It suffices to define μ just on the tight x_n-terms t. For then if t is an x_n-term that is not tight, we may let $\mu t = \mu \tau t$, where $\tau t = f_{x_n}(\eta \hat{y}_1 t \ldots \eta \hat{y}_m t)$. The term τt is tight, since $\hat{y}_i \tau t = \hat{y}_j \tau t$ whenever $\hat{y}_i \tau t \simeq \hat{y}_j \tau t$. Also, $\hat{y}_i \tau t \simeq \hat{y}_i t$ for each $i \le m$. Thus if μ fulfills (b) on τt, then $\hat{y}_i \mu t = \hat{y}_i \mu \tau t \simeq \hat{y}_i \tau t \simeq \hat{y}_i t$, whence μ fulfills (b) on t. Moreover, by $(\beta^*\text{-4})$, $\beta^*(\hat{y}_{i_1}\tau t,\ldots,\hat{y}_{i_p}\tau t) = \beta^*(\hat{y}_{i_1} t,\ldots,\hat{y}_{i_p} t)$ for all numbers $i_1,\ldots,i_p \le m$. Hence if μ fulfills (c) on τt, then μ fulfills (c) on t.

For each tight x_n-term t let $\mu t = f_{x_n}(s_1 \ldots s_m)$, where $\langle s_1,\ldots,s_m \rangle$ is determined from $\langle \hat{y}_1 t,\ldots,\hat{y}_m t \rangle$ thus:

   (μ-1) for each y-set $Y = \{y_{i_1},\ldots,y_{i_p}\}$ in $b(F)$ such that $|\hat{Y}[t]| > 1$, where $i_1 < \cdots < i_p$, let $\langle s_{i_1},\ldots,s_{i_p} \rangle = \beta^*(\hat{y}_{i_1} t,\ldots,\hat{y}_{i_p} t)$;

   (μ-2) for each j such that s_j is not determined by (μ-1), if there is an i such that s_i is so determined and y_j and y_i are hereditarily t-united, then let $s_j = s_i$ for the least such i;

   (μ-3) for each j such that s_j is determined by neither (μ-1) nor (μ-2), let $s_j = \eta \hat{y}_j t$.

From $(\beta^*\text{-3})$ it immediately follows that $\hat{y}_i t \simeq s_i$, $1 \le i \le m$. Hence (b) is fulfilled. Moreover, since β^* preserves permutations and repetitions, by reasoning just as in the amenability proof for Class 2.7 we obtain an analogue to the weak identity preservation property: if y_i and y_j are t-united, then

ISBN 0-201-02540-X

$\hat{y}_i \mu t = \hat{y}_j \mu t$. Hence, to show that μt is tight, it suffices to show that μt-$\approx$-unitedness and t-unitedness coincide. But since t is tight, this follows from (b) and (T-3).

It remains to verify condition (c). This is immediate from clause $(\mu$-1) for those y-sets Y in $b(F)$ such that $|\hat{Y}[t]| > 1$. And for each $i \leq m$ we have $\mathfrak{p}_{\mathfrak{A}}(\beta^*(\hat{y}_i t)) = \mathfrak{p}_{\mathfrak{A}}(\hat{y}_i t) = \mathfrak{p}_{\mathfrak{A}}(\hat{y}_i \mu t)$, since by $(\beta^*$-3$)$ $\beta^*(\hat{y}_i t) \simeq \langle \hat{y}_i t \rangle$ and by (b) $\hat{y}_i t \simeq \hat{y}_i \mu t$. This yields condition (c) for one-membered sets $\{y_i\}$. Finally, if Y is a y-set of cardinality >1 in $b(F)$, but $|\hat{Y}[t]| = 1$, then, again reasoning as in the amenability proof, we have $|\hat{Y}[\mu t]| = 1$. Hence condition (c) for such y-sets follows from (c) for one-membered y-sets. □

Construction of δ. Our procedure here is somewhat simpler than that of the previous two amiability proofs: property $(\beta^*$-4$)$ enables us to do without a new equivalence relation $\sim$. Let $\delta 1 = 1$; for each x_n-term t and each $k \leq n$ let $\delta \hat{x}_k t = \eta \hat{x}_k \mu f_{x_n}(\delta \hat{y}_1 t \ldots \delta \hat{y}_m t)$. Thus the range of δ is included in that of η, whence $|\delta[D(F, \omega)]| \leq r(2rm^4 + 1)^{m^2}$. It remains to verify condition (2) of the Amiability Lemma. Let t be an x_n-term and let $\langle v_1, \ldots, v_p \rangle$ be an arrangement of a set in $b(F)$; we must show that $\mathfrak{p}_{\mathfrak{A}}(\beta^*(\delta \hat{v}_1 t, \ldots, \delta \hat{v}_p t))$ $= \mathfrak{p}_{\mathfrak{A}}(\hat{v}_1 \delta t, \ldots, \hat{v}_p \delta t)$. Let $s = f_{x_n}(\delta \hat{y}_1 t \ldots \delta \hat{y}_m t)$. Since $\delta t \simeq \mu s$, $\mathfrak{p}_{\mathfrak{A}}(\hat{v}_1 \delta t, \ldots, \hat{v}_p \delta t) = \mathfrak{p}_{\mathfrak{A}}(\hat{v}_1 \mu s, \ldots, \hat{v}_p \mu s)$, and it suffices to show that $\mathfrak{p}_{\mathfrak{A}}(\beta^*(\delta \hat{v}_1 t, \ldots, \delta \hat{v}_p t)) = \mathfrak{p}_{\mathfrak{A}}(\hat{v}_1 \mu s, \ldots, \hat{v}_p \mu s)$.

Suppose $\{v_1, \ldots, v_p\}$ is a y-set in $b(F)$. Then, by the definition of s, $\delta \hat{v}_i t = \hat{v}_i s$ for each $i \leq p$. Hence $\mathfrak{p}_{\mathfrak{A}}(\beta^*(\delta \hat{v}_1 t, \ldots, \delta \hat{v}_p t)) = \mathfrak{p}_{\mathfrak{A}}(\hat{v}_1 \mu s, \ldots, \hat{v}_p \mu s)$ by (c) of the Existence Condition and the permutation property $(\beta^*$-2$)$.

Suppose $\{v_1, \ldots, v_p\}$ is an x-focused set in $b(F)$. By (b) of the Existence Condition $\delta \hat{y}_i t = \hat{y}_i s \simeq \hat{y}_i \mu s$ for each $i \leq m$; moreover, $\delta \hat{x}_k t = \eta \hat{x}_k \mu s \simeq \hat{x}_k \mu s$ for each $k \leq n$. Hence $\langle \delta \hat{v}_1 t, \ldots, \delta \hat{v}_p t \rangle \simeq \langle \hat{v}_1 \mu s, \ldots, \hat{v}_p \mu s \rangle$; whence, by $(\beta^*$-4$)$, $\beta^*(\delta \hat{v}_1 t, \ldots, \delta \hat{v}_p t) = \beta^*(\hat{v}_1 \mu s, \ldots, \hat{v}_p \mu s)$. And finally, since the term μs is tight, by $(\beta^*$-1$)$ $\mathfrak{p}_{\mathfrak{A}}(\beta^*(\hat{v}_1 \mu s, \ldots, \hat{v}_p \mu s)) = \mathfrak{p}_{\mathfrak{A}}(\hat{v}_1 \mu s, \ldots, \hat{v}_p \mu s)$.

We have verified condition (2) of the Amiability Lemma; this concludes the proof of amiability for Class 2.7. □

As we mentioned in Chapter 2, §5, Class 2.7 may be extended by allowing schemata to contain initial x-variables. The finite controllability of Initially-extended Class 2.7 follows from that of unextended Class 2.7; for the reduction we present in Chapter 6 of Initially-extended Class 2.7 to unextended Class 2.7 is conservative.

REMARK. We may also treat initially-extended classes directly, modifying the amiability proofs above in systematic ways to take account of the presence in the domain of indicial constants c_{z_i}. In this manner we would obtain, for example, the following;

Initially-extended Class 2.5. Let F be a prenex schema with prefix $\exists z_1 \cdots \exists z_l \forall y_1 \forall y_2 \exists x_1 \cdots \exists x_n$. Then F is $(49n(n + l + 1)^2 d(F, 2 + l) + l + 1)$-amiable. End of Remark.

ISBN 0-201-02540-X

6. Class 2.8

We start our treatment of Class 2.8 by extending the combinatorial results of §5 to arbitrary normal schemata F. Recall that a normal schema is a prenex schema whose prefix begins with a y-variable and ends with an x-variable; whose y-variables are $y_1,\ldots,y_m$, and whose x-variables are $x_1,\ldots,x_n$, where both lists are in left-to-right order of the occurrence of the variables in the prefix of F. To the terminology introduced in §6 of Chapter 2 for describing a normal schema, we add: for each k, $1 \le k \le n$, the kth y-*cluster* is the set of all y-variables governed by x_{k-1} and governing x_k. (When $k = 1$, the kth y-cluster consists of all the y-variables governing x_1; when $k > 1$ but $\exists x_k$ immediately succeeds $\exists x_{k-1}$ in the prefix of the schema, the kth y-cluster is empty.) For each $j > 0$ let κ_j be as in Combinatorial Lemma 1, page 90.

Generalized Combinatorial Lemma 1. There is a function $\alpha\colon D(F, \omega) \to \{1,\ldots,\kappa_{m+n-1}\}$ such that, for all terms s and t and all variables v governing $\chi(t)$ and w governing $\chi(s)$, if $\alpha(s) = \alpha(\hat{v}t)$, then $\alpha(t) \ne \alpha(\hat{w}s)$.

Proof. See §2 of the Appendix. Note that this lemma implies that $\alpha(t) \ne \alpha(\hat{v}t)$ for each term t and each variable v governing $\chi(t)$. $\square$

An equivalence relation $\simeq$ on $D(F, \omega)$ is *faithful* iff, for all terms s and t, $s \simeq t$ only if $\alpha(s) = \alpha(t)$ and $\chi(s) = \chi(t)$.

Generalized Faithfulness Lemma 1. Let $\simeq$ be a faithful equivalence relation on $D(F, \omega)$, let s and t be x_n-terms, and let V and W be x-focused sets such that $\hat{V}[s] \simeq \hat{W}[t]$. Then V and W have the same focus x_k, and $\hat{x}_k s \simeq \hat{x}_k t$.

Proof. Let x_j be the focus of V and let x_k be the focus of W. Since $\hat{V}[s] \simeq \hat{W}[t]$, there are variables $v \in V$ and $w \in W$ such that $\hat{x}_j s \simeq \hat{w}t$ and $\hat{v}s \simeq \hat{x}_k t$.

Since x_j and x_k are the foci of V and W, v dominates x_j and w dominates x_k. Thus $\hat{v}s = \hat{v}\hat{x}_j s$ and $\hat{w}t = \hat{w}\hat{x}_k t$. But then $\hat{x}_j s \simeq \hat{w}\hat{x}_k t$ and $\hat{v}\hat{x}_j s \simeq \hat{x}_k t$, whence $\alpha(\hat{x}_j s) = \alpha(\hat{w}\hat{x}_k t)$ and $\alpha(\hat{v}\hat{x}_j s) = \alpha(\hat{x}_k t)$. By the properties of α, either w does not govern x_k or v does not govern x_j; that is, either $w = x_k$ or $v = x_j$. But if either $w = x_k$ or $v = x_j$, then $\hat{x}_j s \simeq \hat{x}_k t$, whence $x_j = \chi(\hat{x}_j s) = \chi(\hat{x}_k t) = x_k$. $\square$

Generalized Combinatorial Lemma 2. Let E be a faithful equivalence relation of finite order on $D(F, \omega)$. Then there is a faithful equivalence relation $\simeq$ on $D(F, \omega)$, also of finite order, such that

(1) if $s \simeq t$ then sEt;

(2) if $s \simeq t$ and $\hat{v}s \simeq \hat{w}t$, for some variables v and w governing $\chi(s)$, then $\hat{v}s \simeq \hat{v}t \simeq \hat{w}s \simeq \hat{w}t$.

Proof. See the Appendix, §2. We note here that if E has order $\le r$, then $\simeq$ has order $\le r \cdot (2r \cdot (m + n - 1)^4 + 1)^{(m+n-1)^2}$. $\square$

ISBN 0-201-02540-X

Class 2.8. Let F be a normal schema whose basis contains only x-focused sets and y-sets, and that possesses the following property: for each x_k-term t in $D(F, \omega)$, $1 \le k \le n$, and for each y-set Y in $b(F)$ all of whose members govern x_k, if $|\hat{Y}[t]| > 1$ and if a member of Y is hereditarily t-x_k-united with a variable v, then

(1) v dominates no x-variable that governs any member of Y;

(2) v belongs to no y-set Y' in $b(F)$ such that $|\hat{Y}'[t]| > 1$ and $\hat{Y}'[t] \ne \hat{Y}[t]$.

Then F is $(r \cdot (2r \cdot (m + n - 1)^4 + 1)^{(m+n-1)^2})$ -amiable,   where $r = n \cdot \kappa_{m+n-1} \cdot d(F, m + n) + 1$.

Proof.   Let $\mathfrak{A}$ be any truth-assignment on $\bar{L}(F, \omega)$. Define a faithful equivalence relation E on $D(F, \omega)$ thus: sEt iff either $s = t = \dagger$ or else $\alpha(s) = \alpha(t)$, $\chi(s) = \chi(t)$, and $\mathfrak{P}_{\mathfrak{A}}(s) = \mathfrak{P}_{\mathfrak{A}}(t)$. Then let $\simeq$ be the faithful equivalence relation on $D(F, \omega)$ given by Generalized Combinatorial Lemma 2, and let η be the mapping that carries each term s to its earliest $\simeq$-equivalent. The range of η has cardinality at most $r \cdot (2r \cdot (m + n - 1)^4 + 1)^{(m+n-1)^2}$.

Let t be an x_n-term. Two variables v and w are t-x_n-$\simeq$-*united* iff $\hat{v}t \simeq \hat{w}t$ and there are x-focused sets V and W in $b(F)$ such that $\hat{V}[t] \simeq \hat{W}[t]$, $v \in V$, and $w \in W$. Two variables v and w are *hereditarily* t-x_n-$\simeq$-*united* iff either $v = w$ or else v and w stand in the ancestral relation of t-x_n-$\simeq$-united.

Suppose hereditarily t-x_{l+1}-$\simeq$-united is defined for each x_{l+1}-term t, $1 \le l < n$, and let s be an x_l-term. Two variables v and w are s-x_l-$\simeq$-*united* iff v and w govern x_l and there is an x_{l+1}-term t such that $s = \hat{x}_l t$ and v and w are hereditarily t-x_{l+1}-$\simeq$-united. Two variables v and w are *hereditarily* s-x_l-$\simeq$-*united* iff v and w govern x_l and either $v = w$ or else v and w stand in the ancestral relation of s-x_l-$\simeq$-unitedness.

Note that if v and w are hereditarily t-x_k-$\simeq$-united, $1 \le k \le m$, then $\hat{v}t \simeq \hat{w}t$. Another crucial property is this:

(∇)   Let t and s be x_k-terms, $1 \le k \le n$, such that $\hat{v}t \simeq \hat{v}s$ for all variables v governing x_k. Then t-x_k-$\simeq$-unitedness and s-x_k-$\simeq$-unitedness coincide.

To prove (∇) we shall prove the following: let t and s be x_k-terms, $1 \le k \le n$, such that $\hat{v}t \simeq \hat{w}t$ only if $\hat{v}s \simeq \hat{w}s$ for all variables v and w governing x_k; then two variables are t-x_k-$\simeq$-united only if they are s-x_k-$\simeq$-united. Two applications of this result, one in each direction, clearly yield (∇).

Suppose first that $k = n$, and let v_1 and w_1 be t-x_n-$\simeq$-united. Thus $\hat{v}_1 t \simeq \hat{w}_1 t$ and there are x-focused sets V and W in $b(F)$ such that $\hat{V}[t] \simeq \hat{W}[t]$, $v_1 \in V$, and $w_1 \in W$. Now for each pair $\langle v, w \rangle$ of variables in $V \times W$, $\hat{v}t \simeq \hat{w}t$ only if $\hat{v}s \simeq \hat{w}s$: this holds by hypothesis if neither v nor w is x_n; and if one of v and w is x_n then, by the properties of α, $\hat{v}t \simeq \hat{w}t$ only if the other of v and w is also x_n. Hence $\hat{V}[s] \simeq \hat{W}[s]$, and v_1 and w_1 are s-x_n-$\simeq$-united.

Suppose next that $k < n$, and that the result holds for all x_{k+1}-terms. Let v_1 and w_1 be t-x_k-$\simeq$-united. By the definition of t-x_k-$\simeq$-unitedness, there is an x_{k+1}-term t' such that $t = \hat{x}_k t'$ and v_1 and w_1 are hereditarily t'-x_{k+1}-$\simeq$-united. It suffices to define an x_{k+1}-term s' such that $s = \hat{x}_k s'$ and, for all variables v and w governing x_{k+1}, $\hat{v}s' \simeq \hat{w}s'$ only if $\hat{v}t' \simeq \hat{w}t'$. For then, by the induction hypothesis, two variables are t'-x_{k+1}-$\simeq$-united only if they are s'-x_{k+1}-$\simeq$-united; whence two variables are hereditarily t'-x_{k+1}-$\simeq$-united only if they are hereditarily s'-x_{k+1}-$\simeq$-united. Consequently, v_1 and w_1 are hereditarily s'-x_{k+1}-$\simeq$-united; by definition, v_1 and w_1 are then s-x_k-$\simeq$-united.

We define s' by first setting $\hat{x}_k s' = s$ and then, for each y-variable y_j in the $(k + 1)$st y-cluster, setting $\hat{y}_j s'$ thus: if $\hat{y}_j t \simeq \hat{v}t$ for some variable v dominating x_k, then let $\hat{y}_j s' = \hat{v}s$ for the earliest such variable v, and otherwise let $\hat{y}_j s' = \hat{y}_j t'$. This definition and the fact that $t \simeq \hat{v}t$ for no variable v governing x_k quickly yield that $\hat{v}t' \simeq \hat{w}t'$ only if $\hat{v}s' \simeq \hat{w}s'$ for all variables v and w governing x_{k+1}. This completes the proof of (∇).

An x_k-term t is *tight* iff $\hat{v}t = \hat{w}t$ whenever v and w are hereditarily t-x_k-$\simeq$-united, $1 \le k \le n$.

Tightness Properties.   (T-1)   Let t be a tight x_n-term. If V and W are x-focused sets in $b(F)$ such that $\hat{V}[t] \simeq \hat{W}[t]$, and if $v \in V$ and $w \in W$ are variables such that $\hat{v}t \simeq \hat{w}t$, then $\hat{v}t = \hat{w}t$, and hence $\hat{V}[t] = \hat{W}[t]$.

(T-2)   Let t be an x_k-term, $1 \le k \le n$, such that $\hat{x}_{k-1}t$ is tight—a vacuous condition if $k = 1$. Then there is a tight x_k-term u such that (i) $\hat{x}_{k-1}u = \hat{x}_{k-1}t$, and (ii) if v is in the kth y-cluster then $\hat{v}u \simeq \hat{v}t$.

(T-3)   Let t be a tight x_k-term. Then t-x_k-$\simeq$-unitedness and t-x_k-unitedness coincide, whence hereditary t-x_k-$\simeq$-unitedness and hereditary t-x_k-unitedness coincide.

Proof.   (T-1)   Property (T-1) is an immediate consequence of the definitions of t-x_n-$\simeq$-unitedness and of tightness.

(T-2)   Let u be defined as follows. First let $\hat{x}_{k-1}u = \hat{x}_{k-1}t$. Second, for each y-variable y_j in the kth y-cluster, if y_j is hereditarily t-x_k-$\simeq$-united with some variable v dominating x_{k-1}, then let $\hat{y}_j u = \hat{v}t$ for the earliest such variable v, and otherwise let $\hat{y}_j u = \hat{y}_j t$. Clearly $\hat{y}_j u \simeq \hat{y}_j t$ for each such y-variable y_j; thus $\hat{v}u \simeq \hat{v}t$ for each v governing x_k. By (∇), u-x_k-$\simeq$-unitedness and t-x_k-$\simeq$-unitedness coincide, whence hereditary u-x_k-$\simeq$-unitedness and hereditarily t-x_k-$\simeq$-unitedness coincide. From the definition of u and the tightness of $\hat{x}_{k-1}t$ $(=\hat{x}_{k-1}u)$, it follows that if v and w are hereditarily t-x_k-$\simeq$-united, then $\hat{v}u = \hat{w}u$. Hence u is tight.

(T-3)   Property (T-3), in the case that $k = n$, is immediate from (T-1).

ISBN 0-201-02540-X

Suppose now that $k < n$ and that (T-3) holds for all tight x_{k+1}-terms. Let v and w be t-x_k-$\simeq$-united, where t is a tight x_k-term; we must show that v and w are t-x_k-united. By definition there is an x_{k+1}-term s such that $t = \hat{x}_k s$ and v and w are hereditarily s-x_{k+1}-$\simeq$-united. By (T-2) and (∇) we may without loss of generality assume that s is tight. But then, by the induction hypothesis, v and w are hereditarily s-x_{k+1}-united. Hence v and w are t-x_k-united. The converse, that all t-x_k-united variables are t-x_k-$\simeq$-united, is trivial (and holds for all terms t, not just tight ones). $\square$

Properties of β^*. There is a sequence-mapping β^* such that, for all sequences $\langle t_1,\ldots,t_p \rangle$ from $D(F, \omega)$, $p \geq 1$,

(β^*-1) for each tight x_n-term t and each arrangement $\langle v_1,\ldots,v_p \rangle$ of an x-focused set in $b(F)$, $\mathfrak{p}_{\mathfrak{N}}(\beta^*(\hat{v}_1 t,\ldots,\hat{v}_p t)) = \mathfrak{p}_{\mathfrak{N}}(\hat{v}_1 t,\ldots,\hat{v}_p t)$;

(β^*-2) β^* preserves permutations and repetitions;

(β^*-3) $\beta^* (t_1,\ldots,t_p) \simeq \langle t_1,\ldots,t_p \rangle$;

(β^*-4) $\beta^*(s_1,\ldots,s_p) = \beta^*(t_1,\ldots,t_p)$ whenever $\langle s_1,\ldots,s_p \rangle \simeq \langle t_1,\ldots,t_p \rangle$.

Proof. We define β^* just as in the amiability proof for Class 2.7: that is, the definition is word for word the same as that on page 99. (By (T-1) the definition is unambiguous.) Properties (β^*-2), (β^*-3), and (β^*-4) are then immediate. To verify (β^*-1), let t be a tight x_n-term and let $\langle v_1,\ldots,v_p \rangle$ be an arrangement of an x-focused set V in $b(F)$. Then $\beta^*(\hat{v}_1 t,\ldots,\hat{v}_p t)$ is determined by clause (i) on page 99: $\beta^*(\hat{v}_1 t,\ldots,\hat{v}_p t) = \langle \hat{w}_1 s,\ldots,\hat{w}_p s \rangle$ for some tight x_n-term s, some x-focused set W in $b(F)$ such that $\hat{V}[t] \simeq \hat{W}[s]$, and some variables $w_1,\ldots,w_p$ in W such that $\hat{v}_i t = \hat{w}_i s$, $1 \leq i \leq p$. It suffices to show that $\mathfrak{p}_{\mathfrak{N}}(\hat{v}_1 t,\ldots,\hat{v}_p t) = \mathfrak{p}_{\mathfrak{N}}(\hat{w}_1 s,\ldots,\hat{w}_p s)$. Let x_k be the focus of V; by Generalized Faithfulness Lemma 1, x_k is also the focus of W and $\hat{x}_k t \simeq \hat{x}_k s$. Suppose that $v \in V$, $w \in W$, and $\hat{v}t \simeq \hat{w}s$. Then either $v = w = x_k$ or else v and w both govern x_k. Hence, by Generalized Combinatorial Lemma 2, $\hat{v}t \simeq \hat{w}t \simeq \hat{w}s \simeq \hat{v}s$. Consequently, $\hat{V}[t] \simeq \hat{W}[t] \simeq \hat{W}[s] \simeq \hat{V}[s]$, whence $\langle \hat{v}_1 t,\ldots,\hat{v}_p t \rangle \simeq \langle \hat{w}_1 t,\ldots,\hat{w}_p t \rangle$. Since t is tight, by (T-1), $\langle \hat{v}_1 t,\ldots,\hat{v}_p t \rangle = \langle \hat{w}_1 t,\ldots,\hat{w}_p t \rangle$. And since $\hat{x}_k t \simeq \hat{x}_k s$, $\mathfrak{p}_{\mathfrak{N}}(\hat{x}_k t) = \mathfrak{p}_{\mathfrak{N}}(\hat{x}_k s)$, whence $\mathfrak{p}_{\mathfrak{N}}(\hat{w}_1 t,\ldots,\hat{w}_p t) = \mathfrak{p}_{\mathfrak{N}}(\hat{w}_1 s,\ldots,\hat{w}_p s)$. Hence $\mathfrak{p}_{\mathfrak{N}}(\hat{v}_1 t,\ldots,\hat{v}_p t) = \mathfrak{p}_{\mathfrak{N}}(\hat{w}_1 s,\ldots,\hat{w}_p s)$, and property ($\beta^*$-1) is verified. $\square$

Existence Condition. There is a term-mapping μ such that, for every x_k-term t, $1 \leq k \leq n$,

(a) μt is a tight x_k-term;

(b) $\mu \hat{x}_j t = \hat{x}_j \mu t$, $1 \leq j < k$;

(c) $\hat{y}_i t \simeq \hat{y}_i \mu t$ for each y-variable y_i governing x_k;

(d) $\mathfrak{p}_{\mathfrak{N}}(\beta^*(\hat{y}_{i_1} t,\ldots,\hat{y}_{i_p} t)) = \mathfrak{p}_{\mathfrak{N}}(\hat{y}_{i_1} \mu t,\ldots,\hat{y}_{i_p} \mu t)$ whenever $Y = \{y_{i_1},\ldots,y_{i_p}\}$ is a y-set in $b(F)$ each member of which governs x_k, $i_1 < \cdots < i_p$, $p \geq 1$.

ISBN 0-201-02540-X

Proof. Let t be an x_k-term, $1 \le k \le n$, and suppose as induction hypothesis that μ is defined on $\hat{x}_{k-1}t$ and fulfills (a)–(d) (a vacuous supposition if $k = 1$). If the kth y-cluster is empty, then let $\mu t = \hat{x}_k \mu \hat{x}_{k-1}t$; clearly the four conditions are fulfilled by μt. Otherwise, suppose $\{y_{a+1},\ldots,y_{a+b}\}$ is the kth y-cluster, $b > 0$. Let τt be the earliest tight x_k-term such that $\hat{x}_{k-1}\tau t = \mu \hat{x}_{k-1}t$ and, for each y_i in the kth y-cluster, $\hat{y}_i\tau t \simeq \hat{y}_i t$. Such a tight term τt exists by (T-2). Then let μt be the x_k-term determined thus:

(μ-1) $\hat{x}_{k-1}\mu t = \hat{x}_{k-1}\tau t$ $(=\mu \hat{x}_{k-1}t)$;

(μ-2) if y_{a+i}, $1 \le i \le b$, is hereditarily τt-x_k-united with some variable v that dominates x_{k-1}, then let $\hat{y}_{a+i}\mu t = \hat{v}\tau t$ $(= \hat{v}\mu \hat{x}_{k-1}t)$ for the earliest such variable v;

(μ-3) if $Y = \{y_{i_1},\ldots,y_{i_p}\}$ is a y-set in $b(F)$ such that $|\hat{Y}[\tau t]| > 1$, where $a < i_1 < \cdots < i_p \le a + b$, then let $\langle \hat{y}_{i_1}\mu t,\ldots,\hat{y}_{i_p}\mu t \rangle = \beta^*(\hat{y}_{i_1}\tau t,\ldots,\hat{y}_{i_p}\tau t)$;

(μ-4) if $\hat{y}_{a+i}\mu t$ is not determined by (μ-3), but y_{a+i} is hereditarily τt-x_k-united with some y_{a+j} such that $\hat{y}_{a+j}\mu t$ is so determined, then let $\hat{y}_{a+i}\mu t = \hat{y}_{a+j}\mu t$ for the least such $a + j$;

(μ-5) if $\hat{y}_{a+i}\mu t$ is determined by none of (μ-2)–(μ-4), then let $\hat{y}_{a+i}\mu t = \eta \hat{y}_{a+i}\tau t$.

The specification of Class 2.8 assures us that no ambiguity arises from clauses (μ-1)–(μ-4); and an argument like that in the amenability proof for Class 2.8 shows us that if two variables v and w are hereditarily τt-x_k-united then $\hat{v}\mu t = \hat{w}\mu t$. Now $\hat{x}_{k-1}\mu t = \hat{x}_{k-1}\tau t$ and, by the construction of μt and property (β^*-3), $\hat{y}_i\mu t \simeq \hat{y}_i\tau t$ for each y-variable y_i in the kth y-cluster. Hence $\hat{v}\mu t \simeq \hat{v}\tau t$ for each variable v governing x_k. Consequently, by (∇), μt-x_k-$\simeq$-unitedness and τt-x_k-$\simeq$-unitedness coincide. Since τt is tight, by (T-3) τt-x_k-$\simeq$-unitedness and τt-x_k-unitedness coincide. Thus; if v and w are hereditarily μt-x_k-$\simeq$-united then $\hat{v}\mu t = \hat{w}\mu t$. Hence μt is tight, and (a) is fulfilled.

From (μ-1), $\hat{x}_{k-1}\mu t = \mu \hat{x}_{k-1}t$. But then, by the induction hypothesis that μ fulfills (b) on $\hat{x}_{k-1}t$, $\mu \hat{x}_j t = \mu \hat{x}_j \hat{x}_{k-1}t = \hat{x}_j \mu \hat{x}_{k-1}t = \hat{x}_j \hat{x}_{k-1}\mu t = \hat{x}_j \mu t$ for each $j < k - 1$. Hence μ fulfills (b) on t. Moreover, we have seen that $\hat{y}_i\mu t \simeq \hat{y}_i\tau t$ for each y-variable y_i governing x_k. By the induction hypothesis that μ fulfills (c) on $\hat{x}_{k-1}t$ and the construction of τ, $\hat{y}_i\tau t \simeq \hat{y}_i t$ for each such y-variable y_i. Hence μ fulfills (c) on t.

It remains only to show (d). Let Y be a y-set contained in the jth y-cluster for some $j < k$. Then (d) is immediate from the induction hypothesis that μ fulfills (d) on $\hat{x}_{k-1}t$. Let Y now be a y-set in the kth y-cluster. If $|\hat{Y}[\tau t]| > 1$, then (d) follows from clause (μ-3). If $|Y| = 1$, then (d) follows since, by property (β^*-3) and (b), we have $\mathfrak{p}_{\mathfrak{A}}(\beta^*(\hat{y}_i t)) = \mathfrak{p}_{\mathfrak{A}}(\hat{y}_i t) = \mathfrak{p}_{\mathfrak{A}}(\hat{y}_i\mu t)$ for each y_i governing x_k. Finally, just as in the amen-

ISBN 0-201-02540-X

ability proof for Class 2.8, if $|Y| > 1$ but $|\hat{Y}[\tau t]| = 1$, then (d) follows from the previous case and the specification of Class 2.8. $\square$

Construction of δ. Let $\delta\mathord{+} = \mathord{+}$. For each x_k-term t, $1 \le k \le n$, let $\delta t = \eta\mu f_{x_k}(\delta\hat{y}_1 t \ldots \delta\hat{y}_j t)$, where $y_1,\ldots,y_j$ are all the y-variables governing x_k. Thus $|\delta[D(F, \omega)]| \le r(2r(m + n - 1)^4 + 1)^{(m+n-1)^2}$. Now let t be an x_n-term, let $\langle v_1,\ldots,v_p\rangle$ be an arrangement of a set in $b(F)$, and let $s = f_{x_n}(\delta\hat{y}_1 t \ldots \delta\hat{y}_m t)$. Since $\delta t \simeq \mu s$, to verify condition (2) of the Amiability Lemma it suffices to show that $\mathfrak{p}_{\mathfrak{A}}(\beta^*(\delta\hat{v}_1 t,\ldots, \delta\hat{v}_p t)) = \mathfrak{p}_{\mathfrak{A}}(\hat{v}_1\mu s,\ldots,\hat{v}_p\mu s)$. If $\{v_1,\ldots,v_p\}$ is a y-set, then this follows just as in the proof for Class 2.7 from (d) of the Existence Condition and the permutation property (β^*-2). If $\{v_1,\ldots,v_p\}$ is x-focused, then $\langle \delta\hat{v}_1 t,\ldots,\delta\hat{v}_p t\rangle \simeq \langle \hat{v}_1\mu s,\ldots,\hat{v}_p\mu s\rangle$: for, by (c) of the Existence Condition $\delta\hat{y}_i t = \hat{y}_i s \simeq \hat{y}_i\mu s$ for each $i \le m$; and, by the construction of δ and (b) of the Existence Condition, $\delta\hat{x}_k t = \eta\mu\hat{x}_k s = \eta\hat{x}_k\mu s \simeq \hat{x}_k\mu s$. Hence $\beta^*(\delta\hat{v}_1 t,\ldots,\delta\hat{v}_p t) = \beta^*(\hat{v}_1\mu s,\ldots,\hat{v}_p\mu s)$ by (β^*-4), and $\mathfrak{p}_{\mathfrak{A}}(\beta^*(\hat{v}_1\mu s,\ldots,\hat{v}_p\mu s)) = \mathfrak{p}_{\mathfrak{A}}(\hat{v}_1\mu s,\ldots,\hat{v}_p\mu s)$ by (β^*-1) and the tightness of μs. Condition (2) is verified, and the proof of amenability is concluded. $\square$

The Maslov Class

The *Maslov Class* contains just those Skolem schemata whose matrices are *Krom,* that is, whose matrices are conjunctions of binary disjunctions of signed atomic formulas. Thus a Maslov schema has the form

$$\forall y_1 \cdots \forall y_m \exists x_1 \cdots \exists x_n \left[\bigwedge_{1 \leq i \leq k} (A_i \vee B_i) \right],$$

where the A_i and the B_i are signed atomic formulas. The Maslov Class is full in the class of all Skolem schemata; hence, by the Corollary to the Theorem of Chapter 2, §7, it is not amenable. Nor is it amiable (see footnote, page 85). Nonetheless, it is both solvable (Maslov, 1964) and finitely controllable (Aanderaa and Goldfarb, 1974).

As is noted by Maslov (1964) and Aanderaa and Goldfarb (1974), these results also hold for the *Initially-extended Maslov Class,* the class of schemata with prefixes $\exists z_1 \cdots \exists z_l \forall y_1 \cdots \forall y_m \exists x_1 \cdots \exists x_n$ and Krom matrices. Indeed, Kalmár's procedure for eliminating initial x-variables furnishes a reduction of the Initially-extended Maslov Class to the Maslov Class: see Chapter 6, §4.

Like the classes of Chapter 2, §4, the quantifier prefix of a Maslov schema is restricted; unlike those classes, however, the basis is not. Rather, the matrix must be of a special truth-functional form. Now Classes 1.2 and 1.3 are specified by the truth-functional form of the matrix, but those classes involve no restriction on the quantifier prefix. Here, in contrast, the class of all prenex schemata with Krom matrices is unsolvable (Krom, 1970; Aanderaa, 1971). Hence the solvability and finite controllability proofs for the Maslov Class must exploit properties of expansions that reflect both the special quantifier prefix and the special truth-functional form of the matrix.‡

---

‡ We shall meet two other classes of this hybrid type: Class 5.5, the $\forall \exists \forall$–Horn Class, and Class 5.6, the $\forall \exists \forall$–Krom Class. These classes are treated in Chapter 5 since they are not finitely controllable.

Burton Dreben and Warren D. Goldfarb, The Decision Problem: Solvable Classes of Quantificational Formulas

ISBN 0-201-02540-X

The requirement that a Maslov schema F be Skolem assures us that each term t of height >1 in $D(F, \omega)$ occurs as the substituent for the x-variable $\chi(t)$ in exactly one Herbrand instance of F. The requirement that a Maslov schema F have Krom matrix, which implies that the conjunction of any number of Herbrand instances of F is itself a Krom formula, allows us to apply the following Lemma.

Truth-Functional Lemma for Krom Formulas. Let N be a Krom formula and C a conjunction of signed atomic formulas. Then $C \wedge N$ is (truth-functionally) inconsistent only if there are signed atomic formulas A_1 and A_2, each of which is a conjunct in C, such that $A_1 \wedge A_2 \wedge N$ is inconsistent.

Proof. By induction on the number of conjuncts in N. If N has none, then the conclusion is immediate. If $N = M \wedge (B_1 \vee B_2)$, then $C \wedge N$ is inconsistent only if both $C \wedge B_1 \wedge M$ and $C \wedge B_2 \wedge M$ are inconsistent. If there are conjuncts A_1 and A_2 in C such that $A_1 \wedge A_2 \wedge M$ is inconsistent, then we are done. Otherwise, since $C \wedge B_1$ and $C \wedge B_2$ are conjunctions of signed atomic formulas and M is a Krom formula containing one fewer conjunct than N, by the induction hypothesis there are conjuncts A_1 and A_2 in C such that $A_1 \wedge B_1 \wedge M$ and $A_2 \wedge B_2 \wedge M$ are inconsistent. But then $A_1 \wedge A_2 \wedge M \wedge (B_1 \vee B_2)$ is inconsistent. $\square$

We shall use the Truth-Functional Lemma in its contrapositive form: $C \wedge N$ is consistent if $A_1 \wedge A_2 \wedge N$ is consistent for every pair A_1, A_2 of signed atomic formulas that are conjuncts in C.

1. Solvability

Class 4.1. Let F be a Maslov schema containing p predicate letters, and let k be the maximum number of argument places of these predicate letters. Then F has a model provided that $E(F, [2p(2k)^k]^2)$ is consistent.

Proof. For encoding purposes we consider all signed atomic formulas constructed from the p predicate letters of F and the formal variables $\delta_1, \ldots, \delta_{2k}$.‡ Let Ξ be the set of all binary disjunctions of such signed atomic formulas; thus $|\Xi| \leq [2p(2k)^k]^2$. If $\mathfrak{A}$ is any truth-assignment on $\bar{L}(F, \omega)$ and q is any positive integer, then let $\Xi(\mathfrak{A}, q)$ be the set of members D of Ξ such that $\mathfrak{A}$ verifies every instance of D over $D(F, q)$. Note that $\Xi(\mathfrak{A}, q + 1) \subseteq \Xi(\mathfrak{A}, q)$ for every $\mathfrak{A}$ and every q. Note too that $\Xi(\mathfrak{A}, q)$ is never empty, since it always contains disjunctions of the form $A \vee -A$.

We now suppose $E(F, [2p(2k)^k]^2)$ is consistent. Let $\mathfrak{A}$ be any truth-assignment verifying this expansion. By the pigeonhole argument there is a positive integer $r \leq [2p(2k)^k]^2$ such that $\mathfrak{A}$ verifies $E(F, r)$ and $\Xi(\mathfrak{A}, r)$

---

‡ The variables δ_i are, just as in the definition of profile, variables of the formal language, but are not to occur in any schema.

ISBN 0-201-02540-X

$= \Xi(\mathfrak{A}, r + 1)$. (In contrast to Chapter 2, the pigeonhole argument will not be used here to obtain a pigeonhole term-mapping.) We shall show the existence, for each $q > r$, of a truth-assignment $\mathfrak{A}_q$ such that

(1) $\mathfrak{A}_q$ verifies $E(F, q - 1)$;
(2) $\Xi(\mathfrak{A}, r) \subseteq \Xi(\mathfrak{A}_q, q)$.

By (1) we shall have shown the consistency of each expansion $E(F, q)$. Hence, by the Expansion Theorem, F has a model.

Let $\mathfrak{A}_{r+1} = \mathfrak{A}$. By the choice of $\mathfrak{A}$, conditions (1) and (2) are fulfilled for $q = r + 1$. Assume that $q \geq r + 1$ and that we have a truth-assignment $\mathfrak{A}_q$ fulfilling (1) and (2); we seek a suitable truth-assignment $\mathfrak{A}_{q+1}$.

We define three conjunctions K, M_1, and M_2 as follows. Let K be the conjunction of all signed atoms in $\tilde{L}(F, q)$ that are verified by $\mathfrak{A}_q$. Let M_1 be the conjunction of all Herbrand instances in $E(F, q) - E(F, q - 1)$, that is, all Herbrand instances in which the substituents for the x-variables have height $q + 1$. And let M_2 be the conjunction of all instances of members of $\Xi(\mathfrak{A}, r)$ over $D(F, q + 1)$. It suffices to show that the conjunction $K \wedge M_1 \wedge M_2$ is consistent. For if $\mathfrak{A}_{q+1}$ is any truth assignment verifying K, then $\mathfrak{A}_{q+1}$ agrees with $\mathfrak{A}_q$ on $\tilde{L}(F, q)$; in particular, $\mathfrak{A}_{q+1}$ verifies $E(F, q - 1)$. If moreover $\mathfrak{A}_{q+1}$ verifies M_1, then $\mathfrak{A}_{q+1}$ verifies $E(F, q)$. Lastly, if $\mathfrak{A}_{q+1}$ verifies M_2, then clearly every member of $\Xi(\mathfrak{A}, r)$ is also a member of $\Xi(\mathfrak{A}_{q+1}, q + 1)$. Hence any truth-assignment verifying $K \wedge M_1 \wedge M_2$ fulfills conditions (1) and (2) for $q + 1$.

Since $M_1 \wedge M_2$ is a Krom formula and K is a conjunction of signed atoms, by the Truth-Functional Lemma the following condition suffices for the consistency of $K \wedge M_1 \wedge M_2$:

For all signed atoms A and B that are conjuncts in K, the conjunction $A \wedge B \wedge M_1 \wedge M_2$ is consistent.

Suppose A and B given; let $s_1, \ldots, s_j$ be the distinct terms occurring in $A \wedge B$. Thus $0 < j \leq 2k$, and each s_i is in $D(F, q)$. Moreover, $A \wedge B$ is truth-functionally equivalent to $-D(\delta_1/s_1, \ldots, \delta_j/s_j)$ for some disjunction D in Ξ; hence $D \notin \Xi(\mathfrak{A}_q, q)$, since $\mathfrak{A}_q$ verifies $A \wedge B$. Hence by (2), $D \notin \Xi(\mathfrak{A}, r)$; that is, there are (not necessarily distinct) terms $u_1, \ldots, u_j$ in $D(F, r)$ such that $\mathfrak{A}$ verifies $-D(\delta_1/u_1, \ldots, \delta_j/u_j)$. Define a term-mapping γ thus:

(a) $\gamma s_i = u_i$, $1 \leq i \leq j$;
(b) $\gamma t = \mathbf{1}$ if $h(t) \neq q + 1$ and γt is not determined by (a);
(c) γ is homomorphic on terms of height $q + 1$.

Let Γ be the γ-canonical formula-mapping. We show that $\mathfrak{A}$ verifies $\Gamma(A \wedge B \wedge M_1 \wedge M_2)$; by the Consistency Lemma (Chapter 1, §3), $A \wedge B \wedge M_1 \wedge M_2$ is then consistent.

ISBN 0-201-02540-X

First, $\Gamma(A \wedge B)$ is, by clause (a) of the definition of γ, truth-functionally equivalent to $-D(\mathfrak{z}_1/u_1,\ldots,\mathfrak{z}_j/u_j)$, which is verified by $\mathfrak{A}$. Second, by clause (c), Γ is Herbrand instance preserving on $E(F, q) - E(F, q - 1)$. Moreover, by clauses (a) and (b), $\gamma[D(F, q)] \subseteq D(F, r)$; hence $\mathfrak{A}$ verifies $\Gamma(M_1)$. Finally, if G is an instance of a disjunction in $\Xi(\mathfrak{A}, r)$ over $D(F, q + 1)$, then $\Gamma(G)$ is an instance of that disjunction over $D(F, r + 1)$. Since $\Xi(\mathfrak{A}, r + 1) = \Xi(\mathfrak{A}, r)$, $\mathfrak{A}$ verifies $\Gamma(G)$. Hence $\mathfrak{A}$ verifies $\Gamma(M_2)$. $\square$

REMARK. In the proof just given each truth-assignment $\mathfrak{A}_{q+1}$ agrees with $\mathfrak{A}_q$ on $\bar{L}(F, q)$. Hence we may view this proof as an inductive construction of a truth-assignment $\mathfrak{A}_\omega$ that verifies $E(F, \omega)$. Moreover, the core of the construction may be formulated thus: Let $\mathfrak{A}$ be a truth-assignment and r a positive integer such that $\mathfrak{A}$ verifies $E(F, r)$ and $\Xi(\mathfrak{A}, r + 1) = \Xi(\mathfrak{A}, r)$. Then for any truth-assignment $\mathfrak{B}$ and any $q > 0$ such that $\Xi(\mathfrak{A}, r) \subseteq \Xi(\mathfrak{B}, q)$, there is a truth-assignment $\mathfrak{B}'$ with the following properties: $\Xi(\mathfrak{A}, r) \subseteq \Xi(\mathfrak{B}', q + 1)$; $\mathfrak{B}'$ agrees with $\mathfrak{B}$ on $\bar{L}(F, q)$; and $\mathfrak{B}'$ verifies $E(F, q) - E(F, q - 1)$. This formulation will be used in the finite controllability proof immediately below. End of Remark.

2. Finite Controllability

Although the Maslov Class is not amiable, the combinatorial machinery developed in Chapter 3, §4 may be exploited to yield its finite controllability. With the help of Combinatorial Lemmas 1 and 2 and Faithfulness Lemma 1, we shall strengthen the inductive construction that underlies the core of the solvability proof in §1. Our goal is to obtain, given a satisfiable Maslov schema F, a truth-assignment $\mathfrak{A}$ that verifies not just $E(F, \omega)$ but also $\Delta[E(F, \omega)]$ for some term-mapping δ with finite range (Δ is the δ-canonical formula-mapping).

Let $F = \forall y_1 \cdots \forall y_m \exists x_1 \cdots \exists x_n F^M$ be a fixed Maslov schema. For each $q > 1$ let S_q be the set of atoms that either lie in $L(F, q - 1)$ or else have the form $P\hat{v}_1 t \cdots \hat{v}_p t$ for some x_n-term t of height q and some x-focused sequence $\langle v_1,\ldots,v_p \rangle$ of variables of F; and let R_q be the set of other atoms in $L(F, q)$. Note that every atom occurring in $E(F, q - 1)$ is a member of S_q; that if an atom A lies in R_q, then A contains at least one term t of height q and one term not an argument of t; and that the R_q are disjoint. We shall construct a sequence $\mathfrak{A}_1, \mathfrak{A}_2, \mathfrak{A}_3,\ldots$ of truth-assignments such that each $\mathfrak{A}_{q+1}$ agrees with $\mathfrak{A}_q$ on S_q. The desired truth-assignment $\mathfrak{A}$ is the "limit" of this sequence: if A is in S_q, then $\mathfrak{A}(A) = \mathfrak{A}_q(A)$ and hence $\mathfrak{A}(A) = \mathfrak{A}_k(A)$ for each $k > q$.

Our construction may be sketched as follows. Suppose that $\mathfrak{A}_q$ verifies $E(F, q - 1)$. As in the solvability proof we construct $\mathfrak{A}_{q+1}$ so as to verify $E(F, q)$. But, unlike the solvability proof, we do not require that $\mathfrak{A}_{q+1}$ agree

ISBN 0-201-02540-X

with $\mathfrak{A}_q$ on all of $\tilde{L}(F, q)$, rather only on all of S_q: to allow the eventual specification of a term-mapping δ that yields a finite model, we must tamper with the truth-values of atoms in R_q. (Since these atoms do not occur in $E(F, q - 1)$, we incur no risk of undoing the work that $\mathfrak{A}_q$ does and hence of falsifying $E(F, q - 1)$.) To determine how atoms in R_q are to be treated, we use equivalence relations $\simeq$ on $D(F, \omega)$ like those of Chapter 3. These equivalence relations must be defined relative to truth-assignments. In Chapter 3 this presented no problem, since in amiability proofs we start with a given, fixed truth-assignment. But here we must define $\simeq$ in terms of the very truth-assignment we are inductively constructing. Consequently, it becomes important that the equivalences between terms in a domain $D(F, q)$ be determined by the values of the truth-assignment on S_q. That is, whether or not two terms of height $\leq q$ are equivalent must be settled by $\mathfrak{A}_q$; their equivalence or nonequivalence must not change when we pass to the limit and define the equivalence relation with respect to the truth-assignment $\mathfrak{A}$. Fortunately, the procedures of Chapter 3 for constructing equivalence relations have this kind of stability.

Indeed, given any truth-assignment $\mathfrak{B}$ on $\tilde{L}(F, \omega)$, let $E_{\mathfrak{B}}$ be the equivalence relation of finite order on $D(F, \omega)$ defined thus: $s E_{\mathfrak{B}} t$ iff either $s = t = \dashv$ or else $\alpha(s) = \alpha(t)$, $\chi(s) = \chi(t)$, and

$$\mathfrak{p}_{\mathfrak{B}}(\hat{y}_1 s, \ldots, \hat{y}_m s, \hat{x}_1 s, \ldots, \hat{x}_n s) = \mathfrak{p}_{\mathfrak{B}}(\hat{y}_1 t, \ldots, \hat{y}_m t, \hat{x}_1 t, \ldots, \hat{x}_n t),$$

where α is the function given by Combinatorial Lemma 1. Clearly, if two truth-assignments $\mathfrak{B}$ and $\mathfrak{B}'$ agree on S_q, then the two equivalence relations $E_{\mathfrak{B}}$ and $E_{\mathfrak{B}'}$ agree on $D(F, q)$. From an equivalence relation $E_{\mathfrak{B}}$ we may then construct an equivalence relation $\simeq_{\mathfrak{B}}$ with the properties listed in Combinatorial Lemma 2. Here we need some further information, not provided in our previous statement of this Lemma: the relation $\simeq_{\mathfrak{B}}$ on a domain $D(F, q)$ must be determined by the relation $E_{\mathfrak{B}}$ on $D(F, q)$. An inspection of the proof in the Appendix shows that we in fact have

Amplified Combinatorial Lemma 2. Let $r > 0$. There is an effective procedure that yields, given any equivalence relation E on $D(F, \omega)$ of order $\leq r$, an equivalence relation $\simeq$ on $D(F, \omega)$ of order $\leq r \cdot (2rm^4 + 1)^{m^2}$ that fulfills, for all terms s and t and all y-variables y_i and y_j,

(1) if $s \simeq t$, then $s E t$;
(2) if $s \simeq t$ and $\hat{y}_i s \simeq \hat{y}_j t$, then $\hat{y}_i s \simeq \hat{y}_j s \simeq \hat{y}_i t \simeq \hat{y}_j t$.

Moreover, for each $q > 0$, if E' is any equivalence relation on $D(F, \omega)$ that agrees with E on $D(F, q)$ and that has order $\leq r$, then the equivalence relation $\simeq'$ obtained from E' by this procedure agrees with the equivalence relation $\simeq$ on the domain $D(F, q)$. $\square$

ISBN 0-201-02540-X

Thus our demands will be met if we let $\simeq_\mathfrak{B}$ be the result of applying this procedure to $E_\mathfrak{B}$, where $r = \kappa_m \cdot n \cdot d(F, m + n) + 1$. Since the limit truth-assignment $\mathfrak{A}$ is to agree with $\mathfrak{A}_q$ on S_q, we shall have the following: if s and t have height $\leq q$, then $s \simeq_{\mathfrak{A}_q} t$ iff $s \simeq_\mathfrak{A} t$.

A term t is $\mathfrak{B}$-*tight*, where $\mathfrak{B}$ is a truth-assignment, iff $h(t) > 1$ and, for all i and j, $\hat{y}_i t \simeq_\mathfrak{B} \hat{y}_j t$ only if $\hat{y}_i t = \hat{y}_j t$. The whole point of our combinatorial machinery is vested in the following property of tight terms.

Property T. Let t be a $\mathfrak{B}$-tight x_n-term, let u be an x_n-term, and let $\langle v_1, \ldots, v_p \rangle$ and $\langle w_1, \ldots, w_p \rangle$ be x-focused sequences such that $\langle \hat{v}_1 t, \ldots, \hat{v}_p t \rangle \simeq_\mathfrak{B} \langle \hat{w}_1 u, \ldots, \hat{w}_p u \rangle$. Then $\mathfrak{p}_\mathfrak{B}(\hat{v}_1 t, \ldots, \hat{v}_p t) = \mathfrak{p}_\mathfrak{B}(\hat{w}_1 u, \ldots, \hat{w}_p u)$.

Proof. By the definition of $E_\mathfrak{B}$ and by clause (1) of Combinatorial Lemma 2, the equivalence relation $\simeq_\mathfrak{B}$ is faithful. Hence, by Faithfulness Lemma 1, for each $i \leq p$ either $v_i = w_i = $ some x-variable or else v_i and w_i are both y-variables. Since $\langle v_1, \ldots, v_p \rangle$ is x-focused, some v_j is an x-variable x_k, and so $\hat{x}_k t \simeq_\mathfrak{B} \hat{x}_k u$. Hence, if v_i and w_i are both y-variables, say $v_i = y_a$ and $w_i = y_b$, then $\hat{y}_a \hat{x}_k t \simeq_\mathfrak{B} \hat{y}_b \hat{x}_k u$. Hence, by clause (2) of Combinatorial Lemma 2, $\hat{y}_a \hat{x}_k t \simeq_\mathfrak{B} \hat{y}_b \hat{x}_k t$. Hence $\hat{y}_a t \simeq_\mathfrak{B} \hat{y}_b t$; and since t is $\mathfrak{B}$-tight, $\hat{y}_a t = \hat{y}_b t$, that is, $\hat{v}_i t = \hat{w}_i t$. Thus $\langle \hat{v}_1 t, \ldots, \hat{v}_p t \rangle = \langle \hat{w}_1 t, \ldots, \hat{w}_p t \rangle$. But since $\hat{x}_k t \simeq_\mathfrak{B} \hat{x}_k u$, $\mathfrak{p}_\mathfrak{B}(\hat{w}_1 t, \ldots, \hat{w}_p t) = \mathfrak{p}_\mathfrak{B}(\hat{w}_1 u, \ldots, \hat{w}_p u)$. Hence $\mathfrak{p}_\mathfrak{B}(\hat{v}_1 t, \ldots, \hat{v}_p t) = \mathfrak{p}_\mathfrak{B}(\hat{w}_1 u, \ldots, \hat{w}_p u)$. $\square$

For each truth-assignment $\mathfrak{B}$, let $\Lambda_\mathfrak{B}$ be the formula-mapping defined thus: for each atom $A = Pt_1 \cdots t_p$, let $\Lambda_\mathfrak{B}(A) = A$ unless A is in some set R_q and there are a $\mathfrak{B}$-tight x_n-term s in $D(F, q)$ and an x-focused sequence $\langle w_1, \ldots, w_p \rangle$ such that $\langle t_1, \ldots, t_p \rangle \simeq_\mathfrak{B} \langle \hat{w}_1 s, \ldots, \hat{w}_p s \rangle$; in that case, let $\Lambda_\mathfrak{B}(A) = P\hat{w}_1 s \ldots \hat{w}_p s$ for the earliest such s and $\langle w_1, \ldots, w_p \rangle$. Note that if $\Lambda_\mathfrak{B}(A) \neq A$, then $\Lambda_\mathfrak{B}(A)$ is in S_q. Note further two important properties of $\Lambda_\mathfrak{B}$.

(Λ-1)   Let $A = Pt_1 \cdots t_p$ be an atom such that $\Lambda_\mathfrak{B}(A) \neq A$, and suppose $\langle t_1, \ldots, t_p \rangle \simeq_\mathfrak{B} \langle \hat{v}_1 t, \ldots, \hat{v}_p t \rangle$ for some x_n-term t and some x-focused sequence $\langle v_1, \ldots, v_p \rangle$. Then $\mathfrak{B}(\Lambda_\mathfrak{B}(A)) = \mathfrak{B}(P\hat{v}_1 t \cdots \hat{v}_p t)$.

(Λ-2)   Suppose $\mathfrak{B}'$ is a truth-assignment that agrees with $\mathfrak{B}$ on S_q. Then $\Lambda_\mathfrak{B}$ and $\Lambda_{\mathfrak{B}'}$ agree on $L(F, q)$.

Property (Λ-1) is an immediate consequence of Property T. Property (Λ-2) follows from the fact that, if $\mathfrak{B}$ and $\mathfrak{B}'$ agree on S_q, then $\simeq_\mathfrak{B}$ and $\simeq_{\mathfrak{B}'}$ agree on $D(F, q)$; in particular, the $\mathfrak{B}$-tight terms in $D(F, q)$ are the $\mathfrak{B}'$-tight terms in $D(F, q)$.

We need one final preliminary step. We take $L(F, \omega)$ to be so ordered that: (I) for each $q > 1$, every member of S_q precedes every member of R_q; and (II) if the terms occurring in an atom A form a proper subset of those occurring in an atom B, then A precedes B. (Thus if $\Lambda_\mathfrak{B}(A) \neq A$, then $\Lambda_\mathfrak{B}(A)$ precedes A.)

ISBN 0-201-02540-X

Assume that the schema F has a model. Then, just as in the solvability proof, there is a truth-assignment $\mathfrak{A}_1$ and a positive integer r_1 such that $\mathfrak{A}_1$ verifies $E(F, r_1)$ and $\Xi(\mathfrak{A}_1, r_1) = \Xi(\mathfrak{A}_1, r_1 + 1)$. We now construct the sequence $\mathfrak{A}_1, \mathfrak{A}_2,\ldots$; for each q we shall have

(a) $\mathfrak{A}_q$ verifies $E(F, q - 1)$;

(b) $\Xi(\mathfrak{A}_1, r_1) \subseteq \Xi(\mathfrak{A}_q, q)$.

Suppose that either $q = 1$ or else $q > 1$ and we have found a truth-assignment $\mathfrak{A}_q$ fulfilling (a) and (b). (If $q = 1$, then (a) is vacuous and (b) is trivial.) Let K_0 be the conjunction of all signed atoms in $\tilde{S}_q$ that are verified by $\mathfrak{A}_q$, and let N be the conjunction of all instances of members of $\Xi(\mathfrak{A}_1, r_1)$ over $D(F, q)$. By (b), $K_0 \wedge N$ is consistent. We define a sequence $K_0, K_1,\ldots,K_k$ of conjunctions, where $k = |R_q|$. Suppose K_j is defined, $0 \leq j < k$, and let A be the $(j + 1)$st member of R_q. If the conjunction $K_j \wedge (A \Leftrightarrow \Lambda_{\mathfrak{A}_q}(A)) \wedge N$ is consistent, then let $K_{j+1} = K_j \wedge (A \Leftrightarrow \Lambda_{\mathfrak{A}_q}(A))$. Otherwise, let $K_{j+1} = K_j$.

Obviously $K_k \wedge N$ is consistent. Hence there is a truth-assignment $\mathfrak{B}$ verifying it. Since $\mathfrak{B}$ verifies N, $\Xi(\mathfrak{A}_1, r_1) \subseteq \Xi(\mathfrak{B}, q)$. Hence, by the basic construction of the solvability proof (see the Remark at the end of §1), there is a truth assignment $\mathfrak{A}_{q+1}$ such that $\Xi(\mathfrak{A}_1, r_1) \subseteq \Xi(\mathfrak{A}_{q+1}, q + 1)$, $\mathfrak{A}_{q+1}$ verifies every Herbrand instance in $E(F, q) - E(F, q - 1)$, and $\mathfrak{A}_{q+1}$ agrees with $\mathfrak{B}$ on $\tilde{L}(F, q)$. In particular, $\mathfrak{A}_{q+1}$ verifies K_k. Hence $\mathfrak{A}_{q+1}$ verifies K_0, and so $\mathfrak{A}_{q+1}$ agrees with $\mathfrak{A}_q$ on S_q, and verifies $E(F, q - 1)$. It follows that $\mathfrak{A}_{q+1}$ fulfills conditions (a) and (b) for $q + 1$, that is, (a) $\mathfrak{A}_{q+1}$ verifies $E(F, q)$, and (b) $\Xi(\mathfrak{A}_1, r_1) \subseteq \Xi(\mathfrak{A}_{q+1}, q + 1)$.

This completes the construction. We let $\mathfrak{A}$ be the limit of the sequence $\mathfrak{A}_1, \mathfrak{A}_2, \mathfrak{A}_3, \ldots$ in the sense given at the beginning of this section: if A is in S_q, then $\mathfrak{A}(A) = \mathfrak{A}_q(A)$; and if A is in R_q, then $\mathfrak{A}(A) = \mathfrak{A}_{q+1}(A)$. Clearly $\mathfrak{A}$ verifies $E(F, \omega)$, and $\Xi(\mathfrak{A}_1, r_1) \subseteq \Xi(\mathfrak{A}, q)$ for each q. Henceforth we shall write "$\simeq$" for "$\simeq_{\mathfrak{A}}$", "tight" for "$\mathfrak{A}$-tight", and "Λ" for "$\Lambda_{\mathfrak{A}}$".

To show the existence of a finite model for F we require a term-mapping δ with finite range such that $\mathfrak{A}$ verifies $\Delta[E(F, \omega)]$. We shall so construct δ that, for all terms $t_1,\ldots,t_m$ and each $k \leq n$, $\delta f_{x_k}(t_1 \cdots t_m) \simeq f_{x_k}(\delta t_1 \cdots \delta t_m)$. But we shall pick $\delta f_{x_k}(t_1 \cdots t_m)$ not to be the *earliest* $\simeq$-equivalent of $f_{x_k}(\delta t_1 \cdots \delta t_m)$, but rather to be a $\simeq$-equivalent of suitably large height. Our aim in so doing is to assure that if the arguments of an atom A that lies in some R_q are among $\delta t_1,\ldots,\delta t_m$, $\delta f_{x_1}(t_1 \cdots t_m),\ldots,\delta f_{x_n}(t_1 \cdots t_m)$, then $\Lambda(A) \neq A$.

Call a term t *advanced* iff $h(t) > 1$ and, for each tight term s with $s \simeq t$, there is a tight term s' such that $h(s') \leq h(t)$ and $\hat{v}s' \simeq \hat{v}s$ for each variable v of F. Since $\simeq$ has finite order, every term of height >1 possesses

ISBN 0-201-02540-X

an advanced $\simeq$-equivalent. Hence we may define δ thus:

$$\delta \dashv = \dashv;$$
$$\delta f_{x_k}(t_1 \cdots t_m) = \text{the earliest advanced term } t \text{ such that }$$
$$t \simeq f_{x_k}(\delta t_1 \cdots \delta t_m).\ddagger$$

Thus the range of δ contains, aside from $\dashv$, only advanced terms, and no distinct terms in the range are $\simeq$-equivalent. Therefore, the range of δ is finite; moreover, any term of the form $f_{x_k}(\delta t_1 \cdots \delta t_m)$ is tight.

Let t be any x_n-term, and let $H = F^*(y_1/\hat{y}_1 t, \ldots, y_m/\hat{y}_m t)$. We must show that $\mathfrak{A}$ verifies $\Delta(H)$. Since $\mathfrak{A}$ verifies $E(F, \omega)$, it would suffice if $\mathfrak{A}(\Delta(\langle H \rangle_i))$ $= \mathfrak{A}(\langle H' \rangle_i)$ for each i, where $H' = F^*(y_1/\delta \hat{y}_1 t, \ldots, y_m/\delta \hat{y}_m t)$. That is, it would suffice to show:

(†) for each atom $P \delta \hat{v}_1 t \cdots \delta \hat{v}_p t$ in $\Delta(H)$, $\mathfrak{A}(P \delta \hat{v}_1 t \cdots \delta \hat{v}_p t)$ $=$ $\mathfrak{A}(P \hat{v}_1 s, \ldots, \hat{v}_p s)$, where $s = f_{x_n}(\delta \hat{y}_1 t \cdots \delta \hat{y}_m t)$.

(Note that s is tight, that $\delta \hat{v} t \simeq \hat{v} s$ for each variable v of F, and indeed that $\delta \hat{y}_i t = \hat{y}_i s$, $1 \le i \le m$.) Call a term-mapping γ *good* iff $\gamma \delta \hat{v} t = \hat{v} s$ for each variable v of F, that is, iff γ takes each $\delta \hat{y}_j t$ to itself and each $\delta \hat{x}_k t$ to $\hat{x}_k s$. Call a formula-mapping Γ *good* iff it is γ-canonical for some good term-mapping γ. Now for each k, $1 \le k \le n$, the terms $\delta \hat{x}_k t$ are distinct; for, since $\delta \hat{x}_k t \simeq \hat{x}_k s$, $\chi(\delta \hat{x}_k t) = x_k$. Hence good term-mappings exist, and each good formula-mapping carries $P \delta \hat{v}_1 t \cdots \delta \hat{v}_p t$ to $P \hat{v}_1 s \cdots \hat{v}_p s$. Thus (†) becomes

(†′) for each atom in $\Delta(H)$ there is a good formula-mapping Γ such that $\mathfrak{A}(A) = \mathfrak{A}(\Gamma(A))$.

However, to show (†′) we must look at atoms in $L(F, \omega)$ that precede those occurring in $\Delta(H)$ (see the last paragraph of Case 2, Subcase b below). Hence we show the more general

(#) for every atom A in $L(F, \omega)$ there is a good formula-mapping Γ such that $\mathfrak{A}(A) = \mathfrak{A}(\Gamma(A))$.

Our proof is by induction on the ordering of $L(F, \omega)$: we assume (#) holds for all atoms preceding A, and show it holds for A. There are three cases.

Case 1. The atom A contains no term $\delta \hat{x}_k t$. Then, obviously, there are good formula-mappings Γ such that $\Gamma(A) = A$, and condition (#) is trivial.

Case 2. $A = P \delta \hat{v}_1 t \cdots \delta \hat{v}_p t$, where at least one v_j is an x-variable.

Subcase a. For no $q > 1$ does the atom A lie in the set R_q. Then $A = P \hat{w}_1 u \cdots \hat{w}_p u$ for some x_n-term u and some x-focused sequence

---

‡ The notion of advanced is not effective; hence the term-mapping δ need not be effective. However, a more detailed analysis would allow us to replace these definitions with constructive ones.

ISBN 0-201-02540-X

$\langle w_1,\ldots,w_p\rangle$. But then $\langle \hat{w}_1 u,\ldots,\hat{w}_p u\rangle \simeq \langle \hat{v}_1 s,\ldots,\hat{v}_p s\rangle$. By Property T, $\mathfrak{p}_\mathfrak{A}(\hat{w}_1 u,\ldots,\hat{w}_p u) = \mathfrak{p}_\mathfrak{A}(\hat{v}_1 s,\ldots,\hat{v}_p s)$, and so $\mathfrak{A}(A) = \mathfrak{A}(P\hat{v}_1 s\cdots\hat{v}_p s)$. Hence, for any good formula-mapping Γ, $\mathfrak{A}(A) = \mathfrak{A}(\Gamma(A))$.

Subcase b. For some $q > 1$ the atom A lies in R_q. We show first that $\Lambda(A) \neq A$. Let x_k be any x-variable among the v_i; then, by definition, $\delta\hat{x}_k t$ is an advanced $\simeq$-equivalent of the tight term $\hat{x}_k s$, and $h(\delta\hat{x}_k t) \le q$. Hence there is a tight term u such that $h(u) \le q$ and $\hat{v}\hat{x}_k s \simeq \hat{v}u$ for each variable v of F. Let $t' = \hat{x}_n u$; then t' is a tight x_n-term in $D(F, q)$ such that $\langle \delta\hat{v}_1 t,\ldots,\delta\hat{v}_p t\rangle \simeq \langle \hat{v}_1 t',\ldots,\hat{v}_p t'\rangle$. But then Λ takes A to some atom in S_q, and so $\Lambda(A) \neq A$. Hence, by (Λ-1), $\mathfrak{A}(\Lambda(A)) = \mathfrak{A}(P\hat{v}_1 s\cdots\hat{v}_p s)$, and it suffices to show that $\mathfrak{A}(A) = \mathfrak{A}(\Lambda(A))$.

Since $\mathfrak{A}$ agrees with $\mathfrak{A}_q$ on S_q, by (Λ-2) we have $\Lambda(A) = \Lambda_{\mathfrak{A}q}(A)$. Moreover, $\mathfrak{A}(\Lambda(A)) = \mathfrak{A}_q(\Lambda(A))$ and $\mathfrak{A}(A) = \mathfrak{A}_{q+1}(A)$. Hence we have to show that $\mathfrak{A}_q(\Lambda(A)) = \mathfrak{A}_{q+1}(A)$, and this comes to showing that if A is the $(j + 1)$st member of R_q, $j \ge 0$, then the conjunction K_{j+1} used in the construction of $\mathfrak{A}_{q+1}$ contains the conjunct $(A \Leftrightarrow \Lambda(A))$. That is, we have to show the consistency of the conjunction $K_j \wedge (A \Leftrightarrow \Lambda(A)) \wedge N$, where K_j and N are as in that construction. Let Σ be the set of atoms earlier than A, and let K be the conjunction of all signed atoms in $\tilde{\Sigma}$ that are verified by $\mathfrak{A}$. Every atom occurring in K_j is in Σ; in particular, $\Lambda(A)$ is in Σ since $\Lambda(A)$ is in S_q and thus precedes A. Hence we need show only that either

(i) $\mathfrak{A}_q$ verifies $\Lambda(A)$ and $K \wedge A \wedge N$ is consistent; or
(ii) $\mathfrak{A}_q$ falsifies $\Lambda(A)$ and $K \wedge -A \wedge N$ is consistent.

Suppose that $\mathfrak{A}_q$ verifies $\Lambda(A)$; the other case is treated similarly. Since $\mathfrak{A}$ verifies $K \wedge N$, $K \wedge N$ is consistent. Hence, by the Truth-Functional Lemma, it suffices to show the consistency of $B \wedge A \wedge N$ for every signed atom B that is a conjunct in K.

Since B is in $\tilde{\Sigma}$, from the induction hypothesis that (#) holds on all atoms earlier than A it follows that there is a good formula-mapping Γ such that $\mathfrak{A}$ verifies $\Gamma(B)$. (Note that B need not be an atom that occurs in $\Delta(H)$.) Since Γ is good, $\Gamma(A) = P\hat{v}_1 s\cdots\hat{v}_p s$; and since $\mathfrak{A}(P\hat{v}_1 s\cdots\hat{v}_p s) = \mathfrak{A}(\Lambda(A))$, $\mathfrak{A}$ verifies $\Gamma(A)$. Finally, $\mathfrak{A}$ verifies $\Gamma(N)$, since $\Gamma(N)$ is—like N itself—a conjunction of instances of members of $\Xi(\mathfrak{A}_1, r_1)$. Hence $\mathfrak{A}$ verifies $\Gamma(B \wedge A \wedge N)$; by the Consistency Lemma, $B \wedge A \wedge N$ is consistent.

Case 3. The atom A contains at least one term $\delta\hat{x}_k t$ and at least one term not among the $\delta\hat{v}t$. We show that (#) is forthcoming provided that F possesses the following Syntactic Property.

Syntactic Property. Let l be the maximum number of argument places of the predicate letters of F. For each i, $1 \le i \le l$, and each signed atomic formula D constructed from a predicate letter of F and the variables $y_1,\ldots,y_i$, there is an atomic formula D' constructed from a predicate letter of F and

ISBN 0-201-02540-X

the variables $y_1,\ldots,y_{i-1}$ such that the clauses

$$D \Rightarrow D',$$

$$D' \Rightarrow D(y_i/v)$$

are conjuncts in the matrix of F for some variable v of F.‡

For suppose F has this property. We assume that $\mathfrak{A}$ verifies the atom A; the other case is handled similarly, replacing A in the argument below by $-A$. Now there is a signed atomic formula D constructed from a predicate letter of F and the variables $y_1,\ldots,y_i$, $i > 1$, such that $A = D(y_1/u_1,\ldots,y_i/u_i)$, where $u_1,\ldots,u_i$ are distinct terms and the term $u_i \neq \delta\hat{v}t$ for any variable v. (Naturally, among $u_1,\ldots,u_{i-1}$ is at least one term $\delta\hat{v}t$.) Hence there is a D' containing only the variables $y_1,\ldots,y_{i-1}$ such that $A \Rightarrow D'(y_1/u_1,\ldots,y_{i-1}/u_{i-1})$ is a conjunct in an Herbrand instance of F, and hence $\mathfrak{A}$ verifies $D'(y_1/u_1,\ldots,y_{i-1}/u_{i-1})$. The latter atom precedes A in the ordering of $L(F, \omega)$; hence by the induction hypothesis $\mathfrak{A}$ verifies $D'(y_1/\gamma u_1,\ldots,y_{i-1}/\gamma u_{i-1})$ for some good term-mapping γ. But, by the Syntactic Property again, there is a term u' such that

$$D'(y_1/\gamma u_1,\ldots,y_{i-1}/\gamma u_{i-1}) \Rightarrow D(y_1/\gamma u_1,\ldots,y_{i-1}/\gamma u_{i-1},\ y_i/u')$$

is also a conjunct in an Herbrand instance of F. Hence $\mathfrak{A}$ verifies the consequent of this conditional. If we let γ' be the term-mapping that agrees with γ everywhere but on u_i and takes u_i to u', then γ' is good and $\mathfrak{A}$ verifies $D(y_1/\gamma'u_1,\ldots,y_i/\gamma'u_i)$. That is, $\mathfrak{A}(A) = \mathfrak{A}(\Gamma'(A))$, where Γ' is the γ'-canonical formula-mapping. This establishes (#).

We have just shown that $\mathfrak{A}$ verifies $\Delta[E(F, \omega)]$ provided that F possesses the Syntactic Property. We have thereby established the finite controllability of that subclass of the Maslov Class containing just those schemata possessing this property. But of course not all Maslov schemata fall in this subclass. However, the following lemma bridges the gap, and shows the finite controllability of the entire Maslov Class.

Transformation Lemma. We can effectively find for each Skolem schema G a Skolem schema F such that

(1) F possesses the Syntactic Property;
(2) F and G are satisfiable over the same universes;
(3) F is Maslov if G is Maslov.

Proof.   Let $G = \forall y_1\cdots\forall y_m\exists x_1\cdots\exists x_n G^M$. By adding additional y-variables, if necessary (cf. the footnote on page 48), we may assume that the

ISBN 0-201-02540-X

---

‡ Since $(A \Rightarrow B)$ is by definition just $(-A \vee B)$, every conditional of two signed atomic formulas is equivalent to a binary disjunction. Hence we may allow Krom formulas to contain such conditionals as conjuncts. Note that the Syntactic Property implies that the number of y-variables of F is no less than l.

predicate letters of G are at most m-place. We define a sequence $G_m,\ldots,G_1$ of formulas as follows. Let $G_m = G^M$. Suppose G_i defined, $2 \le i \le m$. Then let G_{i-1} be the conjunction of G_i with all clauses

$$D \Rightarrow P_D y_1 \cdots y_{i-1},$$

$$P_D y_1 \cdots y_{i-1} \Rightarrow \exists x D(y_i/x)$$

for each signed atomic formula D constructed from a predicate letter in G_i and the variables $y_1,\ldots,y_i$, where each P_D is a new and distinct $(i-1)$-place predicate letter. Finally, let F be a rectified prenex form of $\forall y_1 \cdots \forall y_m \exists x_1 \cdots \exists x_n G_1$. Clearly F is Maslov whenever G is. If F is satisfiable over a universe U, then G is also, since F logically implies G. Conversely, a model for F with universe U can be obtained from any model for G with universe U thus: if D is a positively signed atomic formula containing the predicate letter Q, then the interpretation assigned to the new predicate letter P_D is an appropriate projection of the interpretation assigned to Q; if D is a negatively signed atomic formula containing Q, then the interpretation assigned to P_D is an appropriate projection of the complement of the interpretation assigned to Q.

It remains to show that F has the Syntactic Property. Suppose D is a signed atomic formula constructed from a predicate letter Q of F and the variables $y_1,\ldots,y_i$. Then Q is at least i-place. But then Q already occurs in G_i, and so conjuncts as required by the Syntactic Property are added to form G_{i-1}. $\square$

ISBN 0-201-02540-X

CHAPTER 5

Solvability without Finite Controllability

All the solvable classes we have discussed to this point are finitely controllable. In this chapter we present three solvable classes which are not, that is, which contain schemata possessing only infinite models. The solvability of the first of these, Class 5.1, was originally shown by Dreben, Kahr, and Wang (1962). Prior to this there were no known "natural" solvable classes that were not also finitely controllable. (We have no precise definition of "natural". We mean to include classes specified by restrictions on prefix, atomic subformulas, or truth-functional form of the matrix, and to exclude obviously *ad hoc* classes, such as those containing only a finite number of schemata or those obtained by adding a finite number of schemata to some natural class.) Our treatment of solvability is followed by one of docility.

Definition 5.1. A class $\mathfrak{C}$ of schemata is *docile* iff there is an effective method for determining, given any schema F in $\mathfrak{C}$, whether F has a finite model.

Note that any finitely controllable class is automatically docile. Moreover, as we shall see, the three solvable but not finitely controllable classes of this chapter are docile. We shall also show the docility of a class whose solvability is open.

In this chapter we are concerned principally with classes of $\forall\exists\forall$-schemata, that is, prenex schemata with prefixes $\forall y_1 \exists x \forall y_2$. The functional form F^* of an $\forall\exists\forall$-schema F contains one function sign f_x, which has one argument place. Hence the domain $D(F, \omega)$ is $\{+, f_x(+), f_x(f_x(+)),...\}$; in particular, for each positive integer p, there is exactly one term in $D(F, \omega)$ of height p. As a result, we may use a convenient numerical notation for the terms in $D(F, \omega)$: we use the same signs for nonnegative integers as for terms. A sign for an integer p may be used for the unique term of height

Burton Dreben and Warren D. Goldfarb, The Decision Problem: Solvable Classes of Quantificational Formulas

ISBN 0-201-02540-X

$p + 1$. Thus we consider a numeral, say "2", to be at one and the same time a notation for the number two and for the term $f_x(f_x(\maltese))$; 0 is zero and is also the term $\maltese$. In general, we use letters like "m" and "n" to range over integers and over terms. This systematic ambiguity enables us to think of a term-mapping γ as a mapping from integers to integers. Notation like "$\gamma n + 2$" will signify the term $f_x(f_x(s))$, where $s = \gamma n$, that is, s is the image under γ of the unique term in $D(F, \omega)$ of height $n + 1$. In this notation, $D(F, p)$ is simply $\{0, 1,..., p - 1\}$; an Herbrand instance $F^*(y_1/m, y_2/n)$ may also be written $F^M(y_1/m, x/m + 1, y_2/n)$.

Despite the simple structure of domain and expansion, the class of all $\forall\exists\forall$-schemata is unsolvable (see the Introduction); indeed, even some remarkably restricted subclasses are unsolvable (see page 124). The power of $\forall\exists\forall$-schemata arises from the x-variable's governing a y-variable. Each term $m + 1$ occurs as the substituent for the x-variable in an infinite number of Herbrand instances; and each term n occurs as the substituent for y_2 in one of these Herbrand instances. Hence an $\forall\exists\forall$-schema can encode conditions that require, for each term m, that certain relations between m and any term n engender certain other relations between $m + 1$ and n. This effect is illustrated by the following example. Let F be the $\forall\exists\forall$-schema whose matrix F^M is

$$Py_1y_1 \wedge (Py_1y_2 \Rightarrow Qxy_2) \wedge (Qy_1y_2 \Rightarrow Qxy_2) \wedge -Qy_1y_1,$$

and let $\mathfrak{A}$ be a truth-assignment verifying $E(F, \omega)$. The first conjunct in F^M requires that $\mathfrak{A}$ verify each atom Pmm; the second and third that $\mathfrak{A}$ verify $Qm + 1\ n$ whenever it verifies Pmn or Qmn. Hence $\mathfrak{A}$ verifies Qmn if $m > n$. Moreover, $\mathfrak{A}$ verifies Qmn only if $m > n$. For if $m \leq n$ and if $\mathfrak{A}$ verified Qmn, then $\mathfrak{A}$ would have to verify Qnn, which violates the fourth conjunct in F^M. This sort of analysis easily shows that F has no finite models. But F clearly has infinite models: for example, we may take the universe to be the integers, and interpret Q as "greater than" and P as identity. Hence no class that contains F is finitely controllable.

The specifications of our solvable classes serve to limit the sort of conditions that can be encoded by the schemata. This allows us to exploit periodicities in the action of truth-assignments on expansions.

1. $\forall\exists\forall$-Classes Specified by Atomic Subformulas

For the most part in this section we limit ourselves to schemata containing dyadic predicate letters only. To obtain a solvable but not finitely controllable class (Class 5.1), we must go further than restrict the basis: we must consider not only the variables that occur in atomic subformulas, but also their order of occurrence. Thus we are implicitly using a notion of "ordered

ISBN 0-201-02540-X

basis''. Other $\forall\exists\forall$-classes aside from 5.1 can be specified in this manner. After our amenability proof for Class 5.1 we sort these other classes into solvable and unsolvable, and into finitely controllable and not finitely controllable.

Class 5.1. Let F be an $\forall\exists\forall$-schema each of whose atomic subformulas has one of the seven forms Py_1y_1, Pxx, Py_2y_2, Py_1x, Pxy_1, Py_1y_2, or Pxy_2. Then $E(F, 2^{d(F, 2)})$ is adequate.

Proof. Let $\mathfrak{A}$ be any truth-assignment on $\hat{L}(F, \omega)$. We define a formula-mapping Γ and an instance-mapping Ψ by means of both a term-mapping γ and a mapping η from *pairs* of terms to terms.

For each atom Pmn in $L(F, \omega)$ let $\Gamma(Pmn) = P\gamma m\ \eta(m, n)$.

For each Herbrand instance $H = F^*(y_1/m, y_2/n)$ let $\Psi(H) = F^*(y_1/\gamma m, y_2/\eta(m, n))$.

We require that Ψ fulfill the boundedness condition

$$\Psi[E(F, \omega)] \subseteq E(F, 2^{d(F, 2)}),$$

and that Γ and Ψ fulfill the central condition

$$\mathfrak{A}(\Gamma(\langle H\rangle_i)) = \mathfrak{A}(\langle \Psi(H)\rangle_i)$$

for each Herbrand instance H of F and each number i. Now the boundedness condition will follow provided that, for all m and n,

(1) $\gamma m < 2^{d(F, 2)}$;
(2) $\eta(m, n) < 2^{d(F, 2)}$.

Moreover, as we shall now show, the central condition is assured if we have, for all m and n,

(3) $\mathfrak{p}_\mathfrak{A}(\gamma(m +1), \eta(m + 1, n)) = \mathfrak{p}_\mathfrak{A}(\gamma m +1, \eta(m, n))$;
(4) $\eta(m, m) = \gamma m$ and $\eta(m, m + 1) = \gamma m + 1$.

For let $H = F^*(y_1/m, y_2/n)$ be any Herbrand instance. If $\langle F\rangle_i = Py_1y_2$, then $\Gamma(\langle H\rangle_i) = P\gamma m\ \eta(m, n) = \langle\Psi(H)\rangle_i$. The central condition is immediate, with no recourse to (3) or (4). If $\langle F\rangle_i = Pxy_2$, then $\Gamma(\langle H\rangle_i) = P\gamma(m + 1)\ \eta(m + 1, n)$, but $\langle\Psi(H)\rangle_i = P\gamma m + 1\ \eta(m, n)$. Here the central condition follows directly from (3); indeed, this case furnishes the motivation underlying that requirement on γ and η. Now suppose $\langle F\rangle_i$ is Py_1y_1 or Py_1x. Then $\Gamma(\langle H\rangle_i) = P\gamma m\ \eta(m, m)$ or $P\gamma m\ \eta(m, m + 1)$, and $\langle\Psi(H)\rangle_i$ is $P\gamma m\ \gamma m$ or $P\gamma m\ \gamma m + 1$, respectively. Hence condition (4) yields $\Gamma(\langle H\rangle_i) = \langle\Psi(H)\rangle_i$. These cases furnish the motivation for condition (4).

We have three further cases. If $\langle F\rangle_i = Pxy_1$, then $\Gamma(\langle H\rangle_i) = P\gamma(m + 1)\ \eta(m + 1, m)$ and $\langle\Psi(H)\rangle_i = P\gamma m + 1\ \gamma m$. Since $\gamma m = \eta(m, m)$ by condition (4), the central condition follows from condition (3). If $\langle F\rangle_i = Pxx$, then $\Gamma(\langle H\rangle_i) = P\gamma(m + 1)\ \gamma(m + 1)$ by condition (4), and $\langle\Psi(H)\rangle_i =$

ISBN 0-201-02540-X

$P\gamma m + 1$ $\gamma m + 1$. Since (3) implies that $\mathfrak{p}_{\mathfrak{A}}(\gamma(m + 1)) = \mathfrak{p}_{\mathfrak{A}}(\gamma m + 1)$, the central condition follows. Finally, if $\langle F \rangle_i = Py_2y_2$, then $\Gamma(\langle H \rangle_i) = P\gamma n$ $\eta(n, n) = P\gamma n \; \gamma n$, while $\langle \Psi(H) \rangle_i = P\eta(m, n) \; \eta(m, n)$. But condition (3) implies that, for each fixed term n, the profile $\mathfrak{p}_{\mathfrak{A}}(\eta(m, n))$ is constant as m ranges over $D(F, \omega)$. Thus $\mathfrak{p}_{\mathfrak{A}}(\eta(m, n)) = \mathfrak{p}_{\mathfrak{A}}(\eta(n, n))$ and so, by condition (4), $\mathfrak{p}_{\mathfrak{A}}(\eta(m, n)) = \mathfrak{p}_{\mathfrak{A}}(\gamma n)$ for all m and n. Again the central condition is fulfilled. This exhausts the seven cases for $\langle F \rangle_i$.

Thus we need only construct mappings γ and η fulfilling (1)–(4). For each m let

$$\xi_{\mathfrak{A}}(m) = \{\mathfrak{p}_{\mathfrak{A}}(m, n) \mid n < 2^{d(F, 2)}\}.$$

Note that the value $\xi_{\mathfrak{A}}(m)$ is a finite nonempty *set* of profiles. Hence there are at most $2^{d(F,2)} - 1$ members in the range of $\xi_{\mathfrak{A}}$. By the pigeonhole argument there are q and r, $0 \leq q < r < 2^{d(F,2)}$, such that $\xi_{\mathfrak{A}}(q) = \xi_{\mathfrak{A}}(r)$. Let γ be the term-mapping that carries each m to $q + k$, where k is the least nonnegative remainder of m (mod $r - q$).‡ Hence $\gamma m < 2^{d(F,2)} - 1$ for each m, and condition (1) is fulfilled. Moreover, $\gamma(m + 1) = \gamma m + 1$ unless $\gamma m = r - 1$, in which case $\gamma(m + 1) = q$. Hence $\xi_{\mathfrak{A}}(\gamma(m + 1)) = \xi_{\mathfrak{A}}(\gamma m + 1)$ for each m.

The mapping η is now defined by induction on the numerical difference between its two arguments.

(η-1) $\eta(m, m) = \gamma m$ and $\eta(m, m + 1) = \gamma m + 1$ for each m. Note that these values are less than $2^{d(F,2)}$.

(η-2)   If $m + 1 < n$, then $\eta(m, n)$ is the least p such that $\mathfrak{p}_{\mathfrak{A}}(\gamma m + 1, p) = \mathfrak{p}_{\mathfrak{A}}(\gamma(m + 1), \eta(m + 1, n))$. On the assumption that $\eta(m + 1, n) < 2^{d(F,2)}$, there will be such a p and it will also be less than $2^{d(F,2)}$, since $\xi_{\mathfrak{A}}(\gamma m + 1) = \xi_{\mathfrak{A}}(\gamma(m + 1))$.

(η-3)   If $n \leq m$, then $\eta(m + 1, n)$ is the least p such that $\mathfrak{p}_{\mathfrak{A}}(\gamma(m + 1), p) = \mathfrak{p}_{\mathfrak{A}}(\gamma m + 1, \eta(m, n))$. On the assumption that $\eta(m, n) < 2^{d(F,2)}$, there will be such a $p < 2^{d(F,2)}$, again since $\xi_{\mathfrak{A}}(\gamma(m + 1)) = \xi_{\mathfrak{A}}(\gamma m + 1)$.

We have thus defined a mapping η on pairs of terms. Clearly $\eta(m, n) < 2^{d(F,2)}$ for all m and n. Moreover, by clause (η-1), η fulfills condition (4). Condition (3) is immediate from clauses (η-2) and (η-3), except in the special case $n = m + 1$. Here we must have $\mathfrak{p}_{\mathfrak{A}}(\gamma(m + 1), \eta(m + 1, m + 1)) = \mathfrak{p}_{\mathfrak{A}}(\gamma m + 1, \eta(m, m + 1))$. By ($\eta$-1) this condition amounts to $\mathfrak{p}_{\mathfrak{A}}(\gamma(m + 1), \gamma(m + 1)) = \mathfrak{p}_{\mathfrak{A}}(\gamma m + 1, \gamma m + 1)$; and this is forthcoming since $\xi_{\mathfrak{A}}(\gamma(m + 1)) = \xi_{\mathfrak{A}}(\gamma m + 1)$.

This concludes the amenability proof for Class 5.1.   □

---

‡ Here we are exploiting fully the ambiguity of our numerical notation for terms, and treating terms as though they were integers.

ISBN 0-201-02540-X

Class 5.1 contains the satisfiable schema F that possesses no finite models given on page 120. The atomic subformulas of F have the forms Py_1y_2, Pxy_2, and Py_1y_1. Another satisfiable schema that has no finite models, this one containing atomic subformulas of the forms Py_1y_2, Pxy_2, and Pxy_1, is the $\forall\exists\forall$-schema with matrix

$$Pxy_1 \wedge (Py_1y_2 \Rightarrow Qxy_2) \wedge (Qy_1y_2 \Rightarrow Qxy_2) \wedge -Qxy_1.$$

Indeed, we can find such an example using any one form in addition to Py_1y_2 and Pxy_2. However, as the following amiability proof shows, we can obtain no such example by using the forms Py_1y_2 and Pxy_2 alone.

Class 5.2. Let F be an $\forall\exists\forall$-schema each of whose atomic subformulas has one of the forms Py_1y_2 or Pxy_2. Then F is $d(F, 2)$-amiable.

Proof. Let $\mathfrak{A}$ be any truth-assignment on $\hat{L}(F, \omega)$. By the pigeonhole argument there are q and r, $0 \le q < r \le d(F, 2)$, such that $\mathfrak{p}_{\mathfrak{A}}(q, 0) = \mathfrak{p}_{\mathfrak{A}}(r, 0)$. Let δ be the term-mapping that carries each m to $q + k$, where k is the least nonnegative remainder of $m \pmod{r - q}$. Thus $|\delta[D(F, \omega)]| = r - q \le d(F, 2)$. Moreover, $\delta(m + 1) = \delta m + 1$ unless $\delta m = r - 1$, in which case $\delta(m + 1) = q$. Hence $\mathfrak{p}_{\mathfrak{A}}(\delta(m+1), 0) = \mathfrak{p}_{\mathfrak{A}}(\delta m + 1, 0)$ for each m.

Let Γ be the formula-mapping that carries each atom Pmn to $Pm0$. (Note, incidentally, that Γ is Herbrand instance preserving on $E(F, \omega)$.) Finally, let Ψ be the instance-mapping that carries each Herbrand instance $F^*(y_1/m, y_2/n)$ to $F^*(y_1/\delta m, y_2/0)$. Suppose $H = F^*(y_1/m, y_2/n)$ is an arbitrary Herbrand instance of F. If $\langle F \rangle_i = Py_1y_2$, then $\Gamma(\Delta(\langle H \rangle_i)) = P\delta m\,0 = \langle \Psi(H) \rangle_i$. If $\langle F \rangle_i = Pxy_2$, then $\Gamma(\Delta(\langle H \rangle_i)) = P\delta(m + 1)\,0$ and $\langle \Psi(H) \rangle_i = P\delta m + 1\,0$. Since $\mathfrak{p}_{\mathfrak{A}}(\delta m + 1, 0) = \mathfrak{p}_{\mathfrak{A}}(\delta(m+1), 0)$, $\mathfrak{A}(\Gamma(\Delta(\langle H \rangle_i))) = \mathfrak{A}(\langle \Psi(H) \rangle_i)$. □

Note that we may extend Class 5.2 and retain amiability by allowing F to contain atomic subformulas of the forms Qy_1 and Qx, where Q is a monadic predicate letter (we need merely let the formula-mapping Γ carry each atom Qm to itself). However, as we have seen, if we allow atomic subformulas of the form Py_1y_1 or of the form Pxx, where P is dyadic, then we can construct a schema possessing infinite models only. Hence essentially monadic atomic subformulas here have a different effect from monadic atomic subformulas.

We may also consider the mirror-images of Classes 5.1 and 5.2, that is, the classes obtained by replacing the forms Py_1y_2 and Pxy_2 in the specifications of 5.1 and 5.2 with the forms Py_2y_1 and Py_2x. The proofs here are analogous to those above, requiring changes solely in the definition of the formula-mapping Γ: for the mirror-image of Class 5.1 we let $\Gamma(Pmn) = P\eta(n, m)\,\gamma n$, and for the mirror-image of Class 5.2 we let $\Gamma(Pmn) =$

ISBN 0-201-02540-X

$P0n$. Thus the mirror-image of Class 5.1 is amenable (but not finitely controllable), and the mirror-image of Class 5.2 is amiable.

These arguments do not apply, however, if we replace the forms Py_1y_2 and Pxy_2 with the forms Py_1y_2 and Py_2x, or with the forms Py_2y_1 and Pxy_2. Indeed, if we allow atomic subformulas of the forms Py_1y_2, Py_2x, and *any* single other (dyadic) form, the resulting class of schemata is unsolvable; similarly if we allow Py_2y_1, Pxy_2, and any single other form. (This remarkable fact is shown by Aanderaa (1966) and Aanderaa and Lewis (1974). See also Lewis (1979). We obtain unsolvability even if we allow just the forms Py_1y_2, Py_2x, and Qy_1, where Q is a monadic predicate letter.) In contrast, if we allow just Py_1y_2 and Py_2x, or just Py_2y_1 and Pxy_2, then we obtain an amiable class.

Class 5.3. Let F be an $\forall\exists\forall$-schema each of whose atomic subformulas has one of the forms Py_1y_2 or Py_2x. Then F is $2d(F, 1)$-amiable.

Proof. Let $\mathfrak{A}$ be any truth-assignment on $\dot{L}(F, \omega)$. By the pigeonhole argument there are q and r, $0 \leq q < r \leq 2d(F, 1)$, such that $\mathfrak{p}_\mathfrak{A}(q) = \mathfrak{p}_\mathfrak{A}(r)$ and $q \equiv r \pmod 2$. Let δ carry each m to $q + k$, where k is the least nonnegative remainder of $m \pmod{r - q}$. Then $|\delta[D(F, \omega)]| = r - q \leq 2d(F, 1)$ and, for each m, $\mathfrak{p}_\mathfrak{A}(\delta m + 1) = \mathfrak{p}_\mathfrak{A}(\delta(m + 1))$ and $\delta m + 1 \equiv \delta(m + 1) \pmod 2$. For each atom Pmn in $L(F, \omega)$, let

$$\Gamma(Pmn) = \begin{cases} Pnn & \text{if } m \equiv n \pmod 2, \\ Pm\ m + 1 & \text{if } m \not\equiv n \pmod 2. \end{cases}$$

(Again, Γ is Herbrand instance preserving on $E(F, \omega)$.) Finally, for each m and n, let

$$\Psi(F^*(y_1/m, y_2/n)) = \begin{cases} F^*(y_1/\delta n, y_2/\delta n) & \text{if } \delta m \equiv \delta n \pmod 2, \\ F^*(y_1/\delta m, y_2/\delta m + 1) & \text{if } \delta m \not\equiv \delta n \pmod 2. \end{cases}$$

Suppose $H = F^*(y_1/m, y_2/n)$ is an arbitrary Herbrand instance of F. If $\langle F \rangle_i = Py_1y_2$, then $\Gamma(\Delta(\langle H \rangle_i)) = \langle \Psi(H) \rangle_i$. If $\langle F \rangle_i = Py_2x$ and if $\delta m \equiv \delta n \pmod 2$, then $\delta n \not\equiv \delta(m + 1) \pmod 2$; hence $\Gamma(\Delta(\langle H \rangle_i)) = P\delta n\ \delta n + 1 = \langle \Psi(H) \rangle_i$. If $\langle F \rangle_i = Py_2x$ and if $\delta m \not\equiv \delta n \pmod 2$, then $\delta n \equiv \delta(m + 1) \pmod 2$, whence $\Gamma(\Delta(\langle H \rangle_i)) = P\delta(m + 1)\ \delta(m + 1)$; and $\langle \Psi(H) \rangle_i = P\delta m + 1\ \delta m + 1$. Since $\mathfrak{p}_\mathfrak{A}(\delta(m + 1)) = \mathfrak{p}_\mathfrak{A}(\delta m + 1)$, $\mathfrak{A}(\Gamma(\Delta(\langle H \rangle_i))) = \mathfrak{A}(\langle \Psi(H) \rangle_i)$. $\square$

Again, a symmetrical argument yields the $2d(F, 1)$-amiability of every $\forall\exists\forall$-schema F each of whose atomic subformulas has one of the forms Py_2y_1 or Pxy_2.

Our sorting of the $\forall\exists\forall$-classes specified by forms of (dyadic) atomic subformulas into solvable and unsolvable, and into finitely controllable and not finitely controllable, will be complete once we establish the following results: the class allowing all forms but Pxy_2 and Py_2x, and the class

ISBN 0-201-02540-X

allowing all forms but Py_1y_2 and Py_2y_1, are amiable. But these classes are subsumed by a more general amiable ∀∃∀-class, one specified by basis and allowing k-place predicate letters for arbitrary k. (In particular, schemata in the class may contain $\{y_1, x, y_2\}$-based atomic subformulas. The presence of such atomic subformulas makes the amiability proof somewhat more complex.)

Class 5.4. Let F be an ∀∃∀-schema whose basis either does not contain $\{x, y_2\}$ or does not contain $\{y_1, y_2\}$. Then F is $4d(F, 3)$-amiable.

Proof. We suppose that $b(F)$ does not contain $\{x, y_2\}$; the other case is treated similarly. Let $\mathfrak{A}$ be any truth-assignment on $\dot{L}(F, \omega)$. By the pigeonhole argument there are q and r, $0 \le q$ and $q + 3 < r \le 4d(F, 3)$, such that $\mathfrak{p}_{\mathfrak{A}}(q, q +1, q + 2) = \mathfrak{p}_{\mathfrak{A}}(r, r +1, r + 2)$. Let δ carry each m to $q + k$, where k is the least nonnegative remainder of m (mod $r - q$). Thus $|\delta[D(F, \omega)]| = r - q \le 4d(F, 3)$. Moreover $\delta(m + 1) = \delta m + 1$ unless $\delta m = r - 1$, in which case $\delta(m + 1) = q$. Let γ carry q to r, $q + 1$ to $r + 1$, and every other term to itself. Clearly $\mathfrak{p}_{\mathfrak{A}}(\gamma m, \gamma m +1) = \mathfrak{p}_{\mathfrak{A}}(m, m +1)$ for each m. Let γ' be the term-mapping like γ except for carrying $q + 2$ to $r + 2$. Define a formula-mapping Γ thus:

(Γ-1) Γ is γ-canonical on an atom $Pm_1\cdots m_k$ if $\{m_1,\ldots,m_k\}$ is one of the sets $\{r - 1, q + 1\}$, $\{r - 2, q, q + 1\}$, $\{r - 2, q\}$, or $\{r -1, q, p\}$ for some p (not necessarily distinct from $r - 1$ or q);

(Γ-2) Γ is γ'-canonical on an atom $Pm_1\cdots m_k$ if $\{m_1,\ldots,m_k\} = \{r - 1, q + 1, q + 2\}$;

(Γ-3) Γ carries every other atom to itself.

(Since $r > q + 3$, no conflict can arise between clauses (Γ-1) and (Γ-2).) Finally, let Ψ carry $F^*(y_1/m, y_2/n)$ to $F^*(y_1/\gamma\delta m, y_2/\gamma\delta n)$ if $\delta m = r - 1$ or if $\{\delta m, \delta n\}$ is one of the sets $\{r - 1, q\}$, $\{r - 1, q + 1\}$, or $\{r - 2, q\}$, and to $F^*(y_1/\delta m, y_2/\delta n)$ otherwise. Suppose $H = F^*(y_1/m, y_2/n)$ is an arbitrary Herbrand instance of F.

(a) If $\delta m \ne r - 1$ and $\{\delta m, \delta n\}$ is neither $\{r - 1, q\}$ nor $\{r - 1, q + 1\}$ nor $\{r - 2, q\}$, then $\Gamma(\Delta(H)) = \Delta(H) = \Psi(H)$.

(b) If $\delta m = r - 1$, then $\Gamma(\Delta(\langle H\rangle_i)) = \langle\Psi(H)\rangle_i$ when $\langle F\rangle_i$ is $\{y_1, y_2\}$-based, $\{y_1, x\}$-based, $\{y_1\}$-based, or $\{y_1, x, y_2\}$-based. Moreover, since $\mathfrak{p}_{\mathfrak{A}}(\delta(m +1)) = \mathfrak{p}_{\mathfrak{A}}(\delta m +1)$ and $\mathfrak{p}_{\mathfrak{A}}(\gamma\delta n) = \mathfrak{p}_{\mathfrak{A}}(\delta n)$, $\mathfrak{A}(\Gamma(\Delta(\langle H\rangle_i))) = \mathfrak{A}(\langle\Psi(H)\rangle_i)$ when $\langle F\rangle_i$ is $\{x\}$-based or $\{y_2\}$-based.

(c) If $\delta m \ne r - 1$ but $\{\delta m, \delta n\}$ is either $\{r - 1, q\}$ or $\{r - 1, q + 1\}$ or $\{r - 2, q\}$, then $\Gamma(\Delta(\langle H\rangle_i)) = \langle\Psi(H)\rangle_i$ when $\langle F\rangle_i$ is $\{y_1, y_2\}$-based or $\{y_1, x, y_2\}$-based. Since $\delta(m + 1) = \delta m + 1$, $\mathfrak{p}_{\mathfrak{A}}(\gamma\delta m, \gamma\delta m +1) = \mathfrak{p}_{\mathfrak{A}}(\delta m, \delta m +1)$, and $\mathfrak{p}_{\mathfrak{A}}(\gamma\delta n) = \mathfrak{p}_{\mathfrak{A}}(\delta n)$, $\mathfrak{A}(\Gamma(\Delta(\langle H\rangle_i))) = \mathfrak{A}(\langle\Psi(H)\rangle_i)$ when $\langle F\rangle_i$ is $\{y_1, x\}$-based, $\{y_1\}$-based, $\{x\}$-based, or $\{y_2\}$-based.

ISBN 0-201-02540-X

Hence in all cases $\mathfrak{A}(\Gamma(\Delta(\langle H \rangle_i))) = \mathfrak{A}(\langle \Psi(H) \rangle_i)$, and amiability is established. $\square$

The unsolvability results we have cited imply, in particular, the unsolvability of the class of all $\forall \exists \forall$-schemata with bases $\{\{y_1, y_2\}, \{x, y_2\}\}$. Hence an $\forall \exists \forall$-class specified by basis is solvable if *and only if* it is subsumed by Class 5.4. Thus restrictions on basis cannot yield a solvable but not finitely controllable $\forall \exists \forall$-class. Indeed, *every* known solvable class specified by prefix and basis restrictions alone is also finitely controllable.

2. $\forall \exists \forall$-Classes Specified by Truth-Functional Form

A *Horn formula* is a quantifier-free formula in conjunctive normal form, each of whose conjuncts contains at most one positively signed atomic formula.‡ Class 5.5 is the $\forall \exists \forall$-Horn Class, which comprises all $\forall \exists \forall$-schemata having Horn matrices. Actually, we restrict attention to schemata containing dyadic predicate letters only. This restriction entails no loss of generality, for the class of all $\forall \exists \forall$-schemata can be reduced to the subclass containing only dyadic predicate letters, and this reduction preserves Horn matrices (see Chapter 6, §5). The solvability of Class 5.5 was first shown by Lewis (1974). The example on page 120 of a satisfiable $\forall \exists \forall$-schema with no finite models is in Class 5.5; hence this class is not finitely controllable. Moreover, Class 5.5 is full in the class of all $\forall \exists \forall$-schemata. By the Corollary of Chapter 2, §7, Class 5.5 is not amenable.

As might be expected, the solvability of Class 5.5 rests on a truth-functional property of Horn formulas.

Truth-Functional Lemma for Horn Formulas. Let E be any consistent set (finite or infinite) of Horn formulas. Let $\mathfrak{A}$ be the truth-assignment defined thus: for every atomic formula A, the assignment $\mathfrak{A}$ verifies A iff E truth-functionally implies A. Then $\mathfrak{A}$ verifies every formula in E.

Proof. We consider the conjuncts in members of E; since all the members are Horn, each such conjunct has one of the forms $-A_1 \vee \cdots \vee -A_n$ or $-A_1 \vee \cdots \vee -A_{n-1} \vee A_n$ (up to permutation of disjuncts), $n \geq 1$, where the A_i are atomic formulas. If $-A_1 \vee \cdots \vee -A_{n-1} \vee A_n$ is a conjunct, then either $\mathfrak{A}$ verifies $-A_1 \vee \cdots \vee -A_{n-1}$ or else E implies $A_1 \wedge \cdots \wedge A_{n-1}$; in the latter case, by truth-functional logic, E implies A_n, and then $\mathfrak{A}$ verifies A_n. If $-A_1 \vee \cdots \vee -A_n$ is a conjunct, then either $\mathfrak{A}$ verifies $-A_1 \vee \cdots \vee -A_n$ or else E implies $A_1 \wedge \cdots \wedge A_n$; but in the latter case E is inconsistent. $\square$

---

‡ Thus a Horn formula may be represented as a conjunction of formulas $(A_1 \wedge \cdots \wedge A_n \Rightarrow B)$, where the A_i are atomic formulas, and B is either an atomic formula or else a symbol for falsity. Such conjuncts are well suited for encoding machine-table instructions of the following sort: If certain states are entered, then go to another state; if certain states are entered, then halt. Horn formulas were first considered by Horn (1951).

ISBN 0-201-02540-X

Class 5.5   (∀∃∀-Horn Class).   Let F be an ∀∃∀-Horn schema containing dyadic predicate letters only. Then F has a model iff there is a truth-assignment $\mathfrak{A}$ that verifies $E(F, d(F, 2) + 1)$ and that possesses the following

Invariance Property.   For all predicate letters P of F and all m and n, $0 \leq m, n < d(F, 2)$, $\mathfrak{A}(Pm\ n) = \mathfrak{A}(Pm + 1\ n + 1)$.

(The existence of such an assignment $\mathfrak{A}$ can, of course, be effectively decided.)

Proof.   Suppose F has a model. We show the existence of a truth-assignment $\mathfrak{A}$ that verifies $E(F, \omega)$, not just $E(F, d(F,2) + 1)$, and that possesses the Invariance Property for all m and n, not just $0 \leq m, n < d(F, 2)$. We note first that

(‡)   if $E(F, \omega)$ truth-functionally implies an atom Pmn, then $E(F, \omega)$ truth-functionally implies $Pm + 1\ n + 1$.

For let γ be the term-mapping that carries each m to $m + 1$, and let Γ be the γ-canonical formula-mapping. Then $\Gamma[E(F, \omega)] \subseteq E(F, \omega)$. If $E(F, \omega)$ implies Pmn, then by the Consistency Lemma of Chapter 1, §3, $\Gamma[E(F, \omega)]$ implies $\Gamma(Pmn)$; whence $E(F, \omega)$ implies $\Gamma(Pmn)$, and $\Gamma(Pmn) = Pm + 1\ n + 1$.

Since F has a model, $E(F, \omega)$ is consistent. Let $\mathfrak{B}$ be the truth-assignment that verifies an atom Pmn iff $E(F, \omega)$ truth-functionally implies Pmn. By the Truth-Functional Lemma, $\mathfrak{B}$ verifies $E(F, \omega)$. Now let $\mathfrak{A}$ be the truth-assignment that verifies an atom Pmn iff $\mathfrak{B}$ verifies $Pm + k\ n + k$ for some $k \geq 0$. By (‡), $\mathfrak{A}$ verifies Pmn iff $\mathfrak{B}$ verifies $Pm + k\ n + k$ for cofinitely many integers k, that is, for all but a finite number of integers k. It quickly follows that $\mathfrak{A}(Pmn) = \mathfrak{A}(Pm + 1\ n + 1)$ for all atoms Pmn in $L(F, \omega)$; hence $\mathfrak{A}$ possesses the Invariance Property. Let $H = F^*(y_1/p, y_2/q)$ be any Herbrand instance of F; we show that $\mathfrak{A}$ verifies H. There are cofinitely many integers k such that, for all of the (finitely many) atoms Pmn occurring in H, $\mathfrak{A}(Pmn) = \mathfrak{B}(Pm + k\ n + k)$. But then, for any of these integers k,

$$\mathfrak{A}(\langle H\rangle_i) = \mathfrak{B}(\langle F^*(y_1/p + k, y_2/q + k)\rangle_i)$$

for each i. Since $\mathfrak{B}$ verifies every Herbrand instance of F, $\mathfrak{A}$ verifies H.

Conversely, suppose there is a truth-assignment $\mathfrak{A}$ verifying $E(F, d(F, 2) + 1)$ and possessing the Invariance Property. We show that $E(F, \omega)$ is consistent by constructing a formula-mapping Γ such that the induced truth-assignment $\mathfrak{A}(\Gamma)$ verifies $E(F, \omega)$. Our proof, like a proof of adequacy, uses the central condition; but here, instead of starting with an arbitrary truth-assignment, we start with the special assignment $\mathfrak{A}$. The Invariance Property assures us that the profile $\mathfrak{p}_{\mathfrak{A}}(m, n)$, where $0 \leq m, n \leq d(F, 2)$, depends only on the numerical difference between m and n. The formula-mapping Γ will induce a truth-assignment $\mathfrak{A}(\Gamma)$ such that the profile $\mathfrak{p}_{\mathfrak{A}(\Gamma)}(m, n)$ of *any* m and n also depends only on the numerical difference between m and n.

By the pigeonhole argument there are numbers q and r, $0 \le q < r \le d(F, 2)$, such that $\mathfrak{p}_{\mathfrak{A}}(q, 0) = \mathfrak{p}_{\mathfrak{A}}(r, 0)$. Define a term-mapping γ thus:

$$\gamma n = \begin{cases} n & \text{if } n < r; \\ q + k & \text{if } n \ge r, \text{ where } k \text{ is the least nonnegative} \\ & \qquad \text{remainder of } n - q \ (\text{mod } r - q). \end{cases}$$

It is immediate that γ fulfills the following two conditions for every n:

(1) $\gamma n < d(F, 2)$;
(2) $\mathfrak{p}_{\mathfrak{A}}(\gamma(n + 1), 0) = \mathfrak{p}_{\mathfrak{A}}(\gamma n + 1, 0)$.

For each atom Pmn in $L(F, \omega)$, let

$$\Gamma(Pmn) = \begin{cases} P\gamma(m - n)\,0 & \text{if } m \ge n; \\ P0\,\gamma(n - m) & \text{if } m < n. \end{cases}$$

Thus $\Gamma(Pmn)$ depends only on the numerical difference between m and n. Note that $\gamma 0 = 0$, so that $\Gamma(Pmm) = P\gamma 0\,0 = P00 = P0\,\gamma 0$.

Define an instance-mapping Ψ thus: for each Herbrand instance $F^*(y_1/m, y_2/n)$ of F,

$$\Psi(H) = \begin{cases} F^*(y_1/\gamma(m - n), y_2/0) & \text{if } m \ge n; \\ F^*(y_1/0, y_2/\gamma(n - m - 1) + 1) & \text{if } m < n. \end{cases}$$

By condition (1), the range of Ψ is included in $E(F, d(F, 2) + 1)$; hence the boundedness condition is fulfilled. It suffices to show the central condition

$$\mathfrak{A}(\Gamma(\langle H \rangle_i)) = \mathfrak{A}(\langle \Psi(H) \rangle_i)$$

for each Herbrand instance H and each i. Let $H = F^*(y_1/m, y_2/n)$.

(a) If $\langle F \rangle_i$ is Py_1y_1, Pxx, or Py_2y_2, then $\Gamma(\langle H \rangle_i) = P00$, and $\langle \Psi(H) \rangle_i = Pkk$ for some $k \le d(F, 2)$. Since the Invariance Property implies that $\mathfrak{p}_{\mathfrak{A}}(0, 0) = \mathfrak{p}_{\mathfrak{A}}(k, k)$, the central condition follows.

(b) If $\langle F \rangle_i$ is Py_1x or Pxy_1, then $\Gamma(\langle H \rangle_i) = P0\,\gamma 1$ or $P\gamma 1\,0$. By condition (2), $\mathfrak{A}(P0\,\gamma 1) = \mathfrak{A}(P01)$ and $\mathfrak{A}(P\gamma 1\,0) = \mathfrak{A}(P10)$. Moreover, $\langle \Psi(H) \rangle_i = Pk\,k + 1$ or $Pk + 1\,k$ for some $k < d(F, 2)$. Again the central condition follows from the Invariance Property.

(c) Now suppose $m \ge n$. If $\langle F \rangle_i$ is Py_1y_2 or Py_2y_1, then $\Gamma(\langle H \rangle_i)$ is $P\gamma(m - n)\,0$ or $P0\,\gamma(m - n)$. Hence $\Gamma(\langle H \rangle_i) = \langle \Psi(H) \rangle_i$. If $\langle F \rangle_i$ is Pxy_2

ISBN 0-201-02540-X

or Py_2x, then $\Gamma(\langle H\rangle_i)$ is $P\gamma(m + 1 - n)\,0$ or $P0\,\gamma(m + 1 - n)$, and $\langle\Psi(H)\rangle_i$ is $P\gamma(m - n) + 1\,0$ or $P0\,\gamma(m - n) + 1$. The central condition follows from condition (2).

(d) Suppose $m < n$. If $\langle F\rangle_i$ is Py_1y_2 or Py_2y_1, then $\Gamma(\langle H\rangle_i)$ is $P0\,\gamma(n - m)$ or $P\gamma(n - m)\,0$, and $\langle\Psi(H)\rangle_i$ is $P0\,\gamma(n - m - 1) + 1$ or $P\gamma(n - m - 1) + 1\,0$. The central condition follows from condition (2). If $\langle F\rangle_i$ is Pxy_2 or Py_2x, then $\Gamma(\langle H\rangle_i)$ is $P0\,\gamma(n - m - 1)$ or $P\gamma(n - m - 1)\,0$, and $\langle\Psi(H)\rangle_i$ is $P1\,\gamma(n - m - 1) + 1$ or $P\gamma(n - m - 1) + 1\,1$. The central condition follows from the Invariance Property.

Since (a)–(d) are exhaustive, the central condition holds in all cases. This completes the solvability proof for Class 5.5. □

We now specify our second nonamenable ∀∃∀-class, although we postpone the proof of its solvability to §3 of the Appendix. Recall that a Krom matrix is a conjunction of binary disjunctions of signed atomic formulas.

Class 5.6. The class of all ∀∃∀-schemata whose matrices are Krom and contain dyadic predicate letters only is solvable.

Class 5.6, the ∀∃∀-Krom Class, is not finitely controllable, since it contains—among others—the satisfiable ∀∃∀-schema F that has no finite models given on page 120. Nor is it amenable, since it is full in the class of all ∀∃∀-schemata. As with Class 5.5, the requirement that matrices contain dyadic predicate letters only entails no loss of generality.

The solvability of the ∀∃∀-Krom Class was shown by Aanderaa and Lewis (1973). We defer the proof because of its length: it relies not just on truth-functional properties of Krom formulas, but also on a sequence of syntactic reductions that enable us to restrict attention to Krom matrices in which each conjunct has a special form.

In Chapter 4 we showed the solvability and finite controllability of the Maslov Class, that is, the ∀···∀∃···∃-Krom Class, and pointed out that this implies the solvability and finite controllability of the Initially-extended Maslov Class, which is the ∃···∃∀···∀∃···∃-Krom Class. However, the ∃∀∃∀-Krom, ∀∃∃∀-Krom, ∀∃∀∀-Krom, and ∀∀∃∀-Krom Classes are all unsolvable (Aanderaa, 1971; Aanderaa and Lewis, 1973). Thus the only Krom prefix class whose status is open is ∀∃∀∃···∃. Similarly, the ∃∀∃∀-Horn, ∀∃∃∀-Horn, ∀∃∀∀-Horn, and ∀∀∃∀-Horn Classes are unsolvable (indeed, the intersection of each of these with the corresponding Krom prefix class is unsolvable (Aanderaa and Lewis, 1973)). Moreover, in contrast to the Maslov Class, the class of Skolem schemata with Horn matrices is also unsolvable. Note that the unsolvability of the ∃∀∃∀-Krom and ∃∀∃∀-Horn Classes, in contrast to the solvability of ∀∃∀-Krom and ∀∃∀-Horn, furnishes two examples of solvable classes that become unsolvable once initial x-variables are allowed.

ISBN 0-201-02540-X

3. Docility

Recall that a class $\mathfrak{C}$ of schemata is *docile* iff there is an effective method for determining, given any schema F in $\mathfrak{C}$, whether F has a finite model. There is no general relation between docility and solvability. By various techniques for encoding Turing machines, we can obtain classes of schemata that are solvable but not docile and that are docile but not solvable. These classes, however, are quite artificial. We have no natural examples of such pathology, although in §5 of the Appendix we show the solvability of an extension of Class 5.1 whose docility is open, and below we mention several unsolvable classes whose docility is open and we prove the docility of a class, Class 5.7, whose solvability is open.

We shall show the docility of Classes 5.1, 5.5, and 5.7. These proofs yield primitive recursive functions φ such that a schema F in one of these classes has a finite model only if it has a model of cardinality $\varphi(F)$. Moreover, the docility proofs for Classes 5.1 and 5.5 closely parallel the solvability proofs given in §§1 and 2.

Let F be an $\forall\exists\forall$-schema and let $p > 0$. Then let δ_p be the term-mapping that carries each m to the least nonnegative remainder of m mod p; and let Δ_p be the δ_p-canonical formula-mapping.

Special Finite Model Lemma. Let F be an $\forall\exists\forall$-schema. If $\Delta_p[E(F, \omega)]$ is consistent, then F has a model of cardinality p. If F has a model of cardinality q, then $\Delta_p[E(F, \omega)]$ is consistent for some $p \leq q$.

Proof. The first claim is a special case of the Finite Model Lemma. For the second claim, which is a special case of its converse, we proceed thus: Suppose F has a model $\mathfrak{M}$ of cardinality q. Let $\mathbf{f}$ be an interpretation of the indicial function sign f_x under which the universal closure of F^* is true in $\mathfrak{M}$; and, for each u in the universe of $\mathfrak{M}$, let $\mathbf{f}^0(u) = u$, $\mathbf{f}^{i+1}(u) = \mathbf{f}(\mathbf{f}^i(u))$. Since $\mathfrak{M}$ has cardinality q, there is an element u in the universe of $\mathfrak{M}$ and a number $p \leq q$ such that $\mathbf{f}^0(u), \mathbf{f}^1(u), \ldots, \mathbf{f}^{p-1}(u)$ are all distinct, and $\mathbf{f}^p(u) = u$. Let $\mathfrak{A}$ be the truth-assignment defined thus: for each atom $Pn_1 \cdots n_k$ in $L(F, \omega)$, $\mathfrak{A}$ verifies $Pn_1 \cdots n_k$ iff $P\mathbf{f}^{n_1}(u) \cdots \mathbf{f}^{n_k}(u)$ is true in $\mathfrak{M}$. Since the universal closure of F^* is true in $\mathfrak{M}$ when f_x is interpreted as $\mathbf{f}$, $\mathfrak{A}$ verifies $E(F, \omega)$. Moreover, by the chose of u, $\mathbf{f}^n(u) = \mathbf{f}^m(u)$ whenever $n \equiv m$ (mod p). Hence $\mathfrak{A}(\Delta_p(A)) = \mathfrak{A}(A)$ for each atom A in $L(F, \omega)$. Consequently, $\mathfrak{A}$ verifies $\Delta_p[E(F, \omega)]$. $\square$

Note that the set $\delta_p[D(F, \omega)]$ is simply $D(F, p)$, that is, $\{0, \ldots, p - 1\}$. Moreover, if $\delta_p m < p - 1$ then $\delta_p(m + 1) = \delta_p m + 1$, and if $\delta_p m = p - 1$ then $\delta_p(m + 1) = 0$. Hence $\Delta_p[E(F, \omega)]$ consists of all Herbrand instances $F^*(y_1/m, y_2/n)$ for $m < p - 1$ and $n \leq p - 1$, together with all instances $F^M(y_1/p - 1, x/0, y_2/n)$ for $n \leq p - 1$ (the latter are instances of F, but not Herbrand instances).

ISBN 0-201-02540-X

Thus a truth-assignment $\mathfrak{A}$ verifying $\Delta_p[E(F, \omega)]$ provides a model for F over the universe $D(F, p)$. Indeed, it provides a model for the universal closure of F^* when the indicial function sign f_x is interpreted as the function that carries m to $m + 1$, $m < p - 1$, and carries $p - 1$ to 0.

Class 5.1. Let F be an $\forall\exists\forall$-schema each of whose atomic subformulas has one of the seven forms Py_1y_1, Pxx, Py_2y_2, Py_1x, Pxy_1, Py_1y_2, or Pxy_2. Then F has a finite model only if it has a model of cardinality $2^{d(F,3)}$.

Proof. Let p be a number such that $\Delta_p[E(F, \omega)]$ is consistent. It suffices to show that if $p > 2^{d(F,3)}$, then there is an $r > 0$ such that $\Delta_{p-r}[E(F, \omega)]$ is consistent. For this shows that F has a finite model only if it has a model of cardinality $\leq 2^{d(F,3)}$; and, since F does not contain the identity sign, F has a model of cardinality $\leq q$ only if F has a model of cardinality q.

Let $\mathfrak{A}$ be any truth-assignment verifying $\Delta_p[E(F, \omega)]$. For each m let

$$\xi_{\mathfrak{A}}(m) = \{\mathfrak{p}_{\mathfrak{A}}(0, m, n) \mid n \leq p - 1\}$$

The range of $\xi_{\mathfrak{A}}$ consists of nonempty finite sets of profiles; indeed, the range contains at most $2^{d(F,3)} - 1$ members. Since $p > 2^{d(F,3)}$, by the pigeonhole argument there are numbers q and r, $0 < q < q + r < p$, such that $\xi_{\mathfrak{A}}(q) = \xi_{\mathfrak{A}}(q + r)$. We shall construct a formula-mapping Γ such that the induced truth-assignment $\mathfrak{A}(\Gamma)$ verifies $\Delta_{p-r}[E(F, \omega)]$. Thus $\mathfrak{A}(\Gamma)$ will provide a model for F over $D(F, p - r)$. The terms $0,\ldots,q - 1$ of this model will correspond to the terms $0,\ldots,q - 1$ in the model for F over $D(F, p)$ provided by $\mathfrak{A}$; while the terms $q,\ldots,p - r - 1$ in the model over $D(F, p - r)$ will correspond to the terms $q + r,\ldots,p - 1$ in the given model over $D(F, p)$. Essentially, we are eliminating the r elements $q,\ldots,q + r - 1$ from the given model over $D(F, p)$.

We start by constructing a term-mapping γ and a mapping η on pairs of terms fulfilling the following five conditions for all m and $n \leq p - r - 1$:

(1) $\gamma m \leq p - 1$; if $m < p - r - 1$, then in fact $\gamma m < p - 1$;
(2) $\eta(m, n) \leq p - 1$;
(3) $\mathfrak{p}_{\mathfrak{A}}(0, \gamma(m + 1), \eta(m + 1, n)) = \mathfrak{p}_{\mathfrak{A}}(0, \gamma m + 1, \eta(m, n))$;
(4) $\eta(m, m) = \gamma m$ and $\eta(m, m + 1) = \gamma m + 1$;
(5) $\eta(m, 0) = \gamma 0 = 0$.

(Both γ and η will be everywhere defined; but we are indifferent to their values on arguments $> p - r - 1$.) Let γ be defined thus:

$$\gamma m = \begin{cases} m & \text{if } m < q, \\ m + r & \text{if } m \geq q. \end{cases}$$

Clearly γ fulfills condition (1). Moreover, $\xi_{\mathfrak{A}}(\gamma(m + 1)) = \xi_{\mathfrak{A}}(\gamma m + 1)$ for every m.

ISBN 0-201-02540-X

We define $\eta(m, n)$ by induction on the numerical difference between m and n. Our definition mimics that in the solvability proof for Class 5.1.

(η-1) $\eta(m, m) = \gamma m$ and $\eta(m, m + 1) = \gamma m + 1$ for each m.

(η-2)   If $m + 1 < n \leq p - r - 1$, then $\eta(m, n)$ is the least $k \leq p - 1$ such that $\mathfrak{p}_\mathfrak{A}(0, \gamma m +1, k) = \mathfrak{p}_\mathfrak{A}(0, \gamma(m +1), \eta(m + 1, n))$. On the assumption that $\eta(m+1, n) \leq p - 1$ there will be such a k, since $\xi_\mathfrak{A}(\gamma(m +1)) = \xi_\mathfrak{A}(\gamma m +1)$.

(η-3)   If $n \leq m \leq p - r - 1$, then $\eta(m + 1, n)$ is the least $k \leq p - 1$ such that $\mathfrak{p}_\mathfrak{A}(0, \gamma(m +1), k) = \mathfrak{p}_\mathfrak{A}(0, \gamma m +1, \eta(m, n))$. On the assumption that $\eta(m, n) \leq p - 1$ there will be such a k, again since $\xi_\mathfrak{A}(\gamma(m +1)) = \xi_\mathfrak{A}(\gamma m +1)$.

(η-4)   If $\eta(m, n)$ is not determined by (η-1)–(η-3), then let $\eta(m, n) = 0$.

Conditions (2)–(4) follow just as did the analogous conditions (2)–(4) in the solvability proof. We prove condition (5) by induction on m. By (η-1), $\eta(0, 0) = \gamma 0$; by the construction of γ, $\gamma 0 = 0$. Suppose $\eta(m, 0) = 0$, where $0 \leq m \leq p - r - 1$. By (η-3), $\eta(m + 1, 0)$ is the least k such that $\mathfrak{p}_\mathfrak{A}(0, \gamma(m +1), k) = \mathfrak{p}_\mathfrak{A}(0, \gamma m +1, 0)$. Since $\xi_\mathfrak{A}(\gamma(m +1)) = \xi_\mathfrak{A}(\gamma m + 1)$, $\mathfrak{p}_\mathfrak{A}(0, \gamma(m +1)) = \mathfrak{p}_\mathfrak{A}(0, \gamma m +1)$. It follows that the least k is 0 itself, that is, $\eta(m + 1, 0) = 0$. This establishes condition (5).

Now let Γ carry each atom Pmn in $L(F, \omega)$ to $P\gamma m \, \eta(m, n)$. To show that $\mathfrak{A}(\Gamma)$ verifies $\Delta_{p-r}[E(F, \omega)]$ we must show

(a) $\mathfrak{A}(\Gamma)$ verifies each Herbrand instance $F^*(y_1/m, y_2/n)$, where $m < p - r - 1$ and $n \leq p - r - 1$;

(b) $\mathfrak{A}(\Gamma)$ verifies each instance $F^M(y_1/p - r - 1, x/0, y_2/n)$, where $n \leq p - r - 1$.

To prove (a), let Ψ be the instance-mapping that carries $H = F^*(y_1/m, y_2/n)$ to $F^*(y_1/\gamma m, y_2/\eta(m, n))$. By conditions (1) and (2), if H is a member of $\Delta_{p-r}[E(F, \omega)]$, then $\Psi(H)$ is a member of $\Delta_p[E(F, \omega)]$. Thus it suffices to show the central condition

$$\mathfrak{A}(\Gamma(\langle H \rangle_i)) = \mathfrak{A}(\langle \Psi(H) \rangle_i)$$

for each i. But the central condition follows from conditions (3) and (4) just as it did from the analogous conditions (3) and (4) in the solvability proof for Class 5.1.

To prove (b), let $I = F^M(y_1/p - r - 1, x/0, y_2/n)$, where $n \leq p - r - 1$. Then let $J = F^M(y_1/p - 1, x/0, y_2/\eta(p - r - 1, n))$. Thus J is a member of $\Delta_p[E(F, \omega)]$, and it suffices to show that

$$\mathfrak{A}(\Gamma(\langle I \rangle_i)) = \mathfrak{A}(\langle J \rangle_i)$$

for each i.

ISBN 0-201-02540-X

Since $p - 1 = \gamma(p - r - 1) = \eta(p - r - 1, p - r - 1)$ and $0 = \gamma0 = \eta(m, 0)$ for each $m \le p - r - 1$, we have $\Gamma(\langle I \rangle_i) = \langle J \rangle_i$ when $\langle F \rangle_i$ is Py_1y_2, Py_1y_1, Pxx, or Py_1x. Three cases remain. If $\langle F \rangle_i = Py_2y_2$, then $\Gamma(\langle I \rangle_i) = P\gamma n \; \eta(n, n) = P\gamma n \; \gamma n$ and $\langle J \rangle_i = P\eta(p - r - 1, n) \; \eta(p - r - 1, n)$. If $\langle F \rangle_i = Pxy_1$, then $\Gamma(\langle I \rangle_i) = P\gamma0 \; \eta(0, p - r - 1) = P0 \; \eta(0, p - r - 1)$ and $\langle J \rangle_i = P0 \; p - 1$. If $\langle F \rangle_i = Pxy_2$, then $\Gamma(\langle I \rangle_i) = P\gamma0 \; \eta(0, n) = P0 \; \eta(0, n)$ and $\langle J \rangle_i = P0 \; \eta(p - r - 1, n)$. In these three cases we may infer $\mathfrak{A}(\Gamma(\langle I \rangle_i)) = \mathfrak{A}(\langle J \rangle_i)$ from the following consequence of conditions (3) and (4): for each fixed $k \le p - r - 1$, the profile $\mathfrak{p}_\mathfrak{A}(0, \eta(j, k))$ is constant as j varies from 0 to $p - r - 1$; indeed, $\mathfrak{p}_\mathfrak{A}(0, \eta(j, k)) = \mathfrak{p}_\mathfrak{A}(0, \gamma k)$.

We have proved (b), and thus shown that $\mathfrak{A}(\Gamma)$ verifies $\Delta_{p-r}[E(F, \omega)]$. This concludes the proof of docility. $\square$

Class 5.5.   Let F be an $\forall\exists\forall$-Horn schema containing dyadic predicate letters only. Then F has a finite model only if it has a finite model of cardinality $2d(F, 2) + 3$.

Proof.   Suppose that $\Delta_p[E(F, \omega)]$ is consistent. We show that if $p > 2d(F, 2) + 3$ then there is an $r > 0$ such that $\Delta_{p-2r}[E(F, \omega)]$ is consistent.

First we apply the Truth-Functional Lemma of §2 to $\Delta_p[E(F, \omega)]$. Let $\mathfrak{A}$ be the truth-assignment that verifies an atom Pmn iff $\Delta_p[E(F, \omega)]$ truth-functionally implies Pmn. Since $\Delta_p[E(F, \omega)]$ is a consistent set of Horn formulas, it is verified by $\mathfrak{A}$. Now let γ be the term-mapping that carries each $m < p - 1$ to $m + 1$ and carries $p - 1$ to 0, and let Γ be the γ-canonical formula-mapping. Then Γ carries $\Delta_p[E(F, \omega)]$ to itself. Hence $\Delta_p[E(F, \omega)]$ implies an atom Pmn if it implies $\Gamma(Pmn)$. Consequently, $\mathfrak{A}$ possesses the following

Invariance Property.   For all predicate letters P of F and all m, n, m', $n' \le p - 1$, if $m - n \equiv m' - n' \pmod{p}$, then $\mathfrak{A}(Pmn) = \mathfrak{A}(Pm'n')$.

Now by the pigeonhole argument there are numbers q and r, $0 < q < q + r \le d(F, 2) + 1$, such that $\mathfrak{p}_\mathfrak{A}(0, q) = \mathfrak{p}_\mathfrak{A}(0, q + r)$. From the Invariance Property we have $\mathfrak{p}_\mathfrak{A}(p - q, 0) = \mathfrak{p}_\mathfrak{A}(0, q) = \mathfrak{p}_\mathfrak{A}(0, q + r) = \mathfrak{p}_\mathfrak{A}(p - q - r, 0)$. Also, since $p > 2d(F, 2) + 3$, we have $q < p - q - 2r$. We shall construct a formula-mapping Γ such that the induced truth-assignment $\mathfrak{A}(\Gamma)$ verifies $\Delta_{p-2r}[E(F, \omega)]$; this will prove the desired result. The truth-assignment $\mathfrak{A}(\Gamma)$ will provide a model for F over $D(F, p - 2r)$, a model obtained essentially by deleting the terms $q, \ldots, q + r - 1$ and $p - q - r, \ldots, p - q - 1$ from the model over $D(F, p)$ that the given truth-assignment provides. More precisely, by the Invariance Property the profile $\mathfrak{p}_\mathfrak{A}(m, n)$, where m, $n \le p - 1$, depends just on the numerical difference between m and n. We so define Γ that if $m, n \le p - 2r$, then the profile $\mathfrak{p}_{\mathfrak{A}(\Gamma)}(m, n)$ also depends just on this difference; but if, say, $m > n$, then $\mathfrak{p}_{\mathfrak{A}(\Gamma)}(m, n)$ will be identified with either $\mathfrak{p}_\mathfrak{A}(m, n)$ or $\mathfrak{p}_\mathfrak{A}(m + r, n)$ or $\mathfrak{p}_\mathfrak{A}(m + 2r, n)$, depending on the magnitude of $m - n$.

Let γ be the term-mapping defined thus:

$$\gamma m = \begin{cases} m & \text{if } m < q, \\ m + r & \text{if } q \leq m < p - q - 2r, \\ m + 2r & \text{if } p - q - 2r \leq m. \end{cases}$$

The term-mapping γ fulfills the following conditions for every m:

(1)   if $m < p - 2r - 1$, then $\gamma m < p - 1$; and $\gamma(p - 2r - 1) = p - 1$;
(2) $\mathfrak{p}_{\mathfrak{A}}(\gamma(m + 1), 0) = \mathfrak{p}_{\mathfrak{A}}(\gamma m + 1, 0)$.

Condition (1) is immediate. For condition (2) note that either $\gamma(m + 1) = m + 1$ or $m + 1 = q$ or $m + 1 = p - q - 2r$. If $m + 1 = q$, then $\gamma(m + 1) = q + r$ and $\gamma m + 1 = q$; if $m + 1 = p - q - 2r$ then $\gamma(m + 1) = p - q$ and $\gamma m + 1 = p - q - r$. Hence in these two cases condition (2) follows from the choice of q and r.

We now proceed just as in the solvability proof for Class 5.5. Let Γ be the formula-mapping defined thus: for each atom Pmn in $L(F, \omega)$,

$$\Gamma(Pmn) = \begin{cases} P\gamma(m - n)\, 0 & \text{if } m \geq n, \\ P0\, \gamma(n - m) & \text{if } m < n. \end{cases}$$

To show that $\mathfrak{A}(\Gamma)$ verifies $\Delta_{p-2r}[E(F, \omega)]$ we must show

(a) $\mathfrak{A}(\Gamma)$ verifies each Herbrand instance $F^*(y_1/m, y_2/n)$, where $m < p - 2r - 1$ and $n \leq p - 2r - 1$;
(b) $\mathfrak{A}(\Gamma)$ verifies each instance $F^M(y_1/p - 2r - 1, x/0, y_2/n)$, where $n \leq p - 2r - 1$.

For (a), let Ψ be the instance-mapping that carries $H = F^*(y_1/m, y_2 n)$ to $F^*(y_1/\gamma(m - n), y_2/0)$ if $m \geq n$ and to $F^*(y_1/0, y_2/\gamma(n - m - 1) + 1)$ if $m < n$. By condition (1), if $m < p - 2r - 1$ and $n \leq p - 2r - 1$, then $\Psi(H)$ is a member of $\Delta_p[E(F, \omega)]$. Hence it suffices to show the central condition

$$\mathfrak{A}(\Gamma(\langle H \rangle_i)) = \mathfrak{A}(\langle \Psi(H) \rangle_i)$$

for each i. But the central condition follows from condition (2) and the Invariance Property, just as it did in the solvability proof for Class 5.5.

For (b), let $I = F^M(y_1/p - 2r - 1, x/0, y_2/n)$, where $n \leq p - 2r - 1$. Then let $J = F^M(y_1/p - 1, x/0, y_2/\gamma n)$. Thus J is in $\Delta_p[E(F, \omega)]$, and it suffices to show that

$$\mathfrak{A}(\Gamma(\langle I \rangle_i)) = \mathfrak{A}(\langle J \rangle_i)$$

for each i. Since $\gamma(p - 2r - 1) = p - 1$ and $\gamma 0 = 0$, we have $\Gamma(\langle I \rangle_i) = \langle J \rangle_i$ if $\langle F \rangle_i$ has one of the forms Py_1x, Pxy_1, Pxx, Pxy_2, or Py_2x. If $\langle F \rangle_i = Py_1y_1$ or Py_2y_2, then $\Gamma(\langle I \rangle_i) = P0\,0$ and $\langle J \rangle_i = Pp - 1\, p - 1$ or $P\gamma n\, \gamma n$. In this case the desired equivalence follows from the Invariance Prop-

ISBN 0-201-02540-X

erty. Finally, if $\langle F \rangle_i$ is Py_1y_2 or Py_2y_1, then $\Gamma(\langle I \rangle_i)$ is $P\gamma(p - 2r - 1 - n)0$ or $P0 \, \gamma(p - 2r - 1 - n)$, whereas $\langle J \rangle_i$ is $Pp - 1 \, \gamma n$ or $P\gamma n \, p - 1$. The desired equivalence follows from the Invariance Property, since, we claim, $\gamma(p - 2r - 1 - n) = p - 1 - \gamma n$. To show this we proceed by cases.

(i) $p - 2r - 1 - n < q$. Then $n \geq p - 2r - q$, so that $\gamma(p - 2r - 1 - n) = p - 2r - 1 - n = p - 1 - (n + 2r) = p - 1 - \gamma n$.

(ii) $q \leq p - 2r - 1 - n < p - q - 2r$. Then $q \leq n < p - q - 2r$, so that $\gamma(p - 2r - 1 - n) = p - 2r - 1 - n + r = p - 1 - (n + r) = p - 1 - \gamma n$.

(iii) $p - q - 2r \leq p - 2r - 1 - n$. Then $n < q$, so that $\gamma(p - 2r - 1 - n) = p - 1 - n = p - 1 - \gamma n$.

This completes the proof that $\mathfrak{A}(\Gamma)$ verifies I. Thus $\mathfrak{A}(\Gamma)$ verifies $\Delta_{p-2r}[E(F, \omega)]$, and docility is established. $\square$

Class 5.6, the $\forall\exists\forall$-Krom Class, is also docile, as we shall show in §3 of the Appendix. On the other hand, the docility of the $\forall\exists\forall\forall$-Krom and $\forall\forall\exists\forall$-Krom Classes, both of which are unsolvable, is open. The other unsolvable Krom classes mentioned at the end of §2, however, are not docile (Aanderaa, 1971).

We now turn to a new class, the study of which was initiated by Ackermann. In 1936 Ackermann noted that an $\forall\exists\forall$-schema F has a model iff the schema

$$\forall y \exists x Pyx \wedge \forall y_1 \forall y_2 \forall y_3 (Py_1y_3 \Rightarrow F^M(x/y_3))$$

has a model, where P is a dyadic predicate letter foreign to F. Moreover, this reduction preserves finite models as well. (The reduction is just that given by Skolem in 1920 to reduce any schema to Skolem form.) Since, as Kahr, Moore, and Wang subsequently showed (1961), the $\forall\exists\forall$-Class is neither solvable nor docile, the same is true for this class of Ackermann's. However, Ackermann asked two further questions: Is the class of schemata $\forall y \exists x Pyx \wedge \forall y_1 \cdots \forall y_n G$, where G is quantifier-free and P is the *only* predicate letter in G, solvable? Is it docile? The answer to the first question is still unknown, although later we shall prove some partial results due to Ackermann. The answer to the second, as Ackermann showed, is affirmative. In fact, we prove the docility of an extension of this class, obtained by allowing G to contain in addition to P arbitrarily many *monadic* predicate letters. This extended result was first shown by Denton (see Dreben and Denton, 1963; Denton, 1963) and by Kostyrko (1962).

Since the functional form of a schema $F = \forall y \exists x Pyx \wedge \forall y_1 \cdots \forall y_n G$ contains only one indicial function sign f_x, and f_x has but one argument, the Special Finite Model Lemma applies to F (and we may use numerical

ISBN 0-201-02540-X

notation for terms in $D(F, \omega)$). Here $\Delta_p[E(F, \omega)]$ is truth-functionally equivalent to the set containing the formula

$$P01 \wedge P12 \wedge \cdots \wedge Pp - 2\ p - 1 \wedge Pp - 1\ 0$$

together with all instances $G(y_1/m_1, \ldots, y_n/m_n)$ of G over $D(F, p)$ (that is, where $m_1, \ldots, m_n \leq p - 1$).

Class 5.7. Let F be a schema $\forall y \exists x Pyx \wedge \forall y_1 \cdots Ay_nG$, where G is quantifier-free and contains, aside from the predicate letter P, only p_0 many monadic predicate letters, $p_0 \geq 0$. Then F has a finite model only if it has a model of cardinality $n \cdot 2^{p_0} + 1$.

Proof. Let $\mathfrak{A}$ be any truth-assignment on $\tilde{L}(F, \omega)$. A sequence $\langle m_0, \ldots, m_{r-1} \rangle$ of distinct terms is an $\mathfrak{A}$-*cycle* iff $\mathfrak{A}$ verifies $Pm_0m_1 \wedge \cdots \wedge Pm_{r-2}m_{r-1} \wedge Pm_{r-1}m_0$.

(A) If $\langle m_0, \ldots, m_{r-1} \rangle$ is an $\mathfrak{A}$-cycle and $\mathfrak{A}$ verifies every instance of G over $\{m_0, \ldots, m_{r-1}\}$, then $\Delta_r[E(F, \omega)]$ is consistent.

For let γ carry each k to m_k if $k \leq r - 1$, and carry k to anything if $k > r - 1$. Then $\mathfrak{A}$ verifies $\Gamma(P01 \wedge \cdots \wedge Pr - 2\ r - 1 \wedge Pr - 1\ 0)$, where Γ is the γ-canonical formula-mapping. Moreover, if I is an instance of G over $D(F, r)$, then $\Gamma(I)$ is an instance of G over $\{m_0, \ldots, m_{r-1}\}$; hence $\mathfrak{A}$ verifies $\Gamma(I)$. We have shown that the induced truth-assignment $\mathfrak{A}(\Gamma)$ verifies $\Delta_r[E(F, \omega)]$.

Now suppose $\Delta_p[E(F, \omega)]$ is consistent and $p > n \cdot 2^{p_0} + 1$. We show that $\Delta_r[E(F, \omega)]$ is consistent for some $r < p$. Let $\mathfrak{A}$ be a truth-assignment verifying $\Delta_p[E(F, \omega)]$; and, for each k, let $|k|$ be the least nonnegative remainder of $k \bmod p$. Thus $\mathfrak{A}$ verifies $Pk\ |k + 1|$ for each $k \leq p - 1$. If there is an $\mathfrak{A}$-cycle $\langle m_0, \ldots, m_{r-1} \rangle$ such that $r < p$ and each m_i is in $D(F, p)$, then the result follows from (A). Hence we may assume for all k and m in $D(F, p)$ that $\mathfrak{A}$ verifies Pkm if *and only if* $m = |k + 1|$. In particular $\mathfrak{A}$ falsifies Pmm for each $m \leq p - 1$. Consequently, as m varies there are at most $n \cdot 2^{p_0}$ values of

$$\langle \mathfrak{p}_{\mathfrak{A}}(|m|), \mathfrak{p}_{\mathfrak{A}}(|m + 1|), \ldots, \mathfrak{p}_{\mathfrak{A}}(|m + n|) \rangle .$$

By the pigeonhole argument there are numbers q and r, $q \leq p - 1$ and $0 < r \leq p - 1$ such that $\mathfrak{p}_{\mathfrak{A}}(|q|) = \mathfrak{p}_{\mathfrak{A}}(|q + r|)$, $\mathfrak{p}_{\mathfrak{A}}(|q + 1|) = \mathfrak{p}_{\mathfrak{A}}(|q + r + 1|), \ldots, \mathfrak{p}_{\mathfrak{A}}(|q + n|) = \mathfrak{p}_{\mathfrak{A}}(|q + r + n|)$. Moreover, we may assume $r > n$. For if $p_0 = 0$, that is, there are no monadic predicate letters in F, then simply pick $q = 0$ and $r = p - 1$. If $p_0 > 0$ but yet $r \leq n$, then let $q' = |q + r|$ and $r' = p - r$; it follows that $|q' + r'| = q$, so that q' and r' fulfills the same condition as do q and r. Moreover, since $p > n \cdot 2^{p_0} + 1 > 2n$, if $r \leq n$ then $r' > n$. Hence we may use q' and r' instead of q and r in what follows.

ISBN 0-201-02540-X

Now let $\mathfrak{B}$ be the truth-assignment that verifies $P|q + r - 1|\ q$ and agrees with $\mathfrak{A}$ on all other atoms. Thus $\langle q, |q + 1|,\ldots,|q + r - 1|\rangle$ is a $\mathfrak{B}$-cycle, and by (A) it suffices to show that every instance of G over the set $S = \{q, |q + 1|,\ldots,|q + r - 1|\}$ is verified by $\mathfrak{B}$. Let I be such an instance. If at least one of q and $|q + r - 1|$ does not occur in I, then $\mathfrak{B}(\langle I\rangle_i) = \mathfrak{A}(\langle I\rangle_i)$ for each i. Since $\mathfrak{A}$ verifies I, it follows that $\mathfrak{B}$ does too. Suppose both q and $|q + r - 1|$ occur in I. Since I contains at most n distinct terms, there is a number k, $1 \leq k \leq n$, such that $|q + k|$ does not occur in I. Since $r > n$, also $k < r$. Let γ be the term-mapping defined thus: if j is among $q, |q + 1|,\ldots,|q + k|$ then $\gamma j = |j + r|$; otherwise, $\gamma j = j$. It quickly follows that for all m_1 and m_2 in S distinct from $|q + k|$, $\mathfrak{p}_\mathfrak{B}(m_1) = \mathfrak{p}_\mathfrak{A}(m_1) = \mathfrak{p}_\mathfrak{A}(\gamma m_1)$ and $\mathfrak{p}_\mathfrak{B}(m_1, m_2) = \mathfrak{p}_\mathfrak{A}(\gamma m_1, \gamma m_2)$. This implies that

$$\mathfrak{B}(\langle I\rangle_i) = \mathfrak{A}(\Gamma(\langle I\rangle_i))$$

for each i, where $\cdot\Gamma$ is the γ-canonical formula-mapping. Now $\Gamma(I)$ is an instance of G over $\gamma[S]$, and $\gamma[S] \subseteq D(F, p)$. Hence $\mathfrak{A}$ verifies $\Gamma(I)$, whence $\mathfrak{B}$ verifies I as required. This concludes the docility proof for Class 5.7. $\square$

The solvability of Class 5.7 is open. Indeed, as we mentioned earlier, it is open even for the subclass of 5.7 containing only those schemata in which occur no monadic predicate letters. We do, however, have results for even more restricted subclasses. There are only finitely many distinct schemata in these subclasses, up to truth-functional equivalence of matrix; hence they are trivially solvable. What we present below, following Ackermann, are decision procedures for these subclasses, and descriptions of the types of models which satisfiable schemata in these subclasses possess.

Class 5.8. Let F be a schema $\forall y\exists xPyx \land \forall y_1\cdots\forall y_nG$, where G is quantifier-free and contains no predicate letters aside from P.

(I)   If $n = 3$, then F has a model provided that $E(F, 3)$ is consistent.
(II) If $n = 4$, then F has a model provided that $E(F, 13)$ is consistent.

Proof. Let $\mathfrak{A}$ be any truth-assignment on $\bar{L}(F, \omega)$, and suppose $\langle m_0,\ldots,m_{r-1}\rangle$ is an r-tuple such that $\mathfrak{A}$ verifies every instance of G over $\{m_0,\ldots,m_{r-1}\}$. In the preceding proof we saw

(A)   if $\langle m_0,\ldots,m_{r-1}\rangle$ is an $\mathfrak{A}$-cycle, then $\Delta_r[E(F, \omega)]$ is consistent.

We also have

(B)   if $r = n$ and if, for all j and k, $0 \leq j, k \leq n - 1$, $\mathfrak{A}$ verifies Pm_jm_k iff $j = k + 1$, then $\Delta_{n+1}[E(F, \omega)]$ is consistent;
(C)   If $r = n$ and if, for all j and k, $0 \leq j, k \leq n - 1$, $\mathfrak{A}$ verifies Pm_jm_k iff $j < k$, then $E(F, \omega)$ is consistent.

Recall that n is the number of y-variables of the second conjunct in F. To prove (B), let $\mathfrak{B}$ be the truth-assignment on $\hat{L}(F, \omega)$ that verifies an atom Pjk iff $j + 1 \equiv k \pmod{n + 1}$. Thus $\langle 0,\ldots,n \rangle$ is a $\mathfrak{B}$-cycle, and it suffices to show that $\mathfrak{B}$ verifies each instance of G over $\{0,\ldots,n\}$. Let I be such an instance. Then there is an integer p, $0 \le p \le n$, that does not occur in I. Define $\gamma \colon \{0,\ldots,n\} \to \{m_0,\ldots,m_{n-1}\}$ thus: if $0 \le k \le p$, then $\gamma k = m_{n-p-k}$; and if $p < k \le n$, then $\gamma k = m_{k-p-1}$. Let Γ be the γ-canonical formula-mapping. Then $\mathfrak{A}$ verifies $\Gamma(I)$, since $\Gamma(I)$ is an instance of G over $\{m_0,\ldots,m_{n-1}\}$. Moreover, since p does not occur in I, $\mathfrak{B}(\langle I\rangle_i) = \mathfrak{A}(\Gamma(\langle I\rangle_i))$ for each i. Hence $\mathfrak{B}$ verifies I. This establishes (B). (Note the similarity of this argument and the docility proof for Class 5.7.)

To prove (C), let $\mathfrak{B}$ be the truth-assignment on $\hat{L}(F, \omega)$ that verifies an atom Pjk iff $j < k$. Clearly $\mathfrak{B}$ verifies $Pj\,j + 1$ for each j, and it suffices to show that $\mathfrak{B}$ verifies each instance of G over $D(F, \omega)$. Let $I = G(y_1/ j_1,\ldots,y_n/j_n)$ be such an instance. Then there is a mapping $\eta \colon \{j_1,\ldots,j_n\} \to \{0,\ldots,n - 1\}$ such that $j_i < j_{i'}$ iff $\eta j_i < \eta j_{i'}$. Let γ be a term-mapping that takes each j_i to $m_{\eta j_i}$; we are indifferent to the values of γ on terms not occurring in I. Let Γ be the γ-canonical formula-mapping. Then once again $\mathfrak{A}$ verifies $\Gamma(I)$ and $\mathfrak{B}(\langle I\rangle_i) = \mathfrak{A}(\Gamma(\langle I\rangle_i))$ for each i. This establishes (C).

Our argument has established the following: if the antecedent of (C) is fulfilled, then the interpretation of the predicate letter P as $<$ provides a model for F over the nonnegative integers.

(A), (B), and (C) quickly yield assertion (I). For suppose $n = 3$ and let $\mathfrak{A}$ verify $E(F, 3)$. Then $\mathfrak{A}$ verifies $P01 \wedge P12$. If $\mathfrak{A}$ also verifies Pjk for some j and k, $0 \le k \le j \le 2$, then by (A) F has a finite model. If not, and in addition $\mathfrak{A}$ falsifies $P02$, then by (B) F has a finite model. Finally, if neither of the preceding cases applies, then the antecedent of (C) holds, so that the interpretation of P as $<$ provides a model for F over the nonnegative integers.

A further argument is required for assertion (II). Suppose $n = 4$, and let $\mathfrak{A}$ be a truth-assignment that verifies every instance of G over a set $\{m_0, m_1, m_2, m_3\}$.

(D)   If, for all j and k, $0 \le j$, $k \le 3$, $\mathfrak{A}$ verifies $Pm_j m_k$ iff either $k = j + 1$ or else $j = 0$ and $k = 3$, then $E(F, \omega)$ is consistent.

For let $\mathfrak{B}$ be the truth-assignment on $\hat{L}(F, \omega)$ that verifies an atom Pjk iff $j < k$ and $j \not\equiv k \pmod 2$. Clearly $\mathfrak{B}$ verifies $Pj\,j + 1$ for each j, and it suffices to show that $\mathfrak{B}$ verifies each instance of G over $D(F, \omega)$. Let $I = G(y_1/j_1,\ldots,y_4/j_4)$ be such an instance. Then there is a mapping $\eta \colon \{j_1, j_2, j_3, j_4\} \to \{0, 1, 2, 3\}$ such that $j_i < j_{i'}$ and $j_i \not\equiv j_{i'} \pmod 2$ just in case $\eta j_i < \eta j_{i'}$ and $\eta j_i \not\equiv \eta j_{i'} \pmod 2$. Let γ take each j_i to $m_{\eta j_i}$, and let Γ be

ISBN 0-201-02540-X

the γ-canonical formula-mapping. Again, $\mathfrak{A}$ verifies $\Gamma(I)$ and $\mathfrak{B}(\langle I \rangle_i) = \mathfrak{A}(\Gamma(\langle I \rangle_i))$ for each i. This proves (D).

Our argument in fact establishes that if the antecedent of (D) is fulfilled, then the interpretation of P as "$<$ and incongruent mod 2" provides a model for F over the nonnegative integers.

To prove assertion (II), let $\mathfrak{A}$ verify $E(F, 13)$. Thus $\mathfrak{A}$ verifies $Pj\ j+1$ for each j in $D(F, 13)$, and $\mathfrak{A}$ verifies each instance of G over $D(F, 13)$. We may assume that for all j and k in $D(F, 13)$, if $k \leq j$, then $\mathfrak{A}$ falsifies Pjk. For otherwise, by (A), F has a finite model. By a certain amount of sheer drudgery, we show that there are m_0, m_1, m_2, m_3 in $D(F, 13)$ that fulfill the antecedent of one of (B), (C), or (D). Indeed, suppose there are no such m_0, m_1, m_2, m_3. Then for all m_0, m_1, m_2, m_3, m_4 in $D(F, 13)$,

(i)   if $\mathfrak{A}$ verifies $Pm_0m_1 \wedge Pm_1m_2 \wedge Pm_2m_3 \wedge -Pm_0m_2$, then $\mathfrak{A}$ verifies Pm_1m_3;

(ii)   if $\mathfrak{A}$ verifies $Pm_0m_1 \wedge Pm_1m_2 \wedge Pm_2m_3 \wedge Pm_3m_4 \wedge -Pm_0m_2$, then $\mathfrak{A}$ falsifies Pm_2m_4.

For if (i) is violated, then $\langle m_0, m_1, m_2, m_3 \rangle$ fulfills the antecedent of either (B) or (D). If (ii) is violated, then either $\langle m_1, m_2, m_3, m_4 \rangle$ fulfills the antecedent of (C), or $\langle m_0, m_1, m_2, m_4 \rangle$ fulfills the antecedent of (D), or $\langle m_0, m_1, m_2, m_4 \rangle$ fulfills the antecedent of (B).

Now $\mathfrak{A}$ verifies either $P01 \wedge P12 \wedge -P02$, or $P02 \wedge P23 \wedge -P03$, or $P01 \wedge P13 \wedge -P03$. For all three fail only if $\mathfrak{A}$ verifies $P02 \wedge P13 \wedge P03$, in which case $\langle 0, 1, 2, 3 \rangle$ fulfills the antecedent of (C). Let $\langle i, j, k \rangle$ be $\langle 0, 1, 2 \rangle$, $\langle 0, 2, 3 \rangle$, or $\langle 0, 1, 3 \rangle$, depending on which case holds (if more than one holds, pick any applicable triple). Note that $k \leq 3$, so that $k + 9 \leq 12$; that is, $k + 9$ is in $D(F, 13)$. By applying (ii) to $\langle i, j, k, k+1, k+2 \rangle$ we have that $\mathfrak{A}$ falsifies $Pk\ k+2$. Applying (ii) to $\langle k, k+1, k+2, k+3, k+4 \rangle$, we obtain that $\mathfrak{A}$ falsifies $Pk+2\ k+4$. Repeating this twice more, we infer that $\mathfrak{A}$ falsifies $Pk+4\ k+6$ and $Pk+6\ k+8$. Then, applying (i) to $\langle k, k+1, k+2, k+3 \rangle$, we find that $\mathfrak{A}$ verifies $Pk+1\ k+3$; applying (i) to $\langle k+2, k+3, k+4, k+5 \rangle$, we find that $\mathfrak{A}$ verifies $Pk+3\ k+5$; similarly, $\mathfrak{A}$ verifies $Pk+5\ k+7$ and $Pk+7\ k+9$. Hence (ii) is applicable to $\langle i, j, k, k+1, k+3 \rangle$, so $\mathfrak{A}$ falsifies $Pk\ k+3$; and (ii) is applicable to $\langle k, k+1, k+2, k+3, k+5 \rangle$, so that $\mathfrak{A}$ falsifies $Pk+2\ k+5$. Similarly, $\mathfrak{A}$ falsifies $Pk+4\ k+7$. Then (i) is applicable to $\langle k+4, k+5, k+7, k+9 \rangle$; hence $\mathfrak{A}$ verifies $Pk+5\ k+9$. But (ii) is applicable to $\langle k+2, k+3, k+5, k+7, k+9 \rangle$, so $\mathfrak{A}$ falsifies $Pk+5\ k+9$. This is a contradiction. Hence the antecedent of one of (B), (C), or (D) must be fulfilled.

To recapitulate, we showed that if $n = 3$, then F has either a finite model or else a model over the nonnegative integers in which P is interpreted as

"$<$". If $n = 4$, then F has either a model of one of these sorts or else a model over the nonnegative integers in which P is interpreted as "$<$ and incongruent mod 2". ☐

The sort of models F can have when $n = 5$ is not known. Presumably, an analysis of cases similar to the foregoing would yield a characterization of such models, but the number of cases is multiplied enormously. Finally, solvability is open not only for the general class in which n is not restricted, but also for that subclass of Class 5.7 in which $n = 3$ and G may contain monadic predicate letters.

ISBN 0-201-02540-X

CHAPTER 6

Reductions

Recall that a *reduction* of a class $\mathfrak{C}_1$ to a class of $\mathfrak{C}_2$ is an effective function φ such that, for every schema F in $\mathfrak{C}_1$, $\varphi(F)$ is a schema in $\mathfrak{C}_2$ that has a model iff F has a model. A reduction φ of $\mathfrak{C}_1$ to $\mathfrak{C}_2$ is *conservative* iff, for every schema F in $\mathfrak{C}_1$, $\varphi(F)$ has a finite model iff F has a finite model. Naturally, $\mathfrak{C}_1$ is *reducible* to $\mathfrak{C}_2$ iff there exists a reduction of $\mathfrak{C}_1$ to $\mathfrak{C}_2$, and is *conservatively reducible* to $\mathfrak{C}_2$ iff there exists a conservative reduction of $\mathfrak{C}_1$ to $\mathfrak{C}_2$. We shall sometimes say, if a function φ is a reduction, that *F reduces to* $\varphi(F)$.

Thus if $\mathfrak{C}_1$ is reducible to $\mathfrak{C}_2$, then $\mathfrak{C}_1$ is solvable provided $\mathfrak{C}_2$ is solvable; and if $\mathfrak{C}_1$ is conservatively reducible to $\mathfrak{C}_2$, then $\mathfrak{C}_1$ is finitely controllable provided $\mathfrak{C}_2$ is finitely controllable, and $\mathfrak{C}_1$ is docile provided $\mathfrak{C}_2$ is docile.

In the first section of this chapter we formulate and apply a simple lemma, the Splitting Lemma, that allows certain types of complexity to be eliminated from the bases of schemata. In this way we prove the solvability and finite controllability of a Skolem class generalizing Class 2.7, as well as new classes of prenex schemata with arbitrary prefixes. Section 2 is concerned solely with Skolem classes: we sort Skolem classes into solvable and unsolvable by using both the results of §1 and a new reduction device. The latter yields unsolvability results by exploiting properties of tied *y*-variables. Moreover, it can be extended to schemata with arbitrary prefixes; in §3 we use it to treat various examples left unsupported in Chapter 2, §6. Finally, the last two sections contain two conservative reductions from the literature: one, due to Kalmár, eliminates initial *x*-variables; the other, whose basic idea stems from Wang, eliminates from ∀∃∀-schemata all but dyadic predicate letters.

1. The Splitting Lemma and Its Applications

The results of this section rest on a simple logical observation. An occurrence of a subformula in a formula F is *positive* iff the occurrence lies

Burton Dreben and Warren D. Goldfarb, The Decision Problem: Solvable Classes of Quantificational Formulas

ISBN 0-201-02540-X

within the scopes of an even number (possibly zero) of negation sign occurrences.

Splitting Lemma. Let F be a formula and let G be a subformula of F
that occurs only positively in F and that contains only one free variable v
(G may contain quantifiers). Let P be a monadic predicate letter foreign to
F; let F_1 be the result of substituting Pv for G in F, and let F_2 be $\forall v(Pv$
$\Rightarrow G)$. Then F and $F_1 \wedge F_2$ are satisfiable over the same universes.

Proof. Since all occurrences of G in F are positive, $F_1 \wedge F_2$ logically
implies F. Hence $F_1 \wedge F_2$ is satisfiable over a universe U only if F is.
Conversely, suppose F has a model $\mathfrak{M}$. Consider the following interpretation
of the predicate letter P over the universe of $\mathfrak{M}$: P is true of an element a
in the universe of $\mathfrak{M}$ iff G is true in $\mathfrak{M}$ when v, the only free variable of G,
takes the value a. Clearly the result of adjoining to $\mathfrak{M}$ this interpretation of
P is a model for $F_1 \wedge F_2$. □

The Splitting Lemma allows us to eliminate from F the *quantified* variables in G, and reintroduce them in a separate clause F_2. Thus these variables are "split" from the other variables of F. This sometimes makes it
possible to simplify the basis of the schema F.

Definition 6.1. Let S be a set of sets of variables. A variable v *cleaves*
S into sets S_1 and S_2 iff $S = S_1 \cup S_2$, $S_1 \cap S_2$ is empty, and v is the only
variable that occurs jointly in some set in S_1 and some set in S_2.

For example, if S is $\{\{y_1, y_2\}, \{y_2, y_3\}, \{y_1, x_1\}, \{y_2, x_1\}, \{y_2, x_2\}, \{y_3, x_2\}\}$, then y_2 cleaves S into $\{\{y_1, y_2\}, \{y_1, x_1\}, \{y_2, x_1\}\}$ and $\{\{y_2, y_3\}, \{y_2, x_2\}, \{y_3, x_2\}\}$.

Our first, and most important, application of the Splitting Lemma is to
Skolem schemata. If F is a Skolem schema whose basis is cleaved by a y-
variable y_p into sets S_1 and S_2, we shall reduce F to a conjunction of
Skolem schemata each of whose bases is either $S_1 \cup \{\{y_p\}\}$ or $S_2 \cup \{\{y_p\}\}$.
Consequently, we obtain the solvability of an extension of Class 2.7. Among
the schemata in this extension but not in Class 2.7 itself is, for example,
each schema $\forall y_1 \forall y_2 \forall y_3 \exists x_1 \exists x_2 F^M$ whose basis is the set S of the preceding
paragraph. (Thus we may have solvability even though $b(F)$ contains dis-
tinct but overlapping y-sets of cardinality >1.)

Class 6.1. The class of Skolem schemata $F = \forall y_1 \cdots \forall y_m \exists x_1 \cdots \exists x_n F^M$
fulfilling the following restriction is solvable and finitely controllable: for all
y-sets Y_1 and Y_2 in $b(F)$ either

(1)   for each x_n-term t, if $\hat{Y}_1[t]$ and $\hat{Y}_2[t]$ are distinct sets of terms of
cardinality >1, then no member of Y_1 is hereditarily t-united to any member
of Y_2; or

(2)   there is a y-variable that cleaves $b(F)$ into sets S_1 and S_2 such that
$Y_1 \in S_1$ and $Y_2 \in S_2$.

ISBN 0-201-02540-X

Proof. We show the solvability and finite controllability of a class superficially broader than Class 6.1, namely, the class of formulas that are conjunctions of schemata in Class 6.1. This class is only superficially broader, since any such conjunction can be transformed into a schema in Class 6.1 by rectifying and applying the usual prenexing rules. We devise a conservative reduction of this class to the class of formulas that are conjunctions of schemata in Class 2.7. Since, as we mentioned in Chapter 2, the latter class is in fact only superficially broader than Class 2.7, and since Class 2.7 is solvable and finitely controllable, this reduction yields the desired result.

The *rank* of a schema F in Class 6.1 is 0 if F lies in Class 2.7 and is the number of y-sets in $b(F)$ that contain more than one member if not. Given a conjunction $F_1 \wedge \cdots \wedge F_l$ of schemata in Class 6.1 at least one of which has positive rank, we shall effectively construct another conjunction of schemata in Class 6.1 which is satisfiable over the same universes as is $F_1 \wedge \cdots \wedge F_l$ and which is obtained from $F_1 \wedge \cdots \wedge F_l$ by replacing some conjunct of positive rank with a conjunction of schemata each of which has lower rank than the replaced conjunct. Iteration of this construction yields the reduction.

Suppose then that $F_1 \wedge \cdots \wedge F_l$ is a conjunction of schemata in Class 6.1, and that F_1 has positive rank. Let $F_1 = \forall y_1 \cdots \forall y_m \exists x_1 \cdots \exists x_n F_1^M$, and let Y_1 and Y_2 be distinct y-sets in $b(F_1)$ that fulfill (2) but not (1) of the basis restriction. Without loss of generality we may assume that for some $p \leq m$ and $q \leq n$ the y-variable y_p cleaves $b(F_1)$ into S_1 and S_2, $Y_1 \in S_1$, $Y_2 \in S_2$, every set in S_1 is a subset of $\{y_1, \ldots, y_p, x_1, \ldots, x_q\}$, and every set in S_2 is a subset of $\{y_p, y_{p+1}, \ldots, y_m, x_{q+1}, \ldots, x_n\}$. By transforming F_1^M into disjunctive normal form and shrinking the scopes of the existential quantifiers—using the equivalences $\exists x(J \vee K) \Leftrightarrow (\exists xJ \vee \exists xK)$ and, for J not containing x, $\exists x(J \wedge K) \Leftrightarrow (J \wedge \exists xK)$—we obtain a formula G_1 that is equivalent to F_1 and that has the form

$$\forall y_1 \cdots \forall y_m \left[\bigvee_{1 \leq i \leq k} (\exists x_1 \cdots \exists x_q J_i \wedge \exists x_{q+1} \cdots \exists x_n K_i) \right],$$

where each J_i contains only V-based atomic subformulas for V in S_1, and each K_i contains only V-based atomic subformulas for V in S_2. By transforming the subformula of G_1 that follows the universal quantifiers into conjunctive normal form (treating each subformula of G_1 that begins with an existential quantifier as a truth-functionally primitive part of the formula), and expanding the scopes of the existential quantifiers somewhat, using the equivalences above, we obtain a formula G_2 that is equivalent to F_1 and that has the form

$$\forall y_1 \cdots \forall y_m \left[\bigwedge_{1 \leq j \leq r} (\exists x_1 \cdots \exists x_q J_j' \vee \exists x_{q+1} \cdots \exists x_n K_j') \right],$$

ISBN 0-201-02540-X

where each J_j' is a disjunction of some of the subformulas J_i of F_1 and each K_j' is a disjunction of some of the K_i. Note that $y_{p+1},\ldots,y_m$ occur in G_2 only in the subformulas K_j'. Hence, by using the equivalences $\forall y(J \wedge K) \Leftrightarrow (\forall yJ \wedge \forall yK)$ and, for J not containing y, $\forall y(J \vee K) \Leftrightarrow (J \vee \forall yK)$, we obtain a formula G_3 equivalent to F_1 and having the form

$$\forall y_1 \cdots \forall y_p \left[\bigwedge_{1 \leq j \leq r} (\exists x_1 \cdots \exists x_q J_j' \vee \forall y_{p+1} \cdots \forall y_m \exists x_{q+1} \cdots \exists x_n K_j') \right].$$

Each of the subformulas $\forall y_{p+1} \cdots \forall y_m \exists x_{q+1} \cdots \exists x_n K_j'$ of G_3 occurs positively in G_3 and contains y_p as its only free variable; hence the Splitting Lemma is applicable to these subformulas. Let $P_1,\ldots,P_r$ be monadic predicate letters foreign to $F_1 \wedge \ldots \wedge F_l$. By r applications of the Splitting Lemma we obtain a formula G_4:

$$\forall y_1 \cdots \forall y_p \left[\bigwedge_{1 \leq j \leq r} (\exists x_1 \cdots \exists x_q J_j' \vee P_j y_p) \right] \wedge$$

$$\bigwedge_{1 \leq j \leq r} [\forall y_p (P_j y_p \Rightarrow \forall y_{p+1} \cdots \forall y_m \exists x_{q+1} \cdots \exists x_n K_j')]$$

such that $F_1 \wedge \cdots \wedge F_l$ and $G_4 \wedge F_2 \wedge \cdots \wedge F_l$ are satisfiable over the same universes. Moreover, G_4 is equivalent to the conjunction of the following $2r$ schemata: for each j, $1 \leq j \leq r$,

$$\forall y_1 \cdots \forall y_p \exists x_1 \cdots \exists x_q (J_j' \vee P_j y_p)$$

and

$$\forall y_p \forall y_{p+1} \cdots \forall y_m \exists x_{q+1} \cdots \exists x_n (P_j y_p \Rightarrow K_j').$$

Hence we shall be done if we show that each of these schemata is in Class 6.1 and has lower rank than F_1. But each atomic subformula of J_j' is V-based for some V in S_1, and each atomic subformula of K_j' is V-based for some V in S_2. Thus the basis of each of the first r schemata is $S_1 \cup \{\{y_p\}\}$, while that of each of the second r schemata is $S_2 \cup \{\{y_p\}\}$. Since S_1 and S_2 are subsets of $b(F_1)$, each of these $2r$ schemata is in Class 6.1. (Strictly speaking, the bound variables of the second r schemata would have to be relettered $y_1,\ldots,y_{m-p+1}, x_1,\ldots,x_{n-q}$.) Moreover, since $Y_1 \notin S_2$ and $Y_2 \notin S_1$, each of these schemata has lower rank than F_1. $\square$

In §2 we shall see more examples of schemata that fall within Class 6.1. Class 6.1 may be extended by allowing initial x-variables; an argument similar to that just given reduces this extended class to Initially-extended Class 2.7.

We now turn to a simpler application of the Splitting Lemma. Let F be a prenex schema with prefix $Q_1 v_1 \cdots Q_p v_p \forall w_1 \cdots \forall w_q$, where each Q_i is

ISBN 0-201-02540-X

either $\forall$ or $\exists$. Thus $w_1,\ldots,w_q$ are *final y-variables*. Suppose that, for some $k \leq p$, the only sets in $b(F)$ that contain any of the w_i are $\{w_1\},\ldots,\{w_q\}$, and $\{v_k, w_1,\ldots,w_q\}$. Then by putting the matrix of F into conjunctive normal form and shrinking the scopes of the universal quantifiers $\forall w_1,\ldots,\forall w_q$, we obtain a formula F_1 that is equivalent to F and that has the form

$$Q_1 v_1 \cdots Q_p v_q \left[\bigwedge_i (J_i \vee \forall w_1 \cdots \forall w_q K_i) \right],$$

where each formula K_i contains only the variables v_k, $w_1,\ldots,w_q$. Applying the Splitting Lemma and then expanding the scopes of some universal quantifiers, we obtain a formula F_2 that has the form

$$Q_1 v_1 \cdots Q_p v_q \left[\bigwedge_i (J_i \vee P_i v_k) \right] \wedge \bigwedge_i [\forall v_k \forall w_1 \cdots \forall w_q (P_i v_k \Rightarrow K_i)]$$

and that is satisfiable over the same universes as is F.

Similarly, let F have prefix $Q_1 v_1 \cdots Q_p v_p \exists w_1 \cdots \exists w_q$, and suppose that, for some $k \leq p$, the only sets in $b(F)$ that contain any of the w_i are $\{w_1\},\ldots,\{w_q\}$, and $\{v_k, w_1,\ldots,w_q\}$. Then by putting the matrix of F into disjunctive normal form, shrinking the scopes of the existential quantifiers $\exists w_1,\ldots,\exists w_q$ and applying the Splitting Lemma, we obtain a formula having the form

$$Q_1 v_1 \cdots Q_p v_p \left[\bigvee_i (J_i \wedge P_i v_k) \right] \wedge \bigwedge_i [\forall v_k (P_i v_k \Rightarrow \exists w_1 \cdots \exists w_q K_i)],$$

in which every atomic subformula of K_i is either $\{w_i\}$-based, $1 \leq i \leq q$, or else $\{v_k, w_1,\ldots,w_q\}$-based.

Clearly, these two reductions may be iterated. We obtain thereby an extension of Class 6.1.

Class 6.2. The class obtained from Class 6.1 by closing under the following operation is both solvable and finitely controllable: if a schema F is in the class, then so is every schema F' whose prefix is like that of F but for containing additional final variables $w_1,\ldots,w_q$, where either all the w_i are y-variables or else all the w_i are x-variables, and whose basis is like that of F but for containing, for some variable v of F, the sets $\{v, w_1,\ldots,w_q\}$, $\{w_1\},\ldots,\{w_q\}$.

Proof. By applying repeatedly the final y-variable reduction and the final x-variable reduction given above, we obtain from any schema F' in Class 6.2 a formula $G_1 \wedge G_2 \wedge \cdots \wedge G_r$ that is satisfiable over the same universes as is F', and such that

(1) G_1 is in Class 6.1;
(2) for each $i \geq 2$, G_i has either the form $\forall v \forall w_1 \cdots \forall w_q G$ or the form

$\forall v \exists w_1 \cdots \exists w_q G$, where G is quantifier free and contains only $\{v\}$-based, $\{w_i\}$-based, and $\{v, w_1, \ldots, w_q\}$-based atomic subformulas, $1 \leq i \leq q$.

It follows that $G_1 \wedge G_2 \wedge \cdots \wedge G_r$ has a rectified prenex equivalent that is in Class 6.1. $\square$

Class 6.2, of course, contains schemata not in Skolem form. A vivid subclass is the class of all prenex schemata $Q_1 v_1 \cdots Q_p v_p F^M$ whose bases contain, aside from one-membered sets, just each set $\{v_i, v_{i+1}\}$, $1 \leq i < p$. There are schemata in Class 6.2 that violate restrictions (I) and (II) of Chapter 2, §6: their bases may contain sets that are neither y-sets nor x-focused sets, and may contain y-sets whose members straddle x-variables. Finally, it should be clear that the final y-variable reduction and the final x-variable reduction also yield the solvability and finite controllability of that class obtained from Class 2.8 in just the way Class 6.2 is obtained from Class 6.1.

Class 6.2 includes the Surányi subclass shown finitely controllable by Dreben (1961), that is, the class of prenex schemata $\forall y_1 \forall y_2 \exists x \forall y_3 F^M$ whose bases include, aside from one-membered sets, just $\{y_1, x\}$, $\{y_2, x\}$, $\{y_1, y_2\}$, and $\{y_1, y_3\}$. Dreben's proof is essentially an amiability proof. Subsequently, Aanderaa showed (1966a) that this Surányi subclass is conservatively reducible to the Minimal Gödel–Kalmár–Schütte Class (Class 2.3) by a procedure that employs a special case of, and indeed furnished the inspiration for, the Splitting Lemma.

We do not know if Classes 6.1 and 6.2 are amenable. This is open even for the following subclass $\mathfrak{C}_1$ of 6.1: prenex schemata $\forall y_1 \forall y_2 \forall y_3 \exists x F^M$ whose bases include, aside from one-membered sets, just $\{y_1, y_2\}$, $\{y_2, y_3\}$, $\{y_1, x\}$, and $\{y_2, x\}$. But we can show the amenability of the following subclass $\mathfrak{C}_2$ of 6.2: the class obtained by starting with Class 2.4, rather than with Class 6.1, and then closing under the operation formulated in the specification of 6.2 restricted to the addition of final y-variables. (Class $\mathfrak{C}_2$ includes the Surányi subclass described in the preceding paragraph.) Every schema in $\mathfrak{C}_1$ is logically equivalent to a schema in $\mathfrak{C}_2$; hence if $\mathfrak{C}_1$ is in fact not amenable, then amenability is not preserved by a simple uniform type of quantificational equivalence.

2. Solvable and Unsolvable Skolem Classes

A *basis-set* S is a (finite) set of (finite) sets of variables among $y_1, \ldots, y_m$ and $x_1, \ldots, x_n$ for some m and n. If S is a basis-set, we let $B(S)$ be the class of Skolem schemata F such that

   (1)   every y-variable of F is a variable y_i in $\cup S$;

ISBN 0-201-02540-X

(2)   every x-variable of F is a variable x_k in $\cup S$;

(3)   every set of cardinality >1 in $b(F)$ is in S.‡

We wish to describe those basis-sets S for which $B(S)$ is solvable, and those for which $B(S)$ is unsolvable. (As we mentioned in Chapter 5, all known solvable classes specified just by prefix and basis are finitely controllable.)

Until further notice we restrict our attention to basis-sets S containing two-membered sets only. We saw in §4 of Chapter 2 that such sets may be represented by graphs: a small circular dot represents a variable, with solid black dots (*x-nodes*) for x-variables and unfilled dots (*y-nodes*) for y-variables; an edge connecting two dots represents a (two-membered) set in S. Thus each y-set in S is represented by a *y-edge*, an edge connecting two y-nodes; each x-focused set $\{x_j, x_k\}$ in S by an *x-edge*, an edge connecting two x-nodes; and each x-focused set $\{y_i, x_k\}$ by a *mixed edge*. If $S = \{\{y_1, y_2\}, \{y_1, x_1\}, \{y_2, x_1\}\}$, then S is represented in Figure 6.1.‖ Here $B(S)$ is just the Minimal Gödel–Kalmár–Schütte Class, Class 2.3; hence Figure 6.1 represents a solvable class.

Figure 6.1

An immediate result is this: if the graph of S contains no mixed edges, then $B(S)$ is solvable. For in this case S contains no set $\{y_i, x_k\}$. Hence any schema F in $B(S)$ can be transformed, by truth-functional manipulations and inverse-prenexing rules, into a schema G in which no y-variable governs any x-variable. That is, G lies in Class 1.1.

---

‡ Until now we have been speaking of Skolem schemata as having prefixes $\forall y_1 \cdots \forall y_m \exists x_1 \cdots \exists x_n$ for some m and n. But for the rest of this chapter we allow Skolem schemata to have prefixes $\forall y_{i_1} \cdots \forall y_{i_p} \exists x_{k_1} \cdots \exists x_{k_q}$ for any subscripts $i_1 < i_2 < \cdots < i_p$ and $k_1 < k_2 < \cdots k_q$. This, together with our stricture against vacuous quantification, makes it possible, given a quantifier-free formula G containing variables y_i and x_k, to speak of *the* Skolem schema with matrix G. (In so liberalizing our terminology, we also understand that the specifications of our solvable classes are liberalized to allow schemata obtained from those with prefixes $\forall y_1 \cdots \forall y_m \exists x_1 \cdots \exists x_n$ by relettering variables.)

‖ Thus every basis-set S is represented by a unique graph. But since we do not label the nodes of a graph, the graph of S represents infinitely many other basis-sets, namely, all those obtainable from S by relettering variables (by relettering we mean, of course, that y-variables are relettered as y-variables and x-variables as x-variables). Since these basis-sets differ only trivially, we speak freely of *the* basis-set represented by a given graph.

ISBN 0-201-02540-X

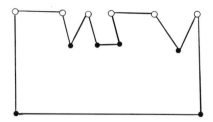

Figure 6.2

As we noted in §4 of Chapter 2, if the graph of S is as in Figure 6.2, then $B(S)$ is a subclass of Class 2.6 and hence is solvable. More generally, for each basis-set S that contains two-membered sets only, $B(S)$ is a subclass of Class 2.6 iff the graph of S possesses the following properties: no y-node lies on two y-edges; and no two y-nodes that lie on different y-edges are connected by a path consisting entirely of mixed edges.

The solvability of Class 6.1 yields additional graphs of solvable classes. For example, we shall soon see that if a graph contains no cycles, then it represents a solvable class. Thus we obtain the solvability of $B(S)$, where the graph of S is as in Figure 6.3. Here S is $\{\{y_1, y_2\}, \{y_2, y_3\}, \{y_3, x_1\}, \{y_4, x_1\}, \{y_4, y_5\}, \{y_5, y_6\}, \{y_6, x_2\}\}$. Note that each y-variable but y_1 cleaves S into two sets.

Figure 6.3

To describe Class 6.1 more fully, we introduce some terminology. Let S be a basis-set and suppose y_i and y_j are in $\cup S$. The basis-set S' obtained from S by *collapsing* y_j into y_i is the set $\{\sigma[V] | V \in S\}$, where σ is the mapping on variables that carries y_j to y_i and carries every other variable to itself. The operation on graphs that corresponds to collapsing is clear: we identify two y-nodes (and omit the y-edge, if any, that connects them). If S_1 and S_2 are basis-sets, then we say a basis-set S' is obtained by *merging* S_1 and S_2 iff S' can be obtained thus:

(1)   the variables in $\cup S_2$ are relettered in such a way that $\cup S_1$ and $\cup S_2$ become disjoint;

(2) S' is the basis-set obtained from the union of S_1 with relettered S_2 by collapsing a y-variable from $\cup S_1$ into a y-variable from $\cup S_2$.

Note that if S' is obtained by merging S_1 and S_2, then some y-variable cleaves S' into S_1 and a relettered S_2. Conversely, if a basis-set S' is cleaved by a y-variable into sets S_1 and S_2, then S' may be obtained by merging S_1 and S_2.

ISBN 0-201-02540-X

A graph obtained by merging two disjoint graphs is simply any graph that results from identifying some y-node in one graph with some y-node in the other graph.

We then have: for all basis-sets S that contain two-membered sets only, $B(S)$ is a subclass of Class 6.1 iff S can be obtained by iterated mergings from basis-sets S_i such that each $B(S_i)$ is a subclass of Class 2.6. This characterization follows from the relation between cleaving and merging mentioned above, together with two easily proved facts. First, if $B(S_1)$ and $B(S_2)$ are subclasses of Class 6.1, and if S' is a basis-set obtained by merging S_1 and S_2, then $B(S')$ is also a subclass of Class 6.1. Second, since we are here concerned just with schemata whose bases contain only one-membered and two-membered sets, clause (1) of the specification of Class 6.1 may be replaced by the following: if Y_1 and Y_2 are distinct y-sets of cardinality >1, then no member of Y_1 is hereditarily tied to any member of Y_2 (see page 60).

This characterization yields the solvability of $B(S)$ when the graph of S contains no cycles. For example, the graph in Figure 6.3 can be obtained by merging ○──○ with ○─○─●─○, merging the result with ○──○, and finally merging the latter result with ○─○─●, where each of these subgraphs represent (quite simple) subclasses of Class 2.6.

A more complex subclass of Class 6.1 is given in Figure 6.4: graphs (a) and (b) represent subclasses of Class 2.6; graph (c), being cycle free, represents a subclass of Class 6.1; and graph (d) is obtained by merging (a) and (b), and then merging the result with (c). Thus (d) represents a subclass of Class 6.1.

We start our discussion of unsolvability by describing two minimal unsolvable Skolem classes. We then obtain additional unsolvable classes by

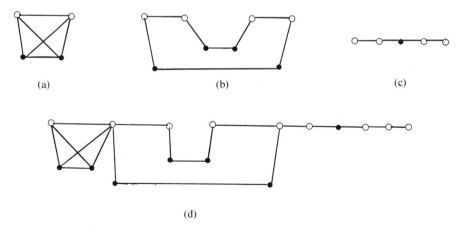

(a)   (b)   (c)

(d)

Figure 6.4

ISBN 0-201-02540-X

two reduction procedures. First, however, we present an unsolvable class
of nonprenex schemata, which is the source of many unsolvability results.

Basic Unsolvable Class.   The following class is unsolvable: schemata
$\forall y \exists x R y x \land \forall y_1 \forall y_2 \forall y_3 F_1$, where R is a dyadic predicate letter, F_1 is
quantifier-free and contains dyadic predicate letters only, and each atomic
subformula of F_1 is either $\{y_1, y_2\}$-based, $\{y_2, y_3\}$-based, or $\{y_1, y_3\}$-based.

Proof.   By means of the reduction we mentioned on page 135, the unsolvability of this class follows from that of the class of $\forall\exists\forall$-schemata
containing no essentially monadic atomic subformulas.   □

From the Basic Unsolvable Class we obtain at once the unsolvability of
$B(S)$ whenever $S = \{\{y_1, y_2\}, \{y_2, y_3\}, \{y_1, y_3\}, \{y_i, x_1\}\}$, where i is any
number: every schema in the Basic Unsolvable Class has a prenex equiv-
alent in $B(S)$. The graph of S is given in Figure 6.5(a) if $i > 3$, and in Figure
6.5(b) if $i \leq 3$.

(a)   (b)

Figure 6.5

The class $B(S)$ is also unsolvable if the graph of S is as in Figure 6.6.
The proof when $\cup S$ contains one x-variable is in Goldfarb and Lewis (1975);
the proof when $\cup S$ contains more than one in Goldfarb (1975). (See also
Lewis (1979). These results are obtained by encoding the unconstrained
domino problem.)

Unsolvable classes $B(S)$ where S has more complex types of cycles can
be obtained by two reduction procecures. Recall that if $B(S')$ reduces to
$B(S)$ and $B(S')$ is unsolvable, then $B(S)$ is unsolvable. (The reductions
below have an opposite point—that of proving *unsolvability*—to those of
§1.)

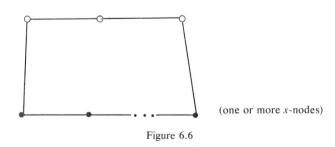

(one or more x-nodes)

Figure 6.6

ISBN 0-201-02540-X

Collapsing Reduction 1.   Let S be a basis-set containing a two-membered y-set $\{y_i, y_j\}$. Let S' be the set obtained from S by collapsing y_j into y_i. Then $B(S')$ is reducible to $B(S)$.

Proof.   Let F be an arbitrary schema in $B(S')$ with matrix F^M. We construct a schema G in $B(S)$ that has a model iff F has a model. By the specification of S there is a quantifier-free formula G_1 such that $G_1(y_j/y_i)$ $= F^M$ and every atomic subformula of G_1 is either V-based for some V in S or else essentially monadic. Let P be a dyadic predicate letter foreign to F, and let G be the Skolem schema with matrix

$$Py_iy_i \wedge (Py_jy_i \Rightarrow G_1).$$

Then G is in $B(S)$ and G logically implies F. Moreover, any model for F becomes one for G if the new predicate letter P is interpreted as identity. $\square$

From the unsolvability of the classes represented in Figures 6.5, we obtain by iterated application of Collapsing Reduction 1 the unsolvability of $B(S)$ whenever the graph of S contains a cycle consisting entirely of y-edges, provided that the graph also contains at least one mixed edge. And from the unsolvability of the classes represented in Figure 6.6 we obtain the unsolvability of each class that can be represented as in Figure 6.7.

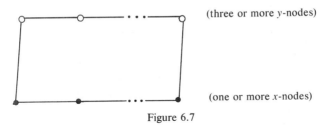

(three or more y-nodes)

(one or more x-nodes)

Figure 6.7

Collapsing Reduction 2.   Let S be a basis-set that contains, for some y_i, y_j, and x_k, the sets $\{y_i, x_k\}$ and $\{y_j, x_k\}$. Let S' be the set obtained from S by collapsing y_j into y_i. Then $B(S')$ reduces to $B(S)$.

Proof.   Let F be an arbitrary schema in $B(S')$, with matrix F^M. Let G_1 be as in the proof of Collapsing Reduction 1, and let Q be a dyadic predicate letter foreign to F. Finally, let G be the Skolem schema with matrix

$$Qy_ix_k \wedge (Qy_jx_k \Rightarrow G_1).$$

Then G is in $B(S)$ and G logically implies F. Now suppose that F has a model. We first transform F into a schema F' that also has a model and that has the same prefix as G. This is easily done: we may let F' be the Skolem schema with matrix $F^M \wedge (Py_j \vee -Py_j)$,‡ where P is any 1-place

---

‡ The point of conjoining F^M with a tautologous formula containing y_j is merely to ensure that y_j is not vacuously quantified in F'.

ISBN 0-201-02540-X

predicate letter. Then $D(F', \omega) = D(G, \omega)$. Let $\mathfrak{A}$ be a truth-assignment on $\hat{L}(F', \omega)$ that verifies $E(F', \omega)$. We define a truth-assignment $\mathfrak{B}$ on $\hat{L}(G, \omega)$ thus: $\mathfrak{B}$ agrees with $\mathfrak{A}$ on all atoms containing predicate letters of F, and $\mathfrak{B}$ verifies an atom Qst iff $s = \hat{y}_i t$. Let H be an arbitrary Herbrand instance of G, and let t be the substituent in H for x_k. It suffices to show that $\mathfrak{B}$ verifies H. Now H has the form

$$Q\hat{y}_i t \, t \wedge (Q\hat{y}_j t \, t \Rightarrow J).$$

By definition $\mathfrak{B}$ verifies the first conjunct in H, and $\mathfrak{B}$ verifies the antecedent of the second conjunct iff $\hat{y}_i t = \hat{y}_j t$. But if $\hat{y}_i t = \hat{y}_j t$, then there is an Herbrand instance of F' having the form $J \wedge (P\hat{y}_j t \vee -P\hat{y}_j t)$; hence $\mathfrak{A}$ verifies J. Since $\mathfrak{A}$ and $\mathfrak{B}$ agree on all subatoms of J, we have shown that $\mathfrak{B}$ verifies H. □

Collapsing Reduction 2 yields the unsolvability of $B(S)$ whenever the graph of S contains a cycle that includes at least two y-edges and at least one mixed edge and that consists of at most one run of x-edges, followed by any combination of y-edges and mixed edges that terminates at the beginning of the run of x-edges. For such a cycle can be collapsed into either a cycle consisting entirely of y-edges or else a cycle of the sort depicted in Figure 6.7. An example of such an unsolvable $B(S)$ is given in Figure 6.8.

Collapsing Reduction 2 makes clear the importance of tied y-variables. Recall that y_i and y_j are tied (relative to a schema F whose basis contains only one- and two-membered sets) iff $\{y_i, x_k\}$ and $\{y_j, x_k\}$ are in $b(F)$ for some x_k. The reduction shows that tied y-variables can function as though they were the same variable; and hence two y-sets that contain hereditarily tied y-variables can function as though they overlapped. The reduction turns on the fact that atomic subformulas $Qy_i x_k$ and $Qy_j x_k$ are coinstantiated in $\langle H, H \rangle$ whenever H is an Herbrand instance in which the substituents for y_i and y_j are identical. In short, the reduction exploits the simultaneous coinstantiability of $Qy_i x_k$ and $Qy_j x_k$. Thus the concerns we had in §4 of Chapter 2 with simultaneously coinstantiable x-focused atomic

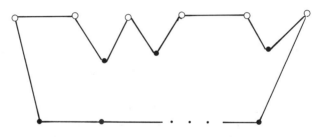

Figure 6.8

ISBN 0-201-02540-X

subformulas, and with the identity preservation requirement on term-mappings, concerns which led to the formulation of the notion of tying, were not adventitious.

The role of tied y-variables is further emphasized by contrasting Figure 6.8 with Figure 6.2. The former represents an unsolvable class, while the latter represents a solvable one.

Collapsing Reduction 2 shows that the presence of any two distinct y-sets of cardinality 2 in a basis-set S leads to the unsolvability of $B(S)$ if all x-focused subsets of $\cup S$ are in S. This result was first shown by Goldfarb and Lewis (1975).

The decision problem for $B(S)$ is open for many basis-sets S that fail to meet the specification of Class 6.1 but nonetheless cannot be shown unsolvable by the two Collapsing Reductions. For example, suppose $\{y_1, y_2\}$ and $\{y_2, y_3\}$ are in S, but the only other path connecting the nodes representing y_1 and y_3 in the graph of S goes by way of several x-nodes and several y-nodes. A graph of a simple set S of this type is given in Figure 6.9. Here we have a cycle; but Collapsing Reduction 2 is not applicable since the y-variable represented by the node in the center is tied to no other y-variable; it is "insulated" from every other y-variable by an x-edge.

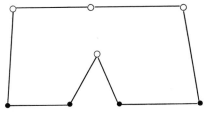

Figure 6.9

Another open problem arises when $\{y_1, y_2\}$ and $\{y_3, y_4\}$ are in S, y_2 and y_3 are tied, but the only path between the nodes representing y_1 and y_4 in the graph of S goes by way of either y_3 or y_2. This is illustrated in Figure 6.10(a). Were we to collapse the two tied y-variables by means of Collapsing Reduction 2 we would obtain Figure 6.10(b). But the latter is a graph that represents a class included in Class 6.1, and hence a solvable class: for it can be obtained by merging two graphs each of which represents a subclass of Class 2.6 (indeed, each represents a subclass of Class 2.5, the Gödel–Kalmár–Schütte Class, since each contains only two y-nodes).

Other open problems are essentially more complicated versions of the two just exhibited. We do not believe, however, that they are intractable: we would hope that the methods presented in Chapter 2 and in this chapter will eventually yield a complete demarcation of those S that give rise to solvable classes $B(S)$ from those that give rise to unsolvable $B(S)$.

ISBN 0-201-02540-X

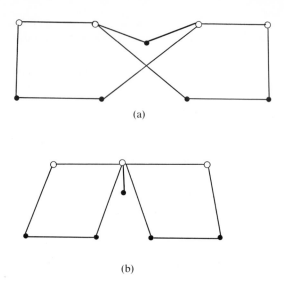

(a)

(b)

Figure 6.10

We now turn our attention to basis-sets containing k-membered sets of variables for $k \geq 2$. The situation here, naturally, is more complex. It is difficult even to describe these sets perspicuously, since the graphical representation used until this point is not adequate. However, using the operations of collapsing and merging, we may characterize the subclasses of Class 6.1 in a manner analogous to that of the more restricted case: for all basis-sets S, $B(S)$ is a subclass of Class 6.1 iff S can be obtained by iterated mergings from basis-sets S_i such that each $B(S_i)$ is a subclass of Class 2.7.

Once k-membered sets occur in basis-sets for $k \geq 2$, the size of these sets can play a role in the solvability of $B(S)$.

Class 6.3. Let S be a basis-set such that the maximum cardinality of the y-sets in S is k, and each x-focused set in S contains at least $k + 1$ variables, of which at least k are x-variables. Then $B(S)$ is solvable.

Proof. Let F be in $B(S)$. If $\langle F \rangle_i$ is Y-based and $\langle F \rangle_j$ is X-based, where Y is a y-set in S and X is an x-focused set in S, then $\langle F \rangle_i$ and $\langle F \rangle_j$ are not coinstantiable. Indeed, for every Herbrand instance H of F, $\langle H \rangle_i$ contains at most k distinct terms and $\langle H \rangle_j$ contains at least $k + 1$ distinct terms. Thus F is in the solvable Skolem class mentioned on page 43. $\square$

Our first new unsolvability result concerns basis-sets containing two y-sets of cardinality ≥ 2, one of which is a proper subset of the other. We shall show that such basis-sets almost always give rise to unsolvable classes. In order to make the strategy of the proof clear, we begin with a simple case. Then we give the general result.

ISBN 0-201-02540-X

Unsolvability Result I. If $S = \{\{y_1, y_2\}, \{y_1, y_2, y_3\}, \{y_i, x_k\}\}$ for any numbers i and k, then $B(S)$ is unsolvable.

Proof. We reduce the Basic Unsolvable Class of page 150 to $B(S)$. Let $F = \forall y \exists x Ryx \wedge \forall y_1 \forall y_2 \forall y_3 F_1$ be any schema in the Basic Unsolvable Class. For each (dyadic) predicate letter P of F let Q_P be a new triadic predicate letter. Let G_1 result from F_1 by replacing atomic subformulas thus:

$$Py_2y_3 \quad \text{by} \quad Q_Py_2y_3y_1$$

$$Py_3y_2 \quad \text{by} \quad Q_Py_3y_2y_1$$

$$Py_1y_3 \quad \text{by} \quad Q_Py_1y_3y_2$$

$$Py_3y_1 \quad \text{by} \quad Q_Py_3y_1y_2.$$

Thus each atomic subformula of G_1 is either $\{y_1, y_2\}$-based or $\{y_1, y_2, y_3\}$-based. Let G_2 be the conjunction of all biconditionals $Py_1y_2 \Leftrightarrow Q_Py_1y_2y_3$. Finally, let G be

$$\forall y \exists x \, Ryx \wedge \forall y_1 \forall y_2 \forall y_3 (G_1 \wedge G_2).$$

Clearly G has a prenex equivalent in $B(S)$. It suffices to show that F has a model iff G has a model.

The schema G logically implies

$$\forall y_1 \forall y_2 \forall y_3 (Py_1y_2 \Leftrightarrow Q_Py_1y_2y_3)$$

for each predicate letter P of F. Hence G logically implies $(G_1 \Rightarrow F_1)$. Consequently, G logically implies F. Conversely, any model for F can be transformed into a model for G (with the same universe) by interpreting each new predicate letter Q_P as true of just those triples $\langle a_1, a_2, a_3 \rangle$ from the universe such that P is true in the given model of the pair $\langle a_1, a_2 \rangle$. $\square$

The proof just given rests on the incorporation into G of clauses that assure the following: the truth of an atomic formula containing a predicate letter Q_P depends only on the first two arguments of the atomic formula. The power of $\{y_2, y_3\}$-based and $\{y_1, y_3\}$-based atomic formulas may then be captured by $\{y_1, y_2, y_3\}$-based atomic formulas. We now apply a more elaborate form of this strategy.

Unsolvability Result II. Let $S = \{\{y_1,\ldots,y_p\}, \{y_1,\ldots,y_q\}, X\}$, where $2 \le p < q$ and X is an x-focused set containing at least one y-variable and at most $q - 1$ x-variables. Then $B(S)$ is unsolvable.

Proof. We consider first the special case in which X contains only one y-variable. Suppose X contains k x-variables, $k \le q - 1$. As before, let F be the schema $\forall y \exists x Ryx \wedge \forall y_1 \forall y_2 \forall y_3 F_1$ in the Basic Unsolvable Class. For each predicate letter P of F let Q_P be a new and distinct q-place

ISBN 0-201-02540-X

predicate letter. Let G_1 result from F_1 by replacing atomic subformula Py_iy_j (where $i \neq j$) with $Q_Py_iy_jv_1\cdots v_{q-2}$, where the v_k are so chosen that $\{y_i, y_j, v_1, \ldots, v_{q-2}\} = \{y_1, \ldots, y_q\}$. Let G_2 be the conjunction of all biconditionals

$$Q_Py_1\cdots y_q \Leftrightarrow Q_Py_1y_2y_qy_3\cdots y_{q-1}$$

and all biconditionals

$$Q_Py_1\cdots y_py_2\cdots y_2 \Leftrightarrow Q_Py_1\cdots y_q,$$

where y_2 occupies the last $q - p$ argument places in the atomic formula on the left side. Finally, let G be

$$\forall y \exists x_1\cdots \exists x_k Q_Ryx_1\cdots x_kx_k\cdots x_k \wedge \forall y_1\cdots \forall y_p(G_1 \wedge G_2),$$

where x_k occupies the last $q - k$ argument places in the atomic formula in the first conjunct. Then G has a prenex equivalent in $B(S)$.

Any model for F can be transformed into a model for G by interpreting each predicate letter Q_P as true of just those q-tuples $\langle a_1, \ldots, a_q \rangle$ from the universe such that P is true in the given model of the pair $\langle a_1, a_2 \rangle$. Conversely, by use of the clauses in G_2 we have that G logically implies

$$\forall y_1\cdots \forall y_q(Q_Py_1y_2y_2\cdots y_2 \Leftrightarrow Q_Py_1\cdots y_q)$$

for each predicate letter P of F. Thus again Q_P depends only on the first two arguments. Hence, given a model for G, we obtain a model for F by interpreting each predicate letter P of F as true of those pairs $\langle a_1, a_2 \rangle$ such that Q_P is true in the given model of $\langle a_1, a_2, a_2, \ldots, a_2 \rangle$. This concludes the proof for the special case.

To handle the general case, in which X contains more than one y-variable, we apply the procedure of the Generalized Collapsing Reduction 1 given three paragraphs below. That is, we may reduce the schema G just constructed to a schema G' that differs from G by containing more y-variables in its first conjunct. This suffices for the result.   □

The restrictions on the x-focused set X given in Unsolvability Result II are essential: if X fails to fulfill them, then $B(S)$ is solvable. For if X contains no y-variables, then each schema in $B(S)$ has an equivalent in which no y-variable governs any x-variable. And if X contains q or more x-variables as well as at least one y-variable, then $B(S)$ falls within Class 6.3.

We now give a generalized version of Collapsing Reduction 1; this enables us to carry out the final part of the proof for Unsolvability Result II, as well as to derive the unsolvability of additional classes $B(S)$.

Generalized Collapsing Reduction 1.   Let S be a basis-set, V a set in S, and y_i and y_j distinct y-variables in V. Let S' be the basis-set obtained from S by collapsing y_j into y_i. Then $B(S')$ is reducible to $B(S)$.

ISBN 0-201-02540-X

Proof. Let F be a schema in $B(S')$. We make the following special assumption on F: every atomic subformula of F that contains y_i contains at least two occurrences of y_i. This assumption entails no loss of generality. For if a given schema in $B(S')$ fails to possess the property, we may transform the schema thus: for each k-place predicate letter P let Q_P be a new $2k$-place predicate letter, and then replace each atomic subformula $Pv_1 \cdots v_k$ of the given schema by $Q_P v_1 \cdots v_k v_1 \cdots v_k$. Clearly this transformation preserves satisfiability.

There is then a quantifier-free formula G_1 such that $G_1(y_j/y_i) = F^M$ and every atomic subformula of G_1 is either essentially monadic or else W-based for some W in S. (The special assumption on F is needed to ensure the existence of G_1. For suppose W is a member of S that contains both y_i and y_j. Then in order to be W-based on atomic subformula $\langle G_1 \rangle_k$ of G_1 might have to be obtained from $\langle F \rangle_k$ by replacing at least one but not all occurrences of y_i by occurrences of y_j.) Let $\langle v_1, \ldots, v_r \rangle$ be an arrangement of $V - \{y_i\}$, and let R be a new $(r + 2)$-place predicate letter. Finally, let G be the Skolem schema with matrix

$$Ry_i y_i v_1 \cdots v_r \wedge (Ry_j y_i v_1 \cdots v_r \Rightarrow G_1).$$

Then G is in $B(S)$, G logically implies F, and any model for F becomes one for G when R is interpreted as true of just those $(r + 2)$-tuples whose first two members are identical. $\square$

Unsolvability Result III. Let $S = \{Y_1, Y_2, X\}$, where Y_1 and Y_2 are distinct y-sets whose intersection has cardinality ≥ 2, and X is an x-focused set containing at least one y-variable and at most $q - 1$ x-variables, where q is the larger of $|Y_1|$ and $|Y_2|$. Then $B(S)$ is unsolvable.

Proof. We suppose that $q = |Y_2| \geq |Y_1|$. Let y_i be the earliest variable in $Y_1 \cap Y_2$, and let S' be the basis-set obtained from S by collapsing y_j into y_i for each y_j in $Y_1 - Y_2$. By Generalized Collapsing Reduction 1, $B(S')$ is reducible to $B(S)$. Moreover, $S' = \{Y_1 \cap Y_2, Y_2, X'\}$, where X' possibly differs from X by having fewer y-variables. Hence $B(S')$ falls under Unsolvability Result II (up to relettering of variables), and $B(S)$ is therefore unsolvable. $\square$

A generalized version of Collapsing Reduction 2 allows us to obtain unsolvability by exploiting y-variables that are t-united for some term t, in much the same way that Collapsing Reduction 2 allows us to exploit tied y-variables.

Generalized Collapsing Reduction 2. Let S be a basis-set and let X_1 and X_2 be x-focused sets in S that share the same x-variables. Let σ be a mapping from $X_1 \cup X_2$ to X_1 such that

(1) σ carries each x-variable to itself;

ISBN 0-201-02540-X

(2) $\sigma[X_1] = \sigma[X_2]$;

(3) $\sigma(\sigma(y_i)) = \sigma(y_i)$ for each y-variable y_i in $X_1 \cup X_2$.

Finally, let S' be the basis-set obtained from S by collapsing y_i into $\sigma(y_i)$ for each y_i in $X_1 \cup X_2$. Then $B(S')$ is reducible to $B(S)$.

Proof.   Let F be a schema in $B(S')$; we assume that each atomic subformula of F that contains a y-variable in $\sigma[X_1]$ contains at least l occurrences of that y-variable, where l is the number of y-variables in $X_1 \cup X_2$. As before, this assumption entails no loss of generality. There is then a quantifier-free formula G_1 such that F^M results from G_1 by replacing each y-variable in $X_1 \cup X_2$ with its image under σ (by property (3) of σ, it makes no difference whether these replacements are done simultaneously or sequentially), and such that every atomic subformula of G_1 is either essentially monadic or else V-based for some V in S.

By the specification of σ there are arrangements $\langle v_1,...,v_p \rangle$ and $\langle w_1,...,w_p \rangle$ of X_1 and X_2 such that

(a) $\sigma(v_k) = \sigma(w_k)$ for each k;

(b) for each y-variable y_i in X_2 there is a k such that $y_i = w_k$ and $\sigma(y_i) = v_k$.

Let Q be a p-place predicate letter foreign to F, and finally let G be the Skolem schema with matrix

$$Qv_1 \cdots v_p \wedge (Qw_1 \cdots w_p \Rightarrow G_1).$$

Then G is in $B(S)$. If in the matrix of G we replace each y-variable in $X_1 \cup X_2$ with its image under σ, then, by (a) and the construction of G_1, we obtain a formula of the form $A \wedge (A \Rightarrow F^M)$. Hence G logically implies F. Conversely, suppose F has a model. Let F' be a schema whose prefix is the same as that of G, and whose matrix is $F^M \wedge N$, where N is some tautologous formula containing all the y-variables in $\cup S - \cup S'$. Then F' has a model, and $D(F', \omega) = D(G, \omega)$. Let $\mathfrak{A}$ be any truth-assignment on $\bar{L}(F', \omega)$ that verifies $E(F', \omega)$. We extend $\mathfrak{A}$ to a truth-assignment $\mathfrak{B}$ on $\bar{L}(G, \omega)$ thus: $\mathfrak{B}$ verifies an atom $Qs_1...s_p$ iff $\langle s_1,...,s_p \rangle = \langle \hat{v}_1 t,...,\hat{v}_p t \rangle$ for some term t of height >1 in $D(G, \omega)$. Let H be an arbitrary Herbrand instance of G, and let t be the substituent in H for some fixed x-variable x_k. We show that $\mathfrak{B}$ verifies H. Now H has the form

$$Q\hat{v}_1 t \cdots \hat{v}_p t \wedge (Q\hat{w}_1 t \cdots \hat{w}_p t \Rightarrow J).$$

Hence $\mathfrak{B}$ verifies the first conjunct in H, and verifies the antecedent of the second conjunct iff $\langle \hat{v}_1 t,...,\hat{v}_p t \rangle = \langle \hat{w}_1 t,...,\hat{w}_p t \rangle$. But by (a) and (b), if $\langle \hat{v}_1 t,...,\hat{v}_p t \rangle = \langle \hat{w}_1 t,...,\hat{w}_p t \rangle$, then $\hat{y}_i t = \hat{y}_j t$, where $y_j = \sigma(y_i)$, for each y-variable y_i in $X_1 \cup X_2$. Hence there is an Herbrand instance of F' having the form $J \wedge N'$ (where N' is tautologous), namely, the Herbrand instance

ISBN 0-201-02540-X

in which t is the substituent for x_k. Consequently, $\mathfrak{A}$ verifies J. And then $\mathfrak{B}$ verifies J too; whence $\mathfrak{B}$ verifies H. □

Corollary. Let S be a basis-set, let F be a schema in $B(S)$, and let t be a term of height >1 in $D(F, \omega)$. For each y-variable y_i of F let $\mu(y_i)$ be the earliest y-variable of F such that y_i and $\mu(y_i)$ are hereditarily t-united. Finally, let S' be the basis-set obtained from S by collapsing y_i into $\mu(y_i)$ for each y-variable y_i in F. Then $B(S')$ is reducible to $B(S)$.

Proof. Any such mapping μ is the composition of several mappings σ which meet the specification given in Generalized Collapsing Reduction 2. □

By means of the Corollary we may extend Unsolvability Result III in the obvious manner. We have thus shown that if S contains two distinct y-sets whose intersection has cardinality ≥ 2 or if S contains two y-sets that can, by means of hereditary t-unitedness, simulate such an intersecting pair of y-sets, then $B(S)$ is unsolvable, provided that S contains one x-focused set of appropriate size. On the other hand, if all y-sets in S are disjoint, and this disjointness is never compromised by hereditary t-unitedness then, of course, $B(S)$ is solvable, since it falls within Class 2.7. We are left then with cases of overlap in which the intersection of two y-sets has cardinality 1. We shall not go into any detail about these. Many analogues to results we gave concerning basis-sets containing only two-membered sets can be obtained; but there are additional open problems, due to considerations of the size of x-focused sets.

We note finally that all the results of this section are unaffected by the presence or absence of essentially monadic atomic formulas. We may restrict the definition of the class $B(S)$ so as to require that $b(F) = S$ for each schema F in $B(S)$: then all basis-sets S that give rise to solvable classes under the old definition of $B(S)$ continue to do so (since under the new definition the class $B(S)$ is smaller); and similarly for unsolvability. (The latter is shown in part by the absence of essentially monadic atomic subformulas from the Basic Unsolvable Class.) We may also expand the definition of $B(S)$ to allow schemata F containing variables aside from those in $\cup S$, provided that these other variables occur solely in essentially monadic atomic subformulas. (Under this expanded definition, schemata in $B(S)$ may have arbitrarily long prefixes.) Again, this affects neither solvability nor unsolvability.

3. Arbitrary Prefixes

In §6 of Chapter 2 we adduced three examples of unsolvable classes of normal schemata that illustrate the difficulties caused by nonfinal x-variables. Here we show these classes unsolvable; the reductions employed in

ISBN 0-201-02540-X

our proofs underscore the role of the nonfinal x-variables. The first two classes in question were

Class A. Prenex schemata with prefixes $\forall y_1 \forall y_2 \exists x_1 \forall y_3 \exists x_2$ and bases $\{\{y_1, y_2, y_3\}, \{x_1\}, \{y_3, x_2\}\}$.

Class B. Prenex schemata with prefixes $\forall y_1 \forall y_2 \exists x_1 \forall y_3 \forall y_4 \exists x_2$ and bases $\{\{y_3, y_4\}, \{x_1\}, \{y_1, x_2\}, \{y_2, x_2\}, \{y_3, x_2\}, \{y_4, x_2\}\}$.

In schemata in these classes the nonfinal x-variable x_1 occurs solely in essentially monadic atomic formulas. Only the position of x_1 plays a role, not any relations between x_1 and other variables that might be encoded by atomic formulas. Indeed, solvable classes result if in the specifications of Classes A and B we either omit x_1, or transform it into a y-variable, or transport it to the end of the prefix.

The presence of nonfinal x-variables like x_1 occasions trouble because any given x_1-term s is the substituent for x_1 in an infinite number of Herbrand instances—in our classes, in all Herbrand instances in which $\hat{y}_1 s$ and $\hat{y}_2 s$ are the substituents for y_1 and y_2. But then "monadic information" about s, information that can be gleaned in one Herbrand instance in which s is the substituent for x_1, can be conveyed to all Herbrand instances in which s is the substituent for x_1. This information can consequently be used to encode information about $\hat{y}_1 s$ and $\hat{y}_2 s$. This idea lies at the root of our two unsolvability proofs, of which the first is similar to that for Unsolvability Result I and of which the second is similar to that for Collapsing Reduction 2.

Unsolvability Proof for Class A. Let F be a schema $\forall y \exists x R y x \wedge \forall y_1 \forall y_2 \forall y_3 F_1$ in the Basic Unsolvable Class. It suffices to construct a schema in Class A that has a model iff F has a model. For each (dyadic) predicate letter P of F let Q_P be a new and distinct triadic predicate letter and let $Q_P{}'$ be a new and distinct monadic predicate letter. Let G_1 result from F_1 by replacing each atomic subformula $P y_i y_j$ (where $i \neq j$) with $Q_P y_i y_j y_k$, where k is the unique number such that $\{i, j, k\} = \{1, 2, 3\}$. Let G_2 be the conjunction of all biconditionals

$$Q_P y_1 y_2 y_3 \Leftrightarrow Q_P{}' x_1.$$

Finally, let G be $\forall y \exists x Q_R y x x \wedge \forall y_1 \forall y_2 \exists x_1 \forall y_3 (G_1 \wedge G_2)$. Then G has a prenex equivalent in Class A. Now G logically implies $\forall y_1 \forall y_2 \exists x_1 \forall y_3 (Q_P y_1 y_2 y_3 \Leftrightarrow Q_P{}' x_1)$ for each predicate letter P of F; this in turn implies $\forall y_1 \forall y_2 \forall y_3 (Q_P y_1 y_2 y_3 \Leftrightarrow Q_P y_1 y_2 y_2)$. (Again Q_P depends just on its first two arguments.) It follows that, given a model for G, we obtain a model for F over the same universe by interpreting each predicate letter P of F as true of those pairs $\langle a_1, a_2 \rangle$ such that Q_P is true in the given model of $\langle a_1, a_2, a_2 \rangle$.

ISBN 0-201-02540-X

Conversely, suppose F has a model. Then so does the schema $F' = \forall y \exists x Ryx \wedge \forall y_1 \forall y_2 \exists x_1 \forall y_3 (F_1 \wedge N)$, where N is any tautologous formula containing the variable x_1. Let $\mathfrak{A}$ be a truth-assignment verifying $E(F', \omega)$. Note that $D(F', \omega) = D(G, \omega)$. We define a truth-assignment $\mathfrak{B}$ on $\tilde{L}(G, \omega)$ thus: for each predicate letter P of F and all terms t_1, t_2, t_3 in $D(G, \omega)$, $\mathfrak{B}(Q_P t_1 t_2 t_3) = \mathfrak{A}(Pt_1 t_2)$; for each predicate letter P of F and all terms s, $\mathfrak{B}$ verifies $Q_P' s$ iff s is an x_1-term and $\mathfrak{A}$ verifies $P\hat{y}_1 s \, \hat{y}_2 s$. It follows quickly that $\mathfrak{B}$ verifies $E(G, \omega)$. Hence G has a model. $\square$

Unsolvability Proof for Class B. We reduce the Skolem class $B(S)$ to Class B, where $S = \{\{y_1, y_2\}, \{y_3, y_4\}, \{y_1, x_2\}, \{y_2, x_2\}, \{y_3, x_2\}, \{y_4, x_2\}\}$. This suffices, since $B(S)$ is unsolvable (see page 152). Let F be a schema in $B(S)$ with matrix F^M. For each $\{y_1, y_2\}$-based atomic subformula A of F let P_A be a new monadic predicate letter. Let G_1 be obtained from F^M by replacing each such A with $P_A x_1$; let G_2 be the conjunction of all biconditionals $(P_A x_1 \Leftrightarrow A(y_1/y_3, y_2/y_4))$; let Q_1 and Q_2 be new dyadic predicate letters; and let G be the schema with prefix $\forall y_1 \forall y_2 \exists x_1 \forall y_3 \forall y_4 \exists x_2$ and matrix $G^M =$

$$G_1 \wedge Q_1 y_1 x_2 \wedge Q_2 y_2 x_2 \wedge (Q_1 y_3 x_2 \wedge Q_2 y_4 x_2 \Rightarrow G_2).$$

Thus G is in Class B.

The schema G logically implies

$$\forall y_1 \forall y_2 \exists x_1 [\forall y_3 \forall y_4 \exists x_2 G_1 \wedge \exists x_2 G^M(y_3/y_1, y_4/y_2)].$$

Moreover, $G^M(y_3/y_1, y_4/y_2)$ truth-functionally implies $G_2(y_3/y_1, y_4/y_2)$, and $G_2(y_3/y_1, y_4/y_2)$ is the conjunction of all biconditionals $P_A x_1 \Leftrightarrow A$. Hence, by the construction of G_1, G logically implies $\forall y_1 \forall y_2 \forall y_3 \forall y_4 \exists x_2 F^M$, that is, G logically implies F.

Conversely, suppose F has a model. Then so does the schema F' whose prefix is like that of G and whose matrix is $F^M \wedge N$, where N is some tautologous formula containing x_1. Note that $D(F', \omega) = D(G, \omega)$ and that $L(G, \omega)$ contains, aside from atoms in $L(F', \omega)$, just atoms $Q_1 st$, $Q_2 st$, and $P_A s$. Let $\mathfrak{A}$ be a truth-assignment verifying $E(F', \omega)$, and extend $\mathfrak{A}$ to a truth-assignment $\mathfrak{B}$ on $\tilde{L}(G, \omega)$ thus: $\mathfrak{B}$ verifies $Q_1 st$ iff $s = \hat{y}_1 t$; $\mathfrak{B}$ verifies $Q_2 st$ iff $s = \hat{y}_2 t$; $\mathfrak{B}$ verifies $P_A s$ iff s is an x_1-term and $\mathfrak{A}$ verifies $A(y_1/\hat{y}_1 s, y_2/\hat{y}_2 s)$.

Let H be any Herbrand instance of G, and let t be the substituent in H for x_2. We show that $\mathfrak{B}$ verifies H. The second and third conjuncts in H are $Q_1 \hat{y}_1 t \, t$ and $Q_2 \hat{y}_2 t \, t$; these are verified by $\mathfrak{B}$. The fourth conjunct has the form $(Q_1 \hat{y}_3 t \, t \wedge Q_2 \hat{y}_4 t \, t \Rightarrow J)$; $\mathfrak{B}$ verifies its antecedent iff $\hat{y}_3 t = \hat{y}_1 t$ and $\hat{y}_4 t = \hat{y}_2 t$. But if $\hat{y}_3 t = \hat{y}_1 t$ and $\hat{y}_4 t = \hat{y}_2 t$, then J is just the conjunction of all biconditionals $P_A \hat{x}_1 t \Leftrightarrow A(y_1/\hat{y}_1 t, y_2/\hat{y}_2 t)$, so that $\mathfrak{B}$ verifies J. Finally, let K be the first conjunct in H, and let K' result from K by

ISBN 0-201-02540-X

replacing each subatom $P_A \hat{x}_1 t$ with $A(y_1/\hat{y}_1 t, y_2/\hat{y}_2 t)$. Thus $\mathfrak{B}(K) = \mathfrak{B}(K')$ $= \mathfrak{A}(K')$. But the Herbrand instance of F' in which t is the substituent for x_2 has the form $K' \wedge N'$, where N' is tautologous. Hence $\mathfrak{A}$ verifies K', whence $\mathfrak{B}$ verifies K. We have shown that $\mathfrak{B}$ verifies H. $\square$

The third example adduced in §6 of Chapter 2 was

Class C.   Prenex schemata with prefixes $\forall y_1 \cdots \forall y_7 \exists x_1 \forall y_8 \exists x_2$ and bases $\{\{y_1, y_2, y_3\}, \{y_4, y_5\}, \{x_1\}, \{y_1, y_2, x_2\}, \{y_4, y_5, x_1, x_2\}, \{y_6, y_7, y_8, x_2\}\}$.

Unsolvability Proof.   Let $S = \{\{y_1, y_2, y_3\}, \{y_1, y_2\}, \{y_1, y_2, x_2\}\}$. Then the Skolem class $B(S)$ is unsolvable (Unsolvability Result II of §2). We reduce $B(S)$ to Class C. Let F be a schema in $B(S)$ with matrix F^M. Let G_1 be the result of replacing each $\{y_1, y_2\}$-based atomic subformula A of F^M with $A(y_1/y_4, y_2/y_5)$. Let Q be a new tetradic predicate letter, and let R_1 and R_2 be new monadic predicate letters. Finally, let G be the schema in Class C with matrix $G^M =$

$$Q y_6 y_7 y_8 x_2 \wedge (Q y_1 y_2 y_2 x_2 \Rightarrow R_1 x_1) \wedge (Q y_4 y_5 x_1 x_2 \Rightarrow R_2 x_1)$$

$$\wedge (R_1 x_1 \wedge R_2 x_1 \Rightarrow G_1).$$

We show first that G logically implies F. Let G_2 be $G^M(y_4/y_1, y_5/y_2, y_6/y_1, y_7/y_2)$. By instantiation, G implies $\forall y_1 \forall y_2 \forall y_3 \exists x_1 \forall y_8 \exists x_2 G_2$, which in turn implies

$$\forall y_1 \forall y_2 \forall y_3 \exists x_1 [\exists x_2 G_2(y_8/y_2) \wedge \exists x_2 G_2(y_8/x_1)].$$

Now $G_2(y_8/y_2)$ truth-functionally implies $R_1 x_1$; and $G_2(y_8/x_1)$ truth-functionally implies

$$R_2 x_1 \wedge (R_1 x_1 \wedge R_2 x_1 \Rightarrow G_1(y_4/y_1, y_5/y_2)).$$

Moreover, $G_1(y_4/y_1, y_5/y_2) = F^M$. Hence G implies

$$\forall y_1 \forall y_2 \forall y_3 \exists x_1 [R_1 x_1 \wedge \exists x_2 (R_2 x_1 \wedge (R_1 x_1 \wedge R_2 x_1 \Rightarrow F^M))].$$

But then G implies $\forall y_1 \forall y_2 \forall y_3 \exists x_2 F^M$.

Conversely, suppose F has a model. Let F' be a schema whose prefix is like that of G and whose matrix is $F^M \wedge N$, where N is some tautologous formula containing $y_4, \ldots, y_8, x_1$. Then F' has a model. Let $\mathfrak{A}$ be a truth-assignment verifying $E(F', \omega)$. Extend $\mathfrak{A}$ to a truth-assignment $\mathfrak{B}$ on $\tilde{L}(G, \omega)$ thus: $\mathfrak{B}$ verifies $Q s_1 s_2 s_3 t$ iff $\chi(t) = x_2$, $s_1 = \hat{y}_6 t$, $s_2 = \hat{y}_7 t$, and $s_3 = \hat{y}_8 t$; $\mathfrak{B}$ verifies $R_1 s$ iff $\chi(s) = x_1$, $\hat{y}_1 s = \hat{y}_6 s$, and $\hat{y}_2 s = \hat{y}_7 s$; and $\mathfrak{B}$ verifies $R_2 s$ iff $\chi(s) = x_1$, $\hat{y}_4 s = \hat{y}_6 s$, and $\hat{y}_5 s = \hat{y}_7 s$. It follows easily that $\mathfrak{B}$ verifies every Herbrand instance of G. Thus G has a model. $\square$

One last unsolvable class, similar to Class C, demonstrates the necessity for the inductive definition of $t\text{-}x_l$-unitedness when $1 \leq l < n$.

Class D.   Prenex schemata with prefixes $\forall y_1 \cdots \forall y_7 \exists x_1 \forall y_8 \forall y_9 \exists x_2 \forall y_{10} \exists x_3$

ISBN 0-201-02540-X

and bases $\{\{y_1, y_2, y_3\}, \{y_4, y_5\}, \{x_1\}, \{x_2\}, \{y_1, y_2, x_3\}, \{y_8, y_9, y_{10}, x_3\}, \{y_6, y_7, x_2, x_3\}, \{y_6, y_7, x_1, x_2\}, \{y_4, y_5, y_8, x_2\}\}$.

Let F be a schema in Class D, and suppose t is an x_3-term in $D(F, \omega)$. Then y_1 and y_8 are t-x_3-united, as are y_2 and y_9, if $\hat{y}_1 t = \hat{y}_8 t$, $\hat{y}_2 t = \hat{y}_9 t$, and either $\hat{y}_{10}t = \hat{y}_1 t$ or $\hat{y}_{10}t = \hat{y}_2 t$; y_6 and y_8 are t-x_3-united, as are y_7 and y_9, if $\hat{y}_6 t = \hat{y}_8 t$, $\hat{y}_7 t = \hat{y}_9 t$, and $\hat{y}_{10}t = \hat{x}_2 t$; and y_6 and y_4 are t-x_3-united, as are y_7 and y_5, if $\hat{y}_6 t = \hat{y}_4 t$, $\hat{y}_7 t = \hat{y}_5 t$, and $\hat{y}_8 t = \hat{x}_1 t$. Moreover, since $\hat{x}_2 t \neq \hat{y}_i t$ for $1 \leq i \leq 9$ and $\hat{x}_1 t \neq \hat{y}_i t$ for $1 \leq i \leq 7$, these three possibilities are mutually exclusive. Now let s be an x_2-term in $D(F, \omega)$. If $\hat{y}_1 s = \hat{y}_6 s = \hat{y}_8 s$ and $\hat{y}_2 s = \hat{y}_7 s = \hat{y}_9 s$, then y_1 and y_6 are hereditarily s-x_2-united, as are y_2 and y_7; if $\hat{y}_6 s = \hat{y}_4 s$, $\hat{y}_7 s = \hat{y}_5 s$, and $\hat{y}_8 s = \hat{x}_1 s$, then y_6 and y_4 are s-x_2-united, as are y_7 and y_5. Again, these two possibilities are mutually exclusive. So far we have come across no violation of the specification of Class 2.8. A violation arises only with hereditary u-x_1-unitedness. For our definition of u-x_1-unitedness in terms of hereditary s-x_2-unitedness yields: if u is an x_1-term such that $\hat{y}_1 u = \hat{y}_4 u = \hat{y}_6 u$ and $\hat{y}_2 u = \hat{y}_5 u = \hat{y}_7 u$, then y_1 and y_4 are hereditarily u-x_1-united, as are y_2 and y_5. Thus the disjointness of $\{y_1, y_2, y_3\}$ and $\{y_4, y_5\}$ is compromised.

Unsolvability Proof. Let $F = \forall y_1 \forall y_2 \forall y_3 \exists x_3 F^M$ be a schema with basis $\{\{y_1, y_2, y_3\}, \{y_1, y_2\}, \{y_1, y_2, x_3\}\}$. It suffices to reduce F to a schema G in Class D.

As before, let G_1 be the result of replacing each $\{y_1, y_2\}$-based atomic subformula A of F^M with $A(y_1/y_4, y_2/y_5)$. Let P and Q be new tetradic predicate letters, and let R_1, R_2, S_1, and S_2 be new monadic predicate letters. Finally, let G be the schema in Class D with matrix $G^M =$

$$Qy_8 y_9 y_{10} x_3 \wedge (Qy_1 y_2 y_2 x_3 \Rightarrow S_1 x_2) \wedge (Qy_6 y_7 x_2 x_3 \Rightarrow S_2 x_2)$$

$$\wedge (S_1 x_2 \wedge S_2 x_2 \Rightarrow R_1 x_1) \wedge Py_4 y_5 y_8 x_2 \wedge (Py_6 y_7 x_1 x_2 \Rightarrow R_2 x_1)$$

$$\wedge (R_1 x_1 \wedge R_2 x_1 \Rightarrow G_1).$$

An argument analogous to that for Class C suffices to show that G logically implies F. Conversely, suppose F has a model. Let F' be a schema whose prefix is like that of G and whose matrix is $F^M \wedge N$, where N is some tautologous formula containing $y_4, \ldots, y_{10}, x_1, x_2$. Then F' has a model. Let $\mathfrak{A}$ be a truth-assignment verifying $E(F', \omega)$. Extend $\mathfrak{A}$ to a truth-assignment $\mathfrak{B}$ on $\hat{L}(G, \omega)$ thus: $\mathfrak{B}$ verifies $Qs_1 s_2 s_3 t$ iff $\chi(t) = x_3$, $s_1 = \hat{y}_8 t$, $s_2 = \hat{y}_9 t$, $s_3 = \hat{y}_{10}t$; $\mathfrak{B}$ verifies $Ps_1 s_2 s_3 t$ iff $\chi(t) = x_2$, $s_1 = \hat{y}_4 t$, $s_2 = \hat{y}_5 t$, $s_3 = \hat{y}_8 t$; $\mathfrak{B}$ verifies $S_1 t$ iff $\chi(t) = x_2$, $\hat{y}_1 t = \hat{y}_8 t$, $\hat{y}_2 t = \hat{y}_9 t$; $\mathfrak{B}$ verifies $S_2 t$ iff $\chi(t) = x_2$, $\hat{y}_6 t = \hat{y}_8 t$, $\hat{y}_7 t = \hat{y}_9 t$; $\mathfrak{B}$ verifies $R_1 t$ iff $\chi(t) = x_1$, $\hat{y}_1 t = \hat{y}_6 t$ and $\hat{y}_2 t = \hat{y}_7 t$; and $\mathfrak{B}$ verifies $R_2 t$ iff $\chi(t) = x_1$, $\hat{y}_4 t = \hat{y}_6 t$ and $\hat{y}_5 t = \hat{y}_7 t$. It follows easily that $\mathfrak{B}$ verifies every Herbrand instance of G. Thus G has a model. $\square$

ISBN 0-201-02540-X

4. Elimination of Initial x-Variables

In 1933 Kalmár showed how to reduce initially-extended Skolem schemata to unextended Skolem schemata. Here we present his reduction, with the modifications needed for the applications of this reduction that we made in earlier chapters. Given an initially-extended Skolem schema F, Kalmár's original reduction yields a Skolem schema G that contains the same number of y-variables as F. We shall see that if the matrix of F is Krom, then so is that of G; this suffices for the application to the initially-extended Maslov Class (see the beginning of Chapter 4). We shall also use Kalmár's technique to reduce F to a conjunction $G_1 \wedge \cdots \wedge G_k$ of Skolem schemata with the following property: for each i there is a set of Y_i of y-variables such that $b(G_i)$ results from $b^p(F)$ by deleting the members of Y_i. This suffices for the application to initially-extended Class 2.7 (see §5 of Chapter 2). Then we shall indicate how to apply the reduction technique to arbitrary normal schemata containing initial x-variables. This enables us to extend Class 2.8.

For the moment we extend our formal language: we allow the schemata to which we reduce initially-extended Skolem schemata to contain 0-place predicate letters, that is, sentential letters. Subsequently we shall show how to eliminate these letters.

Let $F = \exists z_1 \cdots \exists z_l \forall y_1 \cdots \forall y_m \exists x_1 \cdots \exists x_n F^M$ be an initially-extended Skolem schema. A k-mapping, where $k > 0$, is any mapping $\tau: \{1,\ldots,k\} \rightarrow \{0, 1,\ldots,l\}$. If $\langle u_1,\ldots,u_k \rangle$ is a k-tuple and τ is a k-mapping, then the τ-contraction of $\langle u_1,\ldots,u_k \rangle$ is the r-tuple $\langle u_{i_1},\ldots,u_{i_r} \rangle$, where $i_1,\ldots,i_r$ are in ascending order of magnitude the integers which τ carries to 0; r is thus the rank of τ, that is, the number of integers that τ carries to 0. (Note that a k-mapping τ may have rank 0, in which case the τ-contraction of every k-tuple is the empty sequence.)

For each m-mapping σ, let F_σ^M be $F^M(y_1/v_1,\ldots,y_m/v_m)$, where $v_i = z_{\sigma(i)}$ if $\sigma(i) > 0$ and $v_i = y_i$ if $\sigma(i) = 0$. For each k-place predicate letter P of F and each k-mapping τ, $k > 0$, let $Q_{P,\tau}$ be a new r-place predicate letter, where r is the rank of τ. Then, for each m-mapping σ, let G_σ^M be the formula obtained from F_σ^M by replacing each atomic subformula $Pw_1 \cdots w_k$ with the atomic formula $Q_{P,\tau} w_{i_1} \cdots w_{i_r}$, where $\langle w_{i_1},\ldots,w_{i_r} \rangle$ is the τ-contraction of $\langle w_1,\ldots,w_k \rangle$ and τ is the k-mapping such that $\tau(i) = j > 0$ if $w_i = z_j$ and $\tau(i) = 0$ otherwise. For each m-mapping σ, let G_σ be the Skolem schema with matrix G_σ^M. Thus G_σ is $\forall y_{i_1} \cdots \forall y_{i_r} \exists x_1 \cdots \exists x_n G_\sigma^M$, where $\langle y_{i_1},\ldots,y_{i_r} \rangle$ is the σ-contraction of $\langle y_1,\ldots,y_m \rangle$. Finally, let G be the conjunction of all the schemata G_σ. We show that F has a model iff G has a model, and F has a finite model iff G has a finite model.

Suppose F has a model $\mathfrak{M}$ with universe U. Thus there are elements $c_1,\ldots,c_l$ of U such that $\forall y_1 \cdots \forall y_m \exists x_1 \cdots \exists x_n F^M$ is true in $\mathfrak{M}$ when $z_1,\ldots,z_l$

ISBN 0-201-02540-X

take the values $c_1,\ldots,c_l$. Let P be a k-place predicate letter of F and τ a k-mapping of rank r. We interpret the predicate letter $Q_{P,\tau}$ over the universe U thus: $Q_{P,\tau}$ is true of an r-tuple $\langle a_1,\ldots,a_r\rangle$ iff P is true in $\mathfrak{M}$ of the unique k-tuple $\langle b_1,\ldots,b_k\rangle$ such that $\langle a_1,\ldots,a_r\rangle$ is the τ-contraction of $\langle b_1,\ldots,b_k\rangle$ and $b_i = c_{\tau(i)}$ whenever $\tau(i) > 0$. Let $\mathfrak{N}$ be the result of adjoining these interpretations to $\mathfrak{M}$; we show that $\mathfrak{N}$ is a model for each schema G_σ. Note that $\forall y_1\cdots\forall y_m\exists x_1\cdots\exists x_n F^M$ logically implies $\forall y_1\cdots\forall y_m\exists x_1\cdots\exists x_n F_\sigma^M$ for each σ. Hence $\forall y_1\cdots\forall y_m\exists x_1\cdots\exists x_n F_\sigma^M$ is true in $\mathfrak{M}$ (and also in $\mathfrak{N}$) when $z_1,\ldots,z_l$ take the values $c_1,\ldots,c_l$. Moreover, by the constructions of G_σ^M and $\mathfrak{N}$, each biconditional $\langle F_\sigma^M\rangle_i \Leftrightarrow \langle G_\sigma^M\rangle_i$ is true in $\mathfrak{N}$ when $z_1,\ldots,z_l$ take the values $c_1,\ldots,c_l$ and the other variables take any values whatever in U. Hence each G_σ is true in $\mathfrak{N}$. That is, $\mathfrak{N}$ is a model for G with universe U.

Conversely, suppose G has a model $\mathfrak{N}$ with universe U. Let $c_1,\ldots,c_l$ be objects not in U, and let $U' = U \cup \{c_1,\ldots,c_l\}$. We interpret each k-place predicate letter P of F, $k > 0$, over the universe U' thus: P is true of $\langle b_1,\ldots,b_k\rangle$ iff $Q_{P,\tau}$ is true in $\mathfrak{N}$ of the τ-contraction of $\langle b_1,\ldots,b_k\rangle$, where τ is the k-mapping such that $\tau(i) = j > 0$ iff $b_i = c_j$. We show that these interpretations constitute a model $\mathfrak{M}$ for F over U. It suffices to show that $\forall y_1\cdots\forall y_m\exists x_1\cdots\exists x_n F^M$ is true in $\mathfrak{M}$ when $z_1,\ldots,z_l$ take the values $c_1,\ldots,c_l$. Let $\langle b_1,\ldots,b_m\rangle$ be any m-tuple from U'; let σ be the m-mapping such that $\sigma(i) = j > 0$ iff $b_i = c_j$, and let $\langle b_{i_1},\ldots,b_{i_r}\rangle$ be the σ-contraction of $\langle b_1,\ldots,b_m\rangle$. Since $\mathfrak{N}$ is a model for G_σ, there are $d_1,\ldots,d_n$ in U such that G_σ^M is true in $\mathfrak{N}$ when the variables $y_{i_1},\ldots,y_{i_r}$, $x_1,\ldots,x_n$ take the values $b_{i_1},\ldots,b_{i_r}, d_1,\ldots,d_n$. Note that since $d_1,\ldots,d_n$ are in U, they are all distinct from $c_1,\ldots,c_l$. Hence, by the constructions of G_σ^M and $\mathfrak{M}$, $\langle G_\sigma^M\rangle_i$ is true in $\mathfrak{N}$ for these values of the variables iff $\langle F\rangle_i$ is true in $\mathfrak{M}$ for values $c_1,\ldots,c_l$, $b_1,\ldots,b_m, d_1,\ldots,d_n$ of the variables $z_1,\ldots,z_l, y_1,\ldots,y_m, x_1,\ldots,x_n$. Thus F^M is true in $\mathfrak{M}$ for the latter values. Since $\langle b_1,\ldots,b_m\rangle$ is an arbitrary m-tuple from U', it follows that $\forall y_1\cdots\forall y_m\exists x_1\cdots\exists x_n F^M$ is true in $\mathfrak{M}$ when $z_1,\ldots,z_l$ take the values $c_1,\ldots c_l$.

We have shown that if F has a model then G has a model with the same universe; and if G has a model then F has a model whose universe contains just l elements more than that of the given model for G. Thus F is reduced to G; and this reduction is conservative.

The formula G is a conjunction of Skolem schemata G_σ, each of which contains at most m y-variables and exactly n x-variables. For each σ let Y_σ be the set of y-variables y_i such that $\sigma(i) > 0$. Thus Y_σ is the set of y-variables occurring in F but not in G_σ. Moreover, if an atomic subformula $\langle F\rangle_i$ of F is V-based, then $\langle G_\sigma\rangle_i$ is $(V - Y_\sigma - \{z_1,\ldots,z_l\})$-based. Thus the basis $b(G_\sigma)$ of G_σ results from the pruned basis $b^p(F)$ of F by deleting the members of Y_σ.

The formula G also has a prenex equivalent $G' = \forall y_1\cdots\forall y_m\exists x_1\cdots\exists x_{n\cdot p}K$,

ISBN 0-201-02540-X

where p is the number of m-mappings ($p = (l + 1)^m$) and K is a conjunction of formulas resulting from the G_σ^M by relettering the x-variables. Thus G' has the same number of y-variables as F; and the matrix K of G' has the same truth-functional form as a p-fold conjunction of F^M with itself. In particular, if F^M is Krom, then the matrix of G' is Krom.

Finally, we wish to eliminate the 0-place predicate letters from G and from G'. We give two general methods for doing this. Suppose N is a formula containing 0-place predicate letters $P_1, \ldots, P_q$.

(1)   For each truth-assignment $\mathfrak{T}$ on $P_1, \ldots, P_q$ let $N_\mathfrak{T}$ result from N by first replacing each P_i with a conventional symbol for its truth-value under $\mathfrak{T}$, and then eliminating these symbols by means of the usual truth-functional rules. Then N has a model iff at least one formula $N_\mathfrak{T}$ has a model. Note that if we apply this procedure to G and G' above, we no longer obtain a reduction in our sense: what we have, in recursion-theoretic terms, is a truth-table reducibility rather than a many–one reducibility.

(2)   For each 0-place predicate letter P_i let P_i' be a new monadic predicate letter, and let J be the schema

$$\forall y_1 \forall y_2 \left(\bigwedge_i P_i' y_1 \Leftrightarrow P_i' y_2 \right).$$

Let N_1 result from N by replacing each occurrence of P_i with an occurrence of $P_i'v$, where v is any variable (v can be chosen arbitrarily, and need not be the same for each occurrence of P_i). In any model for $N_1 \wedge J$, each predicate letter P_i' is true of all elements or true of none; in short, these new predicate letters act like 0-place letters. Hence N has a model with universe U iff $N_1 \wedge J$ has a model with universe U.

If we apply this second method to G we obtain a formula $G_1 \wedge J$. The new conjunct J is trivially in Class 2.7, since it contains no x-variables; and the basis of each conjunct in G_1 differs from the basis of the homologous conjunct G_σ in G at most by containing new one-membered sets $\{v\}$. The addition of such one-membered sets, of course, does not affect membership in Class 2.7.

If we apply the method to the prenex schema G' we obtain a formula G_1' $\wedge J$; and $G_1' \wedge J$ has a prenex equivalent G'' with the following properties: if the matrix of G' is Krom, then so is the matrix of G''; and if G' contains two or more y-variables, then G'' contains the same number of y-variables.

(If G' has only one y-variable y_1 and we wish to construct a G'' that also has only one y-variable, we may refine method (2) thus: instead of conjoining J to G_1' we conjoin to the matrix of G_1' all biconditionals $P_i'y_1 \Leftrightarrow P_i'x_j$, where $1 \leq i \leq q$ and x_j is an x-variable of G_1'. It follows that if $\mathfrak{A}$ verifies the expansion of the resulting schema, then for each $i \leq q$, $\mathfrak{A}$ assigns the same truth-value to all atoms $P_i't$. Thus we may reduce the

ISBN 0-201-02540-X

Initially-extended Ackermann Class to the unextended Ackermann Class, Class 2.2.)

Our method for eliminating initial x-variables may also be applied to arbitrary prenex schemata. Here the complexity of the reduction grows rapidly with the number of alternations of y-variables and x-variables. Hence we limit ourselves to an example, namely, schemata F that contain two y-clusters. Let F be

$$\exists z_1 \cdots \exists z_l \forall y_1 \cdots \forall y_m \exists x_1 \cdots \exists x_n \forall y_1' \cdots \forall y_p' \exists x_1' \cdots \exists x_q' F^M.$$

For each m-mapping σ and each p-mapping η, let $F_{\sigma,\eta}^M$ be $F^M(y_1/v_1,\ldots,y_m/v_m, y_i'/w_1,\ldots,y_p'/w_p)$, where

(a) for each i, $1 \le i \le m$, $v_i = y_i$ if $\sigma(i) = 0$ and $v_i = z_{\sigma(i)}$ if $\sigma(i) > 0$;

(b) for each i, $1 \le i \le p$, $w_i = y_i'$ if $\eta(i) = 0$ and $w_i = z_{\eta(i)}$ if $\eta(i) > 0$.

For each m-mapping σ and each p-mapping η, let $G_{\sigma,\eta}^M$ result from $F_{\sigma,\eta}^M$ by introducing new predicate letters $Q_{P,\tau}$ and replacing atomic subformulas of $F_{\sigma,\eta}^M$ exactly as in the previous proof. Let $G_{\sigma,\eta}$ be the formula $\forall y_i' \cdots \forall y_p' \exists x_1' \cdots \exists x_q' G_{\sigma,\eta}^M$. For each m-mapping σ, let G_σ be $\forall y_1 \cdots \forall y_m \exists x_1 \cdots \exists x_n (\bigwedge G_{\sigma,\eta})$, where the conjunction is over all p-mappings η. Finally, let G be the conjunction of all the G_σ. Thus G contains no initial x-variables.

Slightly more intricate versions of the arguments given above for Skolem schemata suffice to show the following: if F has a model, then G has a model over the same universe; and if G has a model, then F has a model over a universe containing just l elements more than that of the given model for G.

Our interest in this reduction stems from its employment in showing the solvability and finite controllability of Initially-extended Class 2.8. Suppose then that F is in Initially-extended Class 2.8. That is, Class 2.8 contains (up to relettering of bound variables) some—and hence contains every—prenex schema whose prefix is $\forall y_1 \cdots \forall y_m \exists x_1 \cdots \exists x_n \forall y_1' \cdots \forall y_p' \exists x_1' \cdots \exists x_q'$ and whose basis is the pruned basis $b^p(F)$ of F. We must show that in this case the formula G has a prenex equivalent in Class 2.8. Now G is the conjunction of formulas G_σ, and Class 2.8 is essentially closed under conjunction. Hence it suffices to show that each G_σ has a prenex equivalent in Class 2.8.

Assume that the different p-mappings η have been assigned numbers from 0 to $r - 1$, where $r = (l + 1)^p$. Then G_σ has a prenex equivalent G' of the form

$$\forall y_1 \cdots \forall y_m \exists x_1 \cdots \exists x_n \forall y_1' \cdots \forall y_{r \cdot p}' \exists x_1' \cdots \exists x_{r \cdot q}' (\bigwedge G_{\sigma,\eta}'),$$

where if η has been assigned the number i, then $G'_{\sigma,\eta}$ is the result of replacing in $G^M_{\sigma,\eta}$ all variables y_j' and x_k' by $y'_{i\cdot p+j}$ and $x'_{i\cdot q+k}$, $1 \le j \le p$ and $1 \le k \le q$. (Note that a variable y_j' may not occur in $G^M_{\sigma,\eta}$; this happens if $n(j) > 0$. Hence some y-variables of G' are vacuously quantified.) The basis of G' may be described with the help of mappings ζ_i from variables to variables, $0 \le i < r$: ζ_i carries each y_j and x_k to itself, $1 \le j \le m$ and $1 \le k \le n$, and ζ_i carries y_j' to $y'_{i\cdot p+j}$ and x_k' to $x'_{i\cdot q+k}$, $1 \le j \le p$ and $1 \le k \le q$. Then $b(G')$ is the result of deleting all y-variables that are vacuously quantified in G' from $\{\zeta_i[V] \mid 0 \le i < r$ and $V \in b^p(F)\}$. Since membership in Class 2.8 is preserved by deletion of y-variables, it suffices to show that Class 2.8 contains (up to relettering) some schema G'' whose prefix is identical with that of G' and whose basis is $\{\zeta_i[V] \mid 0 \le i < r$ and $V \in b^p(F)\}$.

Since F is in Initially-extended Class 2.8, each member of $b^p(F)$ is either an x-focused set or a y-set, and each y-set is contained in a single y-cluster. Consequently, the same holds of each member of $b(G'')$. If we let F'' be a schema in Class 2.8 (up to relettering) whose basis is $b^p(F)$, we can show that G'' is in Class 2.8 by a detailed comparison of the relations hereditary t-v-unitedness, where t is a term in $D(G'', \omega)$ and v is an x-variable of G'', with the relations hereditary s-w-unitedness, where s is a term in $D(F'', \omega)$ and w is an x-variable of F''. That is, we can show that if G'' violates the specification of Class 2.8 for some term t in $D(G'', \omega)$ then F'' will violate the specification for some term s in $D(F'', \omega)$. This examination of details is quite long, so we shall not present it here. It relies on the fact that the basis of G'' is the union of copies of the basis of F'' (each copy is $\{\zeta_i[V] \mid V \in b^p(F)\}$ for some i), and the only variables these copies share are those that govern the second y-cluster of G''.

5. Elimination of All but Dyadic Predicate Letters from ∀∃∀-Schemata

The reduction we present in this section is conservative, and preserves Krom matrices and Horn matrices. The basic idea is due to Wang (1962), but his proof has a lacuna.

Let $F = \forall y_1 \exists x \forall y_2 F^M$ be an ∀∃∀-schema. We may assume that F contains no monadic predicate letters. For if it did contain a monadic letter P, we would introduce a new dyadic predicate letter P' and replace each atomic subformula Pv with $P'vv$; this transformation clearly preserves satisfiability. Let Σ be the set of atomic formulas that can be constructed from k-place predicate letters of F for $k > 2$ and from variables of F. For each A in Σ let Q_A be a new dyadic predicate letter, and let G_1 result from F^M by replacing each atomic subformula A in Σ with $Q_A y_1 y_2$.

ISBN 0-201-02540-X

Given any truth-assignment $\mathfrak{A}$ verifying $E(F, \omega)$, we shall so define a truth-assignment $\mathfrak{B}$ that

(‡)   for each A in Σ, $\mathfrak{B}(Q_A mn) = \mathfrak{A}(A(y_1/m, x/m + 1, y_2/n))$.

(We use the numerical notation for terms in $D(F, \omega)$, introduced in Chapter 5.) If $\mathfrak{B}$ also agrees with $\mathfrak{A}$ on atoms containing predicate letters common to F and G_1, $\mathfrak{B}$ will verify each instance $G_1(y_1/m, x/m + 1, y_2/n)$ of G_1. Conversely, given a truth-assignment $\mathfrak{B}$ on atoms $Q_A mn$ we want there to be a truth-assignment $\mathfrak{A}$ on $\tilde{L}(F, \omega)$ that obeys condition (‡). But there will be such an $\mathfrak{A}$ if and only if

(‡‡)   for all A and B in Σ, $\mathfrak{B}(Q_A mn) = \mathfrak{B}(Q_B pq)$ whenever $A(y_1/m, x/m + 1, y_2/n) = B(y_1/p, x/p + 1, y_2/q)$.

To ensure condition (‡‡), we conjoin certain biconditionals to G_1.

Let G_2 be the conjunction of all biconditionals $Q_A y_1 v \Leftrightarrow Q_B w_1 w_2$ such that A and B are in Σ, B is $\{y_1, y_2\}$-based, and $A(y_2/v) = B(y_1/w_1, y_2/w_2)$. (Here and below w_1, w_2, v range over the variables y_1, y_2, x of F. Recall that the substitutions indicated by our substitution notation are carried out simultaneously, not sequentially.)

Moreover, for each A in Σ that contains all three variables of F let R_A be a new dyadic predicate letter. Let G_3 be the conjunction of all biconditionals

$$(Q_A y_2 x \Leftrightarrow R_A y_2 y_1) \wedge (R_A y_1 x \Leftrightarrow Q_B x y_1)$$

such that A and B are members of Σ that contain all three variables and $B = A(y_1/y_2, x/y_1, y_2/x)$.

Finally let $G = \forall y_1 \exists x \forall y_2 (G_1 \wedge G_2 \wedge G_3)$. Thus G contains dyadic predicate letters only.

Suppose F has a model. Let $\mathfrak{A}$ be a truth-assignment verifying $E(F, \omega)$. Let $\mathfrak{B}$ be the truth-assignment on $\tilde{L}(G, \omega)$ such that

(1) $\mathfrak{B}$ agrees with $\mathfrak{A}$ on atoms common to $L(F, \omega)$ and $L(G, \omega)$;

(2) $\mathfrak{B}$ obeys condition (‡) above;

(3)   for each A in Σ that contains all three variables of F, $\mathfrak{B}(R_A mn) = \mathfrak{A}(A(y_1/m, x/m + 1, y_2/n + 1))$.

We show that the truth-assignment $\mathfrak{B}$ verifies $E(G, \omega)$. Let $H = G^*(y_1/m, y_2/n)$ be an arbitrary Herbrand instance of G. Thus $H = H_1 \wedge H_2 \wedge H_3$, where $H_i = G_i(y_1/m, x/m + 1, y_2/n)$.

By the construction of G_1 and condition (‡),

$$\mathfrak{B}(\langle H_1 \rangle_i) = \mathfrak{A}(\langle H' \rangle_i)$$

for each i, where H' is the Herbrand instance $F^*(y_1/m, y_2/n)$ of F. Since $\mathfrak{A}$ verifies H', $\mathfrak{B}$ verifies H_1.

Let σ take y_1 to m, x to $m + 1$, and y_2 to n. Then each conjunct in H_2 has the form

$$Q_A m\, \sigma(v) \Leftrightarrow Q_B \sigma(w_1)\, \sigma(w_2),$$

where A and B are in Σ, B is $\{y_1, y_2\}$-based, and $A(y_2/v) = B(y_1/w_1, y_2/w_2)$. But then $A(y_1/m, x/m + 1, y_2/\sigma(v)) = B(y_1/\sigma(w_1), y_2/\sigma(w_2))$. By condition ($\ddagger$), $\mathfrak{B}$ verifies this conjunct in H_2. Hence $\mathfrak{B}$ verifies H_2.

Each conjunct in H_3 has either the form $Q_A n\ m + 1 \Leftrightarrow R_A nm$ or the form $R_A m\ m + 1 \Leftrightarrow Q_B m + 1\ m$, where A and B are members of Σ that contain all three variables of F and $B = A(y_1/y_2, x/y_1, y_2/x)$. From clause (3) of the definition of $\mathfrak{B}$, $\mathfrak{B}(Q_A n\ m + 1) = \mathfrak{B}(R_A n\ m)$. By the choice of A and B, $A(y_1/m, x/m + 1, y_2/m + 2) = B(y_1/m + 1, x/m + 2, y_2/m)$. Hence $\mathfrak{B}(R_A m\ m + 1)) = \mathfrak{B}(Q_B m + 1\ m)$. We have shown that $\mathfrak{B}$ verifies H_3. This concludes the proof that $\mathfrak{B}$ verifies $E(G, \omega)$.

Suppose G has a model. Let $\mathfrak{B}$ be a truth-assignment verifying $E(G, \omega)$. We show first that $\mathfrak{B}$ obeys condition ($\ddagger\ddagger$). Suppose that A and B are in Σ and that $A(y_1/m, x/m + 1, y_2/n) = B(y_1/p, x/p + 1, y_2/q)$. Let $\sigma(y_1) = m$, $\sigma(x) = m + 1$, and $\sigma(y_2) = n$.

(a)   Assume B is $\{y_1, y_2\}$-based. By the supposition on A and B there are variables w_1, w_2, and v such that $B(y_1/w_1, y_2/w_2) = A(y_2/v)$ and such that $p = \sigma(w_1)$, $q = \sigma(w_2)$, and $n = \sigma(v)$. But then $Q_A y_1 v \Leftrightarrow Q_B w_1 w_2$ is a conjunct in G_2, whence $Q_A mn \Leftrightarrow Q_B pq$ is a conjunct in the Herbrand instance $G^*(y_1/m, y_2/n)$. Since $\mathfrak{B}$ verifies each Herbrand instance of G, condition ($\ddagger\ddagger$) follows.

(b)   Assume B is not $\{y_1, y_2\}$-based, but there are at most two distinct terms in the atom $A(y_1/m, x/m + 1, y_2/n)$. Then there are a $\{y_1, y_2\}$-based C in Σ and j and k such that $A(y_1/m, x/m + 1, y_2/n) = C(y_1/j, y_2/k) = B(y_1/p, x/p + 1, y_2/q)$. By case (a), $\mathfrak{B}(Q_A mn) = \mathfrak{B}(Q_C jk)$ and $\mathfrak{B}(Q_B pq) = \mathfrak{B}(Q_C jk)$. Again ($\ddagger\ddagger$) follows.

(c)   Assume $A(y_1/m, x/m + 1, y_2/n)$ contains three distinct terms. Then either: $A = B$, $m = p$, and $n = q$, so that ($\ddagger\ddagger$) is trivial; or $n = m + 2$, $p = m + 1$, $q = m$, and $B = A(y_1/y_2, x/y_1, y_2/x)$; or $m = n + 1$, $p = n$, $q = m + 1 = n + 2$, and $A = B(y_1/y_2, x/y_1, y_2/x)$. Since the second and third cases are symmetrical, we consider only the second. By the construction of G_3, the Herbrand instance $G^*(y_1/m + 1, y_2/m)$ contains the conjunct $Q_A m\ m + 2 \Leftrightarrow R_A m\ m + 1$; and every Herbrand instance of G in which m is the substituent for y_1 contains the conjunct $R_A m\ m + 1 \Leftrightarrow Q_B m + 1\ m$. Hence $\mathfrak{B}$ verifies $Q_A m\ m + 2 \Leftrightarrow Q_B m + 1\ m$, as required.

Since ($\ddagger\ddagger$) holds, there is a truth-assignment $\mathfrak{A}$ on $\tilde{L}(F, \omega)$ that agrees with $\mathfrak{B}$ on each atom common to $L(G, \omega)$ and $L(F, \omega)$ and that obeys condition ($\ddagger$). Let $H = F^*(y_1/m, y_2/n)$ be any Herbrand instance of F;

ISBN 0-201-02540-X

and let $H_1 = G_1(y_1/m, x/m + 1, y_2/n)$. Then clearly $\mathfrak{A}(\langle H \rangle_i) = \mathfrak{B}(\langle H_1 \rangle_i)$ for each i. Hence $\mathfrak{A}$ verifies H and so F has a model.

This concludes the proof of the reduction. From the construction of G it is clear that G^M is Krom whenever F^M is Krom, and G^M is Horn whenever F^M is Horn. Lastly, to show that the reduction is conservative we would proceed as before, but consider, instead of $E(F, \omega)$ and $E(G, \omega)$, the sets $\Delta_p[E(F, \omega)]$ and $\Delta_p[E(G, \omega)]$ (see Chapter 5, §3). That is, we would show that, for each p, $\Delta_p[E(F, \omega)]$ is consistent iff $\Delta_p[E(G, \omega)]$ is consistent. By the Special Finite Model Lemma, this suffices to show that F has a finite model iff G has a finite model.

ISBN 0-201-02540-X

CHAPTER 7

Solvability and Coinstantiations

We have often noted the pervasive role that coinstantiations play in the decision problem. (Recall that, intuitively, a coinstantiation is simply two occurrences of the same atom in an expansion; technically, a coinstantiation in an expansion $E(F, \omega)$ is a quadruple $\langle H, i, H', j \rangle$, where H and H' are Herbrand instances of F and i and j are integers such that $\langle H \rangle_i = \langle H' \rangle_j$.) In our amenability proofs, various conditions on formula- and instance-mappings were seen to arise from the presence of certain coinstantiations; and the successful construction of such mappings was seen to be possible because of the absence of certain other coinstantiations.

But remarks of this kind depict the relation between solvability and coinstantiations only in a fragmentary manner. One might well ask whether solvability proofs could be obtained by considering coinstantiations directly. The investigation of this question in general quickly engenders syntactic complexities of overwhelming proportions. (It was to avoid such complexities that we first introduced mappings.) Hence we shall here consider in detail only an example: the Gödel–Kalmár–Schütte Class (2.5), which consists of prenex schemata with prefixes $\forall y_1 \forall y_2 \exists x_1 \cdots \exists x_n$. We shall specify a property—which we call the Inevitability Property—of the coinstantiations in the expansions of all such schemata, a property that leads to solvability. Indeed, the Inevitability Property yields an effective criterion for satisfiability couched in terms of the syntactic structure of the matrix. This criterion was central to the original solvability proofs for this class (Gödel, 1932; Kalmár, 1933; Schütte, 1934). We shall also briefly indicate how this approach may be extended to the other classes of Chapter 2.

In the second section of this chapter we derive a windfall from the work of §1. We obtain a straightforward reduction of the Gödel–Kalmár–Schütte Class to the Monadic Class (2.1). A schema F is reduced to a monadic

Burton Dreben and Warren D. Goldfarb, The Decision Problem: Solvable Classes of Quantificational Formulas

ISBN 0-201-02540-X

schema by the encoding of information about the coinstantiations in $E(F, \omega)$. (This reduction may be extended to the other classes of Chapter 2.) Unlike most reductions devised for decision problems, finite models for the monadic endproduct do not correspond to finite models for the schema reduced. We emphasize this feature of our method by providing a similar monadic reduction of Class 5.1, a class that though solvable is not finitely controllable.

A third section is devoted to a different but related issue. We consider classes of schemata specified by the number of atomic subformulas they contain. The class of schemata containing two distinct atomic subformulas is solvable and finitely controllable, facts we prove by analyzing coinstantiability. This contrasts with the unsolvability of the class of schemata containing four distinct atomic subformulas. We also prove a curious result due to Harry Lewis: *Every* class limited by both number of atomic subformulas and number of y-variables is solvable.

1. Coinstantiations Analyzed

We saw in Chapter 1 that, because of coinstantiations, the Herbrand instances of a schema F cannot be treated as independent copies of the matrix of F. If H and H' have an atomic subformula in common, then the value that a truth-assignment $\mathfrak{A}$ takes on H cannot be chosen independently of the value it takes on H'. For if $\langle H \rangle_i = \langle H' \rangle_j$, then of course $\mathfrak{A}(\langle H \rangle_i) = \mathfrak{A}(\langle H' \rangle_j)$.

Thus suppose we are constructing a truth-assignment $\mathfrak{A}$ inductively: we assume $\mathfrak{A}$ is defined on the Herbrand instances earlier than a given H, and we seek to extend $\mathfrak{A}$ to H.‡ In doing this we do not have *carte blanche*. We cannot extend $\mathfrak{A}$ to H by simply choosing for each i, $1 \leq i \leq c(F)$, a value for $\mathfrak{A}(\langle H \rangle_i)$. Coinstantiations give rise to constraints of two sorts on our extension:

(A)   An atomic subformula of H may occur in an earlier Herbrand instance, in which case the value of $\mathfrak{A}$ on the atomic formula is already determined. Our method of extending $\mathfrak{A}$ must not conflict with these previously determined values.

(B)   Atoms may occur multiply in H itself. If $\langle H \rangle_i = \langle H \rangle_j$, then our method of extending $\mathfrak{A}$ must assure that $\mathfrak{A}(\langle H \rangle_i) = \mathfrak{A}(\langle H \rangle_j)$.

In our examination of the Gödel–Kalmár–Schütte Class we shall characterize the coinstantiations that give rise to such constraints. Our concern

ISBN 0-201-02540-X

---

‡ The only fact about the ordering of Herbrand instances we shall exploit is this: each member of $E(F, p)$ precedes each member of $E(F, p + 1) - E(F, p)$.

is thus with atoms that are *old-in-H*, that is, which occur both in H and in at least one Herbrand instance earlier than H, and also with atoms that occur multiply in H itself.

To describe coinstantiations conveniently we introduce the following terminology. Let S be a set of pairs of integers $\leq c(F)$. A pair $\langle H, H' \rangle$ of Herbrand instances *exhibits* S iff, for all pairs $\langle i, j \rangle$ in S, $\langle H \rangle_i = \langle H' \rangle_j$.

Given a Gödel–Kalmár–Schütte schema F we shall specify a number of sets of pairs of integers $\leq c(F)$. These sets have two important properties. First, they are (effectively) specified explicitly in terms of the matrix of F: whether a pair $\langle i, j \rangle$ is in one of the sets depends only on the ith and jth components $\langle F \rangle_i$ and $\langle F \rangle_j$ of F. Second, the sets provide a complete characterization of the coinstantiations that give rise to constraints of sorts (A) and (B). More precisely, for each Herbrand instance H, the constraints that arise for H are determined jointly by the sets exhibited by $\langle H, H' \rangle$ for some Herbrand instance H' earlier than H, and by the sets exhibited by $\langle H, H \rangle$.

Let $F = \forall y_1 \forall y_2 \exists x_1 \cdots \exists x_n F^M$ be a Gödel–Kalmár–Schütte schema. We require that F be *saturated*: every atomic formula that can be constructed from a predicate letter of F and variables of F must occur in F. (There is no loss of generality in requiring saturation, since satisfiability is unaffected when to the matrix of a schema are conjoined tautologies of the form $A \vee -A$.) Let $H = F^*(y_1/t_1, y_2/t_2)$ be an Herbrand instance of F later than the earliest, that is, at least one of t_1, t_2 is not 1. Since F is Skolem, the terms $f_{x_i}(t_1 t_2)$ that occur in H as substituents for the x-variables of F do not occur in any Herbrand instance earlier than H. Hence if $\langle F \rangle_i$ contains at least one x-variable, then $\langle H \rangle_i$ is not old-in-H. On the other hand, since H is not the earliest Herbrand instance of F, each of t_1 and t_2 does occur in at least one earlier Herbrand instance. Hence, by saturation, if $\langle F \rangle_i$ is $\{y_1, y_2\}$-based, then $\langle H \rangle_i$ is old-in-H exactly when both t_1 and t_2 occur together in some Herbrand instance earlier than H (there can be at most two such earlier Herbrand instances), and if $\langle F \rangle_i$ is $\{y_1\}$-based or $\{y_2\}$-based, then $\langle H \rangle_i$ is always old-in-H. Thus there are two cases to consider.

Case 1. There is an Herbrand instance H_1 earlier than H and variables v_1 and v_2 of F such that t_1 and t_2 are the substituents in H_1 for v_1 and v_2, respectively. (We let $v_1 = v_2$ if $t_1 = t_2$.) We say that a triple $\langle H_1, v_1, v_2 \rangle$ consisting of such an Herbrand instance and such variables *covers* H. Let $S(v_1, v_2)$ be the set of pairs of integers

$$\{\langle i, j \rangle \,|\, \langle F \rangle_i \text{ contains } y\text{-variables only and}$$

$$\langle F \rangle_j = \langle F \rangle_i(y_1/v_1, y_2/v_2)\}.\ddagger$$

---

‡ Here and frequently in what follows we leave implicit the bounds $1 \leq i, j \leq c(F)$ in the definitions of the sets of pairs $\langle i, j \rangle$.

ISBN 0-201-02540-X

Then $\langle H, H_1 \rangle$ exhibits $S(v_1, v_2)$. Moreover, by saturation, if $\langle H \rangle_i$ is old-in-H, then there is a j such that $\langle H \rangle_i = \langle H_1 \rangle_j$ and $\langle i, j \rangle$ is in $S(v_1, v_2)$. If $t_1 \neq t_2$ then for all i and j, $\langle H \rangle_i = \langle H \rangle_j$ iff $\langle F \rangle_i = \langle F \rangle_j$; and if $t_1 = t_2$ then $\langle H \rangle_i = \langle H \rangle_j$ iff $\langle F \rangle_i$ and $\langle F \rangle_j$ differ at most by occurrences of one rather than the other y-variable. Thus let $\text{Sim}(v_1, v_2)$ be $\{\langle i, j \rangle \mid \langle F \rangle_i = \langle F \rangle_j\}$ if $v_1 \neq v_2$, and let $\text{Sim}(v_1, v_2)$ be $\{\langle i, j \rangle \mid \langle F \rangle_i(y_2/y_1) = \langle F \rangle_j(y_2/y_1)\}$ if $v_1 = v_2$. Since we have chosen $v_1 = v_2$ just when $t_1 = t_2$, it follows that $\langle H \rangle_i = \langle H \rangle_j$ iff $\langle i, j \rangle$ is in $\text{Sim}(v_1, v_2)$.

Case 2. If H does not fall under Case 1, then there are Herbrand instances H_1 and H_2 earlier than H and variables v_1 and v_2 such that t_1 is the substituent in H_1 for v_1 and t_2 is the substituent in H_2 for v_2. We shall say that any such quadruple $\langle H_1, H_2, v_1, v_2 \rangle$ *covers* H. For each variable v and $\eta = 1$ and 2, let $S_\eta(v)$ be the set

$$\{\langle i, j \rangle \mid \langle F \rangle_i \text{ is } \{y_\eta\}\text{-based and } \langle F \rangle_j = \langle F \rangle_i(y_\eta/v)\}.$$

Then $\langle H, H_1 \rangle$ exhibits $S_1(v_1)$ and $\langle H, H_2 \rangle$ exhibits $S_2(v_2)$. By saturation, if $\langle H \rangle_i$ is old-in-H, then there is a j such that $\langle i, j \rangle$ is in either $S_1(v_1)$ or $S_2(v_2)$. Finally, note that $t_1 \neq t_2$; otherwise H would fall under Case 1. Hence $\langle H \rangle_i = \langle H \rangle_j$ iff $\langle i, j \rangle$ is in Sim_0, where $\text{Sim}_0 = \{\langle i, j \rangle \mid \langle F \rangle_i = \langle F \rangle_j\}$.

We have now analyzed how constraints of sorts (A) and (B) arise for Herbrand instances of F. Indeed, if $\mathfrak{A}$ is a truth-assignment defined on all Herbrand instances preceding a given H, then $\mathfrak{A}$ may be extended to H in any way that assures the following conditions:

(1) If H falls under Case 1, then for some triple $\langle H_1, v_1, v_2 \rangle$ that covers H,

    (a) $\mathfrak{A}(\langle H \rangle_i) = \mathfrak{A}(\langle H_1 \rangle_j)$ whenever $\langle i, j \rangle \in S(v_1, v_2)$; and

    (b) $\mathfrak{A}(\langle H \rangle_i) = \mathfrak{A}(\langle H \rangle_j)$ whenever $\langle i, j \rangle \in \text{Sim}(v_1, v_2)$.

(2) If H falls under Case 2, then for some quadruple $\langle H_1, H_2, v_1, v_2 \rangle$ that covers H,

    (a) $\mathfrak{A}(\langle H \rangle_i) = \mathfrak{A}(\langle H_1 \rangle_j)$ whenever $\langle i, j \rangle \in S_1(v_1)$, and $\mathfrak{A}(\langle H \rangle_i) = \mathfrak{A}(\langle H_2 \rangle_j)$ whenever $\langle i, j \rangle \in S_2(v_2)$; and

    (b) $\mathfrak{A}(\langle H \rangle_i) = \mathfrak{A}(\langle H \rangle_j)$ whenever $\langle i, j \rangle \in \text{Sim}_0$.

That is, in each case the conditions assure: (A) if $\langle H \rangle_i$ is old-in-H, then the extension of $\mathfrak{A}$ agrees with the previously determined value of $\mathfrak{A}$ on $\langle H \rangle_i$; and (B) if $\langle H \rangle_i = \langle H \rangle_j$, then $\mathfrak{A}(\langle H \rangle_i) = \mathfrak{A}(\langle H \rangle_j)$.

The crucial fact about Gödel–Kalmár–Schütte schemata, the fact that leads to solvability, is that the coinstantiations described in our analysis inevitably recur. More precisely, we have the

Inevitability Property.   For all variables v_1 and v_2 of F and all Herbrand instances H_1, there is an Herbrand instance H such that $\langle H, H_1 \rangle$ exhibits $S(v_1, v_2)$ and $\langle H, H \rangle$ exhibits $\text{Sim}(v_1, v_2)$. Moreover, if H_1 is in an expansion $E(F, p)$, then H is in $E(F, p + 1)$. For all variables v_1 and v_2

ISBN 0-201-02540-X

of F and all Herbrand instances H_1 and H_2, there is an Herbrand instance H such that $\langle H, H_1 \rangle$ exhibits $S_1(v_1)$, $\langle H, H_2 \rangle$ exhibits $S_2(v_2)$, and $\langle H, H \rangle$ exhibits Sim_0. Again, if H_1 and H_2 are in $E(F, p)$, then H is in $E(F, p + 1)$.

The Inevitability Property is obvious. Given v_1 and v_2 and H_1, the desired H is that Herbrand instance in which the substituents for y_1 and y_2 are identical with the substituents in H_1 for v_1 and v_2, respectively. Also, no Herbrand instance in an expansion $E(F, p)$ contains terms of height $>$ $p + 1$. Hence if H_1 is in $E(F, p)$, then H is in $E(F, p + 1)$. Similarly, given v_1, v_2, H_1, and H_2, the desired H is that Herbrand instance in which the substituent for y_1 is identical with the substituent in H_1 for v_1 and the substituent for y_2 is identical with the substituent in H_2 for v_2. Again, it is immediate that if H_1 and H_2 are in $E(F, p)$, then H is in $E(F, p + 1)$.

We shall exploit the Inevitability Property to construct inductively well-behaved truth-assignments. To do this we first specify a special "reflection condition" on truth-assignments. The Inevitability Property enters in the inductive step, that is, in extending any given truth-assignment fulfilling the reflection condition to an Herbrand instance on which the assignment is not yet defined. The basis step, as we shall see, amounts simply to an application of the pigeonhole argument. In the end, our proof comes down to showing the amenability of the Gödel–Kalmár–Schütte Class.

Let $\mathfrak{A}$ be a truth-assignment and let H and H' be (not necessarily distinct) Herbrand instances on which $\mathfrak{A}$ is defined. We say that H and H' $\mathfrak{A}$-*match* (or that H $\mathfrak{A}$-matches H') iff $\mathfrak{A}(\langle H \rangle_i) = \mathfrak{A}(\langle H' \rangle_i)$ for every i. A truth-assignment $\mathfrak{A}$ defined on some or all of $E(F, \omega)$ is *r-reflecting*, $r > 0$, iff $\mathfrak{A}$ is defined on every Herbrand instance in $E(F, r + 1)$ and, for each H on which $\mathfrak{A}$ is defined, there is an H' in $E(F, r)$ that $\mathfrak{A}$-matches H.

Now suppose that $\mathfrak{A}$ is an r-reflecting truth-assignment defined on all Herbrand instances earlier than a given Herbrand instance H. We so extend $\mathfrak{A}$ to H that r-reflection is preserved.

(1) H falls under Case 1. Choose a triple $\langle H_1, v_1, v_2 \rangle$ that covers H. Since H_1 is earlier than H, by the induction hypothesis there is an H_1' in $E(F, r)$ that $\mathfrak{A}$-matches H_1. By the Inevitability Property there is an H' in $E(F, r + 1)$ such that $\langle H', H_1' \rangle$ exhibits $S(v_1, v_2)$ and $\langle H', H' \rangle$ exhibits $\mathrm{Sim}(v_1, v_2)$. We claim that $\mathfrak{A}$ may be extended to H be setting $\mathfrak{A}(\langle H \rangle_i) =$ $\mathfrak{A}(\langle H' \rangle_i)$ for each i. For if $\langle i, j \rangle \in S(v_1, v_2)$, then $\langle H' \rangle_i = \langle H_1' \rangle_j$; hence $\mathfrak{A}(\langle H \rangle_i) = \mathfrak{A}(\langle H_1' \rangle_j) = \mathfrak{A}(\langle H_1 \rangle_j)$. And if $\langle i, j \rangle \in \mathrm{Sim}(v_1, v_2)$, then $\langle H' \rangle_i = \langle H' \rangle_j$; hence $\mathfrak{A}(\langle H \rangle_i) = \mathfrak{A}(\langle H \rangle_j)$.

Moreover, this extension of $\mathfrak{A}$ is r-reflecting: it is so defined that H $\mathfrak{A}$-matches H', and by the induction hypothesis there is an Herbrand instance in $E(F, r)$ that $\mathfrak{A}$-matches H'.

ISBN 0-201-02540-X

(2) H falls under Case 2. Choose a quadruple $\langle H_1, H_2, v_1, v_2 \rangle$ that covers H. By the induction hypothesis there are H_1' and H_2' in $E(F, r)$ that $\mathfrak{A}$-match H_1 and H_2, respectively. By the Inevitability Property there is an H' in $E(F, r + 1)$ such that $\langle H', H_1' \rangle$ exhibits $S_1(v_1)$, $\langle H', H_2' \rangle$ exhibits $S_2(v_2)$, and $\langle H', H' \rangle$ exhibits Sim_0. An argument parallel to that just given shows that without violating any constraints we may so extend $\mathfrak{A}$ to H that H $\mathfrak{A}$-matches H'. Again, this extension of $\mathfrak{A}$ is r-reflecting.

Our proof shows in particular that, starting with an r-reflecting truth-assignment $\mathfrak{A}$ defined just on $E(F, r + 1)$, we may proceed step by step to obtain an r-reflecting extension of $\mathfrak{A}$ defined on all of $E(F, \omega)$. Because this extension is r-reflecting it will verify $E(F, \omega)$ provided that $\mathfrak{A}$ verifies $E(F, r)$. Hence if there is an r-reflecting truth-assignment verifying $E(F, r)$ then F has a model.

Furthermore, there is an r-reflecting truth-assignment verifying $E(F, r)$ provided that $E(F, 2^{c(F)})$ is consistent. For let $\mathfrak{B}$ be any truth-assignment verifying $E(F, 2^{c(F)})$. We define a mapping $\xi_{\mathfrak{B}}$ on $E(F, 2^{c(F)})$ by setting

$$\xi_{\mathfrak{B}}(H) = \{i \mid 1 \leq i \leq c(F) \text{ and } \mathfrak{B} \text{ verifies } \langle H \rangle_i\}.$$

Since the range of $\xi_{\mathfrak{B}}$ contains at most $2^{c(F)}$ members, by the pigeonhole argument there is an r, $1 \leq r \leq c(F)$, such that for each H in $E(F, r + 1)$ there is an H' in $E(F, r)$ that $\mathfrak{B}$-matches H. Thus the restriction of $\mathfrak{B}$ to $E(F, r + 1)$ is r-reflecting.

The solvability of the Gödel–Kalmár–Schütte Class is immediate. For we have shown that a (saturated) schema F in this class has a model iff $E(F, 2^{c(F)})$ is consistent.

Indeed, we have in fact shown that $E(F, 2^{c(F)})$ is adequate. For we have shown that, given any truth-assignment $\mathfrak{B}$, there is an $r \leq 2^{c(F)}$ and a truth-assignment $\mathfrak{A}$ on $E(F, \omega)$ that agrees with $\mathfrak{B}$ on $E(F, r)$ and that is r-reflecting. Thus there is an instance-mapping Ψ whose range is included in $E(F, r)$ and such that $\mathfrak{A}(\langle H \rangle_i) = \mathfrak{B}(\langle \Psi(H) \rangle_i)$ for each H and each i. We may take $\mathfrak{A}$ to be an induced truth-assignment $\mathfrak{B}(\Gamma)$ thus: let Γ take each atom A to the atom $\langle \Psi(H) \rangle_i$, where H and i are the earliest Herbrand instance and integer for which $A = \langle H \rangle_i$. We then have $\mathfrak{B}(\Gamma(\langle H \rangle_i)) = \mathfrak{B}(\langle \Psi(H) \rangle_i)$, which is the central condition of the definition of adequacy.

REMARK.   In Chapter 2 we saw that if F is a Gödel–Kalmár–Schütte schema then $E(F, 2d(F, 2))$ is adequate. Now if F is saturated, then $2^{c(F)}$ is of the same order of magnitude as $d(F, 2 + n)$, where n is the number of x-variables of F. Thus the bound obtained immediately above is larger than that obtained in Chapter 2. But the smaller bound can in fact be obtained by the methods of this chapter. Such a refinement of the foregoing proof turns on the fact that an atom $\langle H \rangle_i$ is old-in-H only if $\langle F \rangle_i$ contains

ISBN 0-201-02540-X

no x-variables. Consequently a weaker notion of $\mathfrak{A}$-matching than that employed earlier may be formulated and applied in the inductive construction of a truth-assignment $\mathfrak{A}$. This weaker notion of $\mathfrak{A}$-matching allows a smaller bound to be used in the application of the pigeonhole argument. End of Remark.

The proof we have just given uses both parts of our analysis of coinstantiations: first, that the sets of pairs $S(v_1, v_2)$, $\text{Sim}(v_1, v_2)$, $S_1(v_1)$, $S_2(v_2)$, and Sim_0 provide a complete description of the coinstantiations that put constraints on the inductive construction of truth-assignments; second, that the coinstantiations so described inevitably recur. It is of some interest to formulate a proof that separates the contribution these two parts make to solvability. We shall see that the second part yields a condition on the syntactic structure of matrices necessary for the satisfiability of a (saturated) Gödel–Kalmár–Schütte schema, and the first part shows this condition to be sufficient as well.

Let F be an arbitrary schema. We consider the set $\mathbf{T}$ of truth-assignments on the matrix F^M of F. If $\mathfrak{T}$ is a member of $\mathbf{T}$ and if $\mathfrak{A}$ is a truth-assignment defined on an Herbrand instance H, we say that $\mathfrak{A}$ on H *matches* $\mathfrak{T}$ (on F^M) iff $\mathfrak{A}(\langle H \rangle_i) = \mathfrak{T}(\langle F \rangle_i)$ for each i. Hence for each $\mathfrak{A}$ and each H there is a unique $\mathfrak{T}$ in $\mathbf{T}$ that is matched by $\mathfrak{A}$ on H. And $\mathfrak{A}$ verifies H iff the matched truth-assignment verifies F^M.

Thus every truth-assignment on a set of Herbrand instances generates in a natural manner a mapping from those Herbrand instances to $\mathbf{T}$. The converse does not hold. Not every mapping from Herbrand instances to $\mathbf{T}$ is generated by a truth-assignment on those instances: for if $\langle H_1 \rangle_i = \langle H_2 \rangle_j$ then unless $\mathfrak{T}_1(\langle F \rangle_i) = \mathfrak{T}_2(\langle F \rangle_j)$ there will be no truth-assignment $\mathfrak{A}$ that on H_1 matches $\mathfrak{T}_1$ and on H_2 matches $\mathfrak{T}_2$.‡

We say that a pair $\langle \mathfrak{T}, \mathfrak{T}' \rangle$ of members of $\mathbf{T}$ *fits* a set S of pairs of integers iff $\mathfrak{T}(\langle F \rangle_i) = \mathfrak{T}'(\langle F \rangle_j)$ whenever $\langle i, j \rangle$ is in S. If two members $\mathfrak{T}$ and $\mathfrak{T}'$ of $\mathbf{T}$ are matched by a truth-assignment $\mathfrak{A}$ on two Herbrand instances H and H' and if $\langle H, H' \rangle$ exhibits S, then $\langle \mathfrak{T}, \mathfrak{T}' \rangle$ fits S.

Now let F be a saturated Gödel–Kalmár–Schütte schema. The following condition on $\mathbf{T}$ is necessary and sufficient for the satisfiability of F.

GKS Condition.   There is a subset $\mathbf{U}$ of $\mathbf{T}$ such that

(I)   every member of $\mathbf{U}$ verifies F^M;

(II)   for all variables v_1 and v_2 and every $\mathfrak{T}_1$ in $\mathbf{U}$ there is a $\mathfrak{T}$ in $\mathbf{U}$ such that $\langle \mathfrak{T}, \mathfrak{T}_1 \rangle$ fits $S(v_1, v_2)$ and $\langle \mathfrak{T}, \mathfrak{T} \rangle$ fits $\text{Sim}(v_1, v_2)$;

---

‡ This is another way in which we may view the effect coinstantiations have on the decision problem. For it is the occurrence of nontrivial coinstantiations that prevents an arbitrary mapping from $E(F, \omega)$ to the set of truth-assignments verifying F^M from being generated by a truth-assignment verifying $E(F, \omega)$.

ISBN 0-201-02540-X

(III)   for all variables v_1 and v_2 and all $\mathfrak{T}_1$ and $\mathfrak{T}_2$ in **U** there is a $\mathfrak{T}$ in **U** such that $\langle \mathfrak{T}, \mathfrak{T}_1 \rangle$ fits $S_1(v_1)$ and $\langle \mathfrak{T}, \mathfrak{T}_2 \rangle$ fits $S_2(v_2)$.

Proof of Necessity.   Assume F has a model. Let $\mathfrak{A}$ be a truth-assignment verifying $E(F, \omega)$, and let **U** be the set of members of **T** that are matched by $\mathfrak{A}$ on at least one Herbrand instance. Then every member of **U** verifies F^M. Now let v_1 and v_2 be any variables and let $\mathfrak{T}_1$ be a member of **U**; suppose $\mathfrak{T}_1$ is matched by $\mathfrak{A}$ on H_1. By the Inevitability Property there is an Herbrand instance H such that $\langle H, H_1 \rangle$ exhibits $S(v_1, v_2)$ and $\langle H, H \rangle$ exhibits $\text{Sim}(v_1, v_2)$. If $\mathfrak{T}$ is the member of **T** matched by $\mathfrak{A}$ on H, then $\langle \mathfrak{T}, \mathfrak{T}_1 \rangle$ fits $S(v_1, v_2)$ and $\langle \mathfrak{T}, \mathfrak{T} \rangle$ fits $\text{Sim}(v_1, v_2)$. By the choice of **U**, this $\mathfrak{T}$ is in **U**. The proof that clause (III) of the GKS Condition is fulfilled is analogous. $\square$

Proof of Sufficiency.   Assume F fulfills the GKS Condition. We construct inductively a truth-assignment $\mathfrak{A}$ that verifies $E(F, \omega)$.

Basis.   Let $H = F^*(y_1/\uparrow, y_2/\uparrow)$. No atom is old-in-$H$, since H is the earliest Herbrand instance. Also $\langle H \rangle_i = \langle H \rangle_j$ iff $\langle i, j \rangle \in \text{Sim}(y_1, y_1)$. By (II) of the GKS Condition there is a $\mathfrak{T}$ in **U** such that $\langle \mathfrak{T}, \mathfrak{T} \rangle$ fits $\text{Sim}(y_1, y_1)$. We may define $\mathfrak{A}$ on H to match $\mathfrak{T}$ without violating any constraints.

Induction.   Suppose $\mathfrak{A}$ is so defined on all Herbrand instances earlier than H that $\mathfrak{A}$ on each such Herbrand instance matches some member of **U**.

(1) H falls under Case 1. Choose a triple $\langle H_1, v_1, v_2 \rangle$ that covers H. Let $\mathfrak{T}_1$ be that member of **U** matched by $\mathfrak{A}$ on H_1, and let $\mathfrak{T}$ be a member of **U** such that $\langle \mathfrak{T}, \mathfrak{T}_1 \rangle$ fits $S(v_1, v_2)$ and $\langle \mathfrak{T}, \mathfrak{T} \rangle$ fits $\text{Sim}(v_1, v_2)$. We may so extend $\mathfrak{A}$ to H that $\mathfrak{A}$ on H matches $\mathfrak{T}$. For if $\langle i, j \rangle \in S(v_1, v_2)$, then $\mathfrak{T}(\langle F \rangle_i) = \mathfrak{T}_1(\langle F \rangle_j)$; hence $\mathfrak{A}(\langle H \rangle_i) = \mathfrak{A}(\langle H_1 \rangle_j)$. And if $\langle i, j \rangle \in \text{Sim}(v_1, v_2)$, then $\mathfrak{T}(\langle F \rangle_i) = \mathfrak{T}(\langle F \rangle_j)$; hence $\mathfrak{A}(\langle H \rangle_i) = \mathfrak{A}(\langle H \rangle_j)$.

(2) H falls under Case 2. Choose a quadruple $\langle H_1, H_2, v_1, v_2 \rangle$ that covers H. Let $\mathfrak{T}_1$ and $\mathfrak{T}_2$ be the members of **U** matched by $\mathfrak{A}$ on H_1 and on H_2. Then let $\mathfrak{T}$ be a member of **U** such that $\langle \mathfrak{T}, \mathfrak{T}_1 \rangle$ fits $S_1(v_1)$ and $\langle \mathfrak{T}, \mathfrak{T}_2 \rangle$ fits $S_2(v_2)$. Note that $\langle \mathfrak{T}, \mathfrak{T} \rangle$ fits Sim_0 since $\text{Sim}_0 = \{\langle i, j \rangle \,|\, \langle F \rangle_i = \langle F \rangle_j\}$. Reasoning as in the previous case, we conclude that $\mathfrak{A}$ may be so extended to H that $\mathfrak{A}$ on H matches $\mathfrak{T}$.

We have thus shown how to construct a truth-assignment $\mathfrak{A}$ on $E(F, \omega)$ such that $\mathfrak{A}$ on each Herbrand instance matches some member of **U**. Since each member of **U** verifies F^M, $\mathfrak{A}$ verifies $E(F, \omega)$. Hence if F fulfills the GKS Condition, then F has a model. $\square$

REMARK.   We may reformulate the GKS Condition by using a modification of the notion of profile. Since F is saturated, each member $\mathfrak{T}$ of **T** assigns values to all atomic formulas constructed from the predicate letters and the variables of F. Hence we may use the notation $\mathfrak{p}_{\mathfrak{T}}(v_1, \ldots, v_k)$, where

ISBN 0-201-02540-X

$v_1, \ldots, v_k$ are variables of F: this is just the set of all signed atomic formulas A constructed from predicate letters of F and "indexing variables" $\delta_1, \ldots, \delta_k$ such that $\mathfrak{T}$ verifies $A(\delta_1/v_1, \ldots, \delta_k/v_k)$ (see page 39). Clauses (II) and (III) of the GKS Condition are then equivalent to the following:

(II′)   for all variables v_1 and v_2 and every member $\mathfrak{T}_1$ of U there is a $\mathfrak{T}$ in U such that $\mathfrak{p}_{\mathfrak{T}}(y_1, y_2) = \mathfrak{p}_{\mathfrak{T}_1}(v_1, v_2)$; if moreover $v_1 = v_2$, then $\mathfrak{p}_{\mathfrak{T}}(y_1, y_2, x_1, \ldots, x_n) = \mathfrak{p}_{\mathfrak{T}}(y_1, y_1, x_1, \ldots, x_n) = \mathfrak{p}_{\mathfrak{T}}(y_2, y_2, x_1, \ldots, x_n)$;

(III′)   for all variables v_1 and v_2 and all members $\mathfrak{T}_1$ and $\mathfrak{T}_2$ of U there is a $\mathfrak{T}$ in U such that $\mathfrak{p}_{\mathfrak{T}}(y_1) = \mathfrak{p}_{\mathfrak{T}_1}(v_1)$ and $\mathfrak{p}_{\mathfrak{T}}(y_2) = \mathfrak{p}_{\mathfrak{T}_2}(v_2)$.

This reformulation is close to the original Gödel–Kalmár–Schütte approach, and will play a role in Chapter 8, §4.   End of Remark.

The sort of analysis we have carried out for Gödel–Kalmár–Schütte schemata may be extended to the other classes of Chapter 2. For each class we would first seek characterizations of the coinstantiations that give rise to constraints on the inductive construction of truth-assignments on $E(F, \omega)$. Second we would show that the coinstantiations so characterized inevitably recur, in some sense akin to that provided by the Inevitability Property. A suitable execution of this program would allow us to construct well-behaved truth-assignments on $E(F, \omega)$.

In this generalization of the approach, the Inevitability Property has to be modified. An example will show the reason. Let $F = \forall y_1 \forall y_2 \forall y_3 \exists x F^M$ and let $\{y_1, y_2, y_3\}$ be the only y-set of cardinality >1 in $b(F)$. Hence F is in Class 2.4. Suppose that $H = F^*(y_1/t_1, y_2/t_2, y_3/t_3)$ is an Herbrand instance in which $t_3 = t_2$, and such that t_1, t_2, and t_3 all occur in an Herbrand instance H_1 earlier than H, say as the substituents for y_3, y_2, and y_1. Thus $\langle H \rangle_i$ is old-in-H iff $\langle F \rangle_i$ contains y-variables only. Let $S = \{\langle i, j \rangle \mid \langle F \rangle_i$ contains y-variables only and $\langle F \rangle_j = \langle F \rangle_i(y_1/y_3, y_3/y_1)\}$; and let $\text{Sim}_1 = \{\langle i, j \rangle \mid \langle F \rangle_i(y_3/y_2) = \langle F \rangle_j(y_3/y_2)\}$. Then $\langle H, H_1 \rangle$ exhibits S, and $\langle H \rangle_i = \langle H \rangle_j$ iff $\langle i, j \rangle \in \text{Sim}_1$. Now a straightforward analogue to the Inevitability Property would require for every H_1 an Herbrand instance H such that $\langle H, H_1 \rangle$ exhibits S and $\langle H, H \rangle$ exhibits Sim_1. But this fails. Given H_1, there is such an H if and only if the substituents in H_1 for y_1 and y_2 are identical. Thus let $\text{Sim}_2 = \{\langle i, j \rangle \mid \langle F \rangle_i(y_2/y_1) = \langle F \rangle_j(y_2/y_1)\}$. We then have a

Modified Inevitability Property.   For every Herbrand instance H_1 such that $\langle H_1, H_1 \rangle$ exhibits Sim_2, there is an Herbrand instance H such that $\langle H, H_1 \rangle$ exhibits S and $\langle H, H \rangle$ exhibits Sim_1.

In fact we can prove, for each coinstantiation-characterization needed to describe constraints, either a straightforward analogue to the Inevitability Property or else a Modified Inevitability Property. Truth-assignments $\mathfrak{A}$ on $E(F, \omega)$ may then be inductively constructed. In such a construction we

ISBN 0-201-02540-X

use an induction hypothesis stronger than r-reflection. At each step we require the following: for every H on which $\mathfrak{A}$ is defined there is an H' in $E(F, r)$ such that H and H' $\mathfrak{A}$-match and moreover $\langle H' \rangle_i = \langle H' \rangle_j$ whenever $\langle H \rangle_i = \langle H \rangle_j$. This strengthened induction hypothesis allows us to use a Modified Inevitability Property in the construction. And the attention we give to the sets Sim that describe simultaneous coinstantiations assures us that the induction hypothesis is preserved by the construction. Moreover, the pigeonhole argument still suffices to provide the basis of the induction, although the bound will be larger than $2^{c(F)}$. Finally, such an analysis of coinstantiations would also provide a syntactic criterion for satisfiability along the lines of the GKS Condition, but the criterion would be considerably more complex.

The concern with simultaneous coinstantiations that has arisen here is of a piece with our concern in Chapter 2 that the sequence-mapping γ^* preserve repeated occurrences of terms in sequences and that the term-mapping γ preserve (some) identity of arguments. For as we pointed out in Chapter 2, these conditions assure that $\langle \Psi(H) \rangle_i = \langle \Psi(H) \rangle_j$ whenever $\langle H \rangle_i = \langle H \rangle_j$.

A Modified Inevitability Property can be used for each of Classes 2.4–2.8. But as the classes treated increase in extent, the specification of the coinstantiation-characterizations becomes more and more complex. For this reason we shall not pursue this approach in any further detail.

We close this section by examining how the presence of two overlapping y-sets in the basis of a schema produces coinstantiations that do *not* allow the inductive construction of well-behaved truth-assignments. Let $F = \forall y_1 \forall y_2 \forall y_3 \exists x F^M$ and let $\{y_1, y_2\}$ and $\{y_2, y_3\}$ be in $b(F)$. Suppose $H = F^*(y_1/t_1, y_2/t_2, y_3/t_3)$ is an Herbrand instance such that just the atoms containing t_1 and t_2 and those containing t_2 and t_3 are old-in-H. We may choose H_1 and H_2 earlier than H, and variables v_1, v_2, w_1, w_2 such that t_1 and t_2 are the substituents in H_1 for v_1 and v_2, and t_2 and t_3 are the substituents in H_2 for w_1 and w_2. Thus let $S_1 = \{\langle i, j \rangle | \langle F \rangle_i \text{ is } \{y_1, y_2\}\text{-based and } \langle F \rangle_j = \langle F \rangle_i(y_1/v_1, y_2/v_2)\}$, and let $S_2 = \{\langle i, j \rangle | \langle F \rangle_i \text{ is } \{y_2, y_3\}\text{-based and } \langle F \rangle_j = \langle F \rangle_i(y_2/w_1, y_3/w_2)\}$. Then $\langle H, H_1 \rangle$ exhibits S_1 and $\langle H, H_2 \rangle$ exhibits S_2. But the only analogue to the Inevitability Property that holds for these coinstantiations is a considerably weakened one: for all Herbrand instances H_1 and H_2, *if* the substituent in H_1 for v_2 is identical with the substituent in H_2 for w_1, *then* there is an Herbrand instance H such that $\langle H, H_1 \rangle$ exhibits S_1 and $\langle H, H_2 \rangle$ exhibits S_2. And this analogue of the Inevitability Property is quite unlike that of the previous case. If we wished to exploit it in an inductive construction of a truth-assignment $\mathfrak{A}$, it would not be enough to require an induction hypothesis of the following form: for every Herbrand instance H on which $\mathfrak{A}$ is defined there is an H' in $E(F, r)$ that $\mathfrak{A}$-matches H and, perhaps, that has some further properties

ISBN 0-201-02540-X

in common with H. Rather, we would have to require that if $\mathfrak{A}$ is defined on H_1 and H_2 and the substituent in H_1 for v_2 is identical with that in H_2 for w_1, then there are $H_1{}'$ and $H_2{}'$ in $E(F, r)$ that $\mathfrak{A}$-match H_1 and H_2, respectively, and such that the substituent in $H_1{}'$ for v_2 is identical with the substituent in $H_2{}'$ for w_1. But this is too much to ask: we cannot inductively extend $\mathfrak{A}$ so that this hypothesis is preserved. Moreover, we cannot even obtain a basis for such an induction. Indeed, this problem is at bottom just that pointed out in Chapter 2: we cannot require of the sequence-mapping γ^* that it take overlapping sequences to overlapping sequences.

2. Reductions to Monadic

The analysis of coinstantiations carried out in the previous section enables us to reduce the Gödel–Kalmár–Schütte Class to the Monadic Class (2.1). Given any saturated Gödel-Kalmár-Schütte schema F we shall construct monadic schemata F_1 and F_2 with the following properties:

(1)   Each truth-assignment $\mathfrak{A}$ on $E(F, \omega)$ corresponds in a canonical manner to a truth-assignment $\mathfrak{B}$ on $E(F_1 \wedge F_2, \omega)$ that verifies every instance of the functional form $F_2{}^*$. If moreover $\mathfrak{A}$ verifies $E(F, \omega)$, then $\mathfrak{B}$ verifies every instance of $F_1{}^*$. Hence $F_1 \wedge F_2$ has a model provided that F has a model.

(2)   Each truth-assignment $\mathfrak{B}$ on $E(F_1 \wedge F_2, \omega)$ that verifies every instance of $F_2{}^*$ corresponds in a canonical manner to a truth-assignment $\mathfrak{A}$ on $E(F, \omega)$. If moreover $\mathfrak{B}$ verifies every instance of $F_1{}^*$, then $\mathfrak{A}$ verifies $E(F, \omega)$. Hence F has a model provided that $F_1 \wedge F_2$ has a model.

We may view the schema F_2 as encoding into monadic quantification theory properties of the coinstantiations in $E(F, \omega)$, or, more precisely, properties of the constraints on truth-assignments that those coinstantiations produce. In the proofs of properties (1) and (2) we shall make frequent use both of the sets of pairs $S(v_1, v_2)$, $\mathrm{Sim}(v_1, v_2)$, $S_1(v_1)$, $S_2(v_2)$, and Sim_0 defined earlier on pages 174–175, and of the Inevitability Property.

Let $F = \forall y_1 \forall y_2 \exists x_1 \cdots \exists x_n F^M$ be a saturated Gödel–Kalmár–Schütte schema. The schemata F_1 and F_2 will contain monadic predicate letters P_i for $1 \le i \le c(F)$. If we think of each term t in $D(F_1 \wedge F_2, \omega)$ as being correlated with an Herbrand instance H of F, then each truth-assignment $\mathfrak{A}$ on $E(F, \omega)$ will correspond to the truth-assignment on $E(F_1 \wedge F_2, \omega)$ that verifies $P_i t$ iff $\mathfrak{A}$ verifies $\langle H \rangle_i$.

Let F_1 be the schema $\forall y B$, where B results from the matrix F^M by replacing the ith atomic-formula-occurrence with $P_i y$, $1 \le i \le c(F)$. Thus F^M differs from the formula B in containing $\langle F \rangle_i$ at that place where B

ISBN 0-201-02540-X

contains $P_i y$. (Note that to obtain B we replace occurrences rather than atomic formulas. We may have $\langle F \rangle_i = \langle F \rangle_j$ yet $\langle B \rangle_i = P_i y \neq P_j y = \langle B \rangle_j$.)

To obtain F_2 we first construct schemata C_{v_1, v_2} and D_{v_1, v_2} for all variables v_1 and v_2 of F. For each pair $\langle v_1, v_2 \rangle$ of variables, let C_{v_1, v_2} be

$$\forall y \exists x [\bigwedge (P_i x \Leftrightarrow P_j y) \wedge \bigwedge (P_i x \Leftrightarrow P_j x)],$$

where the first conjunction is over the pairs $\langle i, j \rangle$ in $S(v_1, v_2)$ and the second conjunction is over the pairs $\langle i, j \rangle$ in $\mathrm{Sim}(v_1, v_2)$. Moreover, for each pair $\langle v_1, v_2 \rangle$ of variables, let D_{v_1, v_2} be the schema

$$\forall y_1 \forall y_2 \exists x [\bigwedge (P_i x \Leftrightarrow P_j y_1) \wedge \bigwedge (P_i x \Leftrightarrow P_j y_2)$$
$$\wedge \bigwedge (P_i x \Leftrightarrow P_j x)],$$

where the first conjunction is over the pairs $\langle i, j \rangle$ in $S_1(v_1)$, the second is over the pairs $\langle i, j \rangle$ in $S_2(v_2)$, and the third is over the pairs $\langle i, j \rangle$ in Sim_0.

Finally, let F_2 be (a rectified equivalent of) the conjunction of all C_{v_1, v_2} and D_{v_1, v_2} (that is, for all variables v_1 and v_2 of F). The functional form $F_2{}^*$ contains $(n + 2)^2$ 1-place function signs g_{v_1, v_2} arising from the conjuncts C_{v_1, v_2} and $(n + 2)^2$ 2-place function signs h_{v_1, v_2} arising from the conjuncts D_{v_1, v_2}. Hence the domain $D(F_1 \wedge F_2, \omega)$ is generated from $\dashv$ by these function signs.

Proof of (1). We define a mapping $\theta : D(F_1 \wedge F_2, \omega) \to E(F, \omega)$. That is, θ takes terms to Herbrand instances. Our definition makes crucial use of the Inevitability Property.

$\theta \dashv = F^*(y_1 / \dashv, y_2 / \dashv)$;

$\theta g_{v_1, v_2}(t) = H$, where H is the Herbrand instance of F such that $\langle H, \theta t \rangle$ exhibits $S(v_1, v_2)$ and $\langle H, H \rangle$ exhibits $\mathrm{Sim}(v_1, v_2)$;

$\theta h_{v_1, v_2}(t_1 t_2) = H$, where H is the Herbrand instance of F such that $\langle H, \theta t_1 \rangle$ exhibits $S_1(v_1)$, $\langle H, \theta t_2 \rangle$ exhibits $S_2(v_2)$, and $\langle H, H \rangle$ exhibits Sim_0.

Given a truth-assignment $\mathfrak{A}$ on $E(F, \omega)$, we let $\mathfrak{B}$ be the truth-assignment on $E(F_1 \wedge F_2, \omega)$ such that

$$\mathfrak{B}(P_i t) = \mathfrak{A}(\langle \theta t \rangle_i)$$

for every term t and every i. We show first that $\mathfrak{B}$ verifies every instance of $F_2{}^*$ over $D(F_1 \wedge F_2, \omega)$. Each such instance is merely a conjunction of biconditionals. Hence it is more than enough to show that if $P_i s \Leftrightarrow P_j t$ is a conjunct in an instance of $F_2{}^*$, then $\langle \theta s \rangle_i = \langle \theta t \rangle_j$; for then $\mathfrak{B}(P_i s) = \mathfrak{B}(P_j t)$, and $P_i s \Leftrightarrow P_j t$ is verified. But $\langle \theta s \rangle_i = \langle \theta t \rangle_j$, as we may easily see by examining these biconditionals. For example, among the conjuncts in an

ISBN 0-201-02540-X

instance of $F_2{}^*$ are biconditionals

$$P_i g_{v_1, v_2}(t) \Leftrightarrow P_j t$$

for $\langle i, j \rangle$ in $S(v_1, v_2)$. But $\langle \theta g_{v_1, v_2}(t), \theta t \rangle$ exhibits $S(v_1, v_2)$; hence $\langle \theta g_{v_1, v_2}(t) \rangle_i = \langle \theta t \rangle_j$. The other cases are similar.

Property (1) will be established once we show that if $\mathfrak{A}$ verifies every Herbrand instance of F, then $\mathfrak{B}$ verifies every instance of $F_1{}^*$ over $D(F_1 \wedge F_2, \omega)$. By the construction of F_1 and the definition of $\mathfrak{B}$ we have

$$\mathfrak{B}(\langle F_1{}^*(y/t) \rangle_i) = \mathfrak{A}(\langle \theta t \rangle_i)$$

for every term t in $D(F_1 \wedge F_2, \omega)$ and every integer i. Moreover, the matrices of F_1 and F share the same truth-functional form. The desired result is immediate. $\square$

Proof of (2). We shall define a mapping $\theta: E(F, \omega) \to D(F_1 \wedge F_2, \omega)$. That is, θ takes Herbrand instances to terms. First we select for each Herbrand instance H of F a triple $\langle H_1, v_1, v_2 \rangle$ or quadruple $\langle H_1, H_2, v_1, v_2 \rangle$ that covers H, depending on whether H falls under Case 1 or under Case 2 (page 174). Then we let

$$\theta H = \begin{cases} g_{y_1, y_1}(4) & \text{if } H = F^*(y_1/4, y_2/4); \\ g_{v_1, v_2}(\theta H_1) & \text{if } H \text{ falls under Case 1,} \\ & \text{where } \langle H_1, v_1, v_2 \rangle \text{ is the} \\ & \text{selected triple that covers } H; \\ h_{v_1, v_2}(\theta H_1\ \theta H_2) & \text{if } H \text{ falls under Case 2,} \\ & \text{where } \langle H_1, H_2, v_1, v_2 \rangle \text{ is the} \\ & \text{selected quadruple that covers } H. \end{cases}$$

Now let $\mathfrak{B}$ be a truth-assignment on $E(F_1 \wedge F_2, \omega)$ that verifies every instance of $F_2{}^*$ over $D(F_1 \wedge F_2, \omega)$. We show by induction that there is a truth-assignment $\mathfrak{A}$ on $E(F, \omega)$ such that

$$(\#) \quad \mathfrak{A}(\langle H \rangle_i) = \mathfrak{B}(P_i \theta H)$$

for every Herbrand instance H of F and every integer i. Our argument closely parallels the two inductive constructions of truth-assignments $\mathfrak{A}$ in §1; it uses conditions (1) and (2) stated on page 175 that assure no violation of constraints occurs in such a construction.

Basis. Let $H = F^*(y_1/4, y_2/4)$. Since no atom $\langle H \rangle_i$ is old-in-H, there is a truth-assignment $\mathfrak{A}$ on H fulfilling $(\#)$ provided that $\mathfrak{B}(P_i \theta H) = \mathfrak{B}(P_j \theta H)$ whenever $\langle H \rangle_i = \langle H \rangle_j$. Now $\langle H \rangle_i = \langle H \rangle_j$ iff $\langle i, j \rangle \in \mathrm{Sim}(y_1, y_1)$. By the definition of the schema C_{y_1, y_1} and the choice of θH, for each pair $\langle i, j \rangle$ in $\mathrm{Sim}(y_1, y_1)$ the biconditional $P_i \theta H \Leftrightarrow P_j \theta H$ is a conjunct in some instance of $F_2{}^*$. Thus $\mathfrak{B}(P_i \theta H) = \mathfrak{B}(P_j \theta H)$ for all such pairs $\langle i, j \rangle$.

Induction. Suppose there is a truth-assignment $\mathfrak{A}$ that is defined on all

Herbrand instances earlier than a given Herbrand instance H and that fulfills (#). We seek to show the existence of an extension of $\mathfrak{A}$ to H that continues to fulfill (#).

If H falls under Case 1, then $\theta H = g_{v_1,v_2}(\theta H_1)$ for some triple $\langle H_1, v_1, v_2 \rangle$ that covers H. By condition (1) on page 175, it suffices to show that (a) if $\langle i, j \rangle \in S(v_1, v_2)$, then $\mathfrak{B}(P_i\theta H) = \mathfrak{B}(P_j\theta H_1)$, and (b) if $\langle i, j \rangle \in \mathrm{Sim}(v_1, v_2)$, then $\mathfrak{B}(P_i\theta H) = \mathfrak{B}(P_j\theta H)$. But by the definition of the schema C_{v_1,v_2}, if $\langle i, j \rangle \in S(v_1, v_2)$, then $P_i\theta H \Leftrightarrow P_j\theta H_1$ is a conjunct in some instance of $F_2{}^*$; and if $\langle i, j \rangle \in \mathrm{Sim}(v_1, v_2)$, then $P_i\theta H \Leftrightarrow P_j\theta H$ is a conjunct in the same instance of $F_2{}^*$. Hence, since $\mathfrak{B}$ verifies every instance of $F_2{}^*$, (a) and (b) hold.

If H falls under Case 2, then $\theta H = h_{v_1,v_2}(\theta H_1\, \theta H_2)$ for some quadruple $\langle H_1, H_2, v_1, v_2 \rangle$ that covers H. By condition (2) on page 175 it suffices to show that (a) if $\langle i, j \rangle \in S_1(v_1)$, then $\mathfrak{B}(P_i\theta H) = \mathfrak{B}(P_j\theta H_1)$, and if $\langle i, j \rangle \in S_2(v_2)$, then $\mathfrak{B}(P_i\theta H) = \mathfrak{B}(P_j\theta H_2)$; and (b) if $\langle i, j \rangle \in \mathrm{Sim}_0$, then $\mathfrak{B}(P_i\theta H) = \mathfrak{B}(P_j\theta H)$. Again, from the construction of the schemata D_{v_1,v_2} and the assumption that $\mathfrak{B}$ verifies each instance of $F_2{}^*$, (a) and (b) are seen to hold.

It remains to show that if $\mathfrak{B}$ also verifies every instance of $F_1{}^*$ over $D(F_1 \wedge F_2, \omega)$, then $\mathfrak{A}$ verifies $E(F, \omega)$. Now $\langle F_1{}^* \rangle_i = P_i y$ for each i, and the matrices of F and F_1 share the same truth-functional form. Since we have just established that $\mathfrak{A}(\langle H \rangle_i) = \mathfrak{B}(P_i\theta H)$ for each Herbrand instance H of F and each integer i, the desired result is immediate. $\square$

REMARK. Although each truth-assignment verifying $E(F_1 \wedge F_2, \omega)$ corresponds to a truth-assignment verifying $E(F, \omega)$, a finite model for $F_1 \wedge F_2$ does not correspond to one for F.‡ Indeed, let us consider models $\mathfrak{M}$ for $F_1 \wedge F_2$ that are "small" in the following sense: for no two distinct elements a and b of $\mathfrak{M}$ are just the same predicate letters true in $\mathfrak{M}$ of a and of b. (Every model of a monadic schema contains a finite submodel that is small in this sense. In fact, our finite controllability proof for the Monadic Class yields small models.) If $\mathfrak{M}$ is a small model, then there is an injection from the universe of $\mathfrak{M}$ into the set $\mathbf{T}$ of truth-assignments on the matrix of F: an element of $\mathfrak{M}$ is carried to that truth-assignment $\mathfrak{T}$ that, for each i, verifies $\langle F \rangle_i$ iff P_i is true in $\mathfrak{M}$ of the element. Hence a small model corresponds to a subset $\mathbf{U}$ of $\mathbf{T}$. Moreover, this set $\mathbf{U}$ fulfills the GKS Condition: since $\mathfrak{M}$ is a model for F_1, every truth-assignment in $\mathbf{U}$ verifies F^M; since $\mathfrak{M}$ is a model for the schemata C_{v_1,v_2}, the set $\mathbf{U}$ fulfills clause (II) of the condition; and since $\mathfrak{M}$ is a model for the schemata D_{v_1,v_2}, the set $\mathbf{U}$ fulfills clause (III)

---

‡ Thus our reduction is unlike those of Ackermann (1954, Chapter 7). Ackermann proves the finite controllability of the Ackermann Class (Class 2.2) and the Gödel–Kalmár–Schütte Class by *conservatively* reducing them to the Monadic Class.

of the condition. Conversely, any subset U of T that fulfills the GKS Condition corresponds to a small model for $F_1 \wedge F_2$. The universe of this model may be taken to be U itself, and then each predicate letter P_i is interpreted as true of just those truth-assignments that verify $\langle F \rangle_i$. Clause (I) of the condition ensures that this interpretation provides a model for F_1, and clauses (II) and (III) that it provides a model for F_2.   End of Remark.

Monadic reductions of the other solvable classes of Chapter 2 may also be devised. Such reductions, of course, rely on a complete analysis of coinstantiations of the sort sketched near the end of §1.

More surprisingly, perhaps, we may also reduce Class 5.1 to the Monadic Class, even though Class 5.1 is not finitely controllable. Once again, given a schema F in Class 5.1, we proceed by encoding the constraints on truth-assignments to $E(F, \omega)$ spawned by coinstantiations. But here, finite models for the monadic schema correspond to sets of sets of truth-assignments on F^M that possess certain closure properties.

Class 5.1 contains those $\forall \exists \forall$-schemata each atomic subformula of which has one of the seven forms Py_1y_1, Pxx, Py_2y_2, Py_1x, Pxy_1, Py_1y_2, Pxy_2. We call such a schema F *saturated* if every atomic formula containing a predicate letter of F and having one of the allowable forms actually occurs in F. As before, for each saturated schema F in Class 5.1, we shall construct monadic schemata F_1 and F_2 with the following properties:

(1)   Each truth-assignment $\mathfrak{A}$ on $E(F, \omega)$ corresponds to a truth-assignment $\mathfrak{B}$ on $E(F_1 \wedge F_2, \omega)$ that verifies every instance of $F_2{}^*$. If moreover $\mathfrak{A}$ verifies $E(F, \omega)$, then $\mathfrak{B}$ verifies every instance of $F_1{}^*$.

(2)   Each truth-assignment $\mathfrak{B}$ on $E(F_1 \wedge F_2, \omega)$ that verifies every instance of $F_2{}^*$ corresponds to a truth-assignment $\mathfrak{A}$ on $E(F, \omega)$. If moreover $\mathfrak{B}$ verifies every instance of $F_1{}^*$, then $\mathfrak{A}$ verifies $E(F, \omega)$.

To obtain F_1 and F_2 we first define four sets of pairs of positive integers $\leq c(F)$ that enable us to characterize the coinstantiations in $E(F, \omega)$.

$S_1 = \{\langle i, j \rangle \,|\, \langle F \rangle_i(y_2/y_1) = \langle F \rangle_j(y_2/y_1)\}$. Thus S_1 is exhibited by $\langle H, H \rangle$ whenever H is an Herbrand instance having the form $F^*(y_1/s, y_2/s)$.

$S_2 = \{\langle i, j \rangle \,|\, \langle F \rangle_i(y_2/x) = \langle F \rangle_j(y_2/x)\}$. Thus S_2 is exhibited by $\langle H, H \rangle$ whenever H has the form $F^*(y_1/s, y_2/f_x(s))$.

$S_3 = \{\langle i, i \rangle \,|\, \langle F \rangle_i \text{ does not contain } y_2\}$. Thus $\langle H, H' \rangle$ exhibits S whenever the substituents in H and in H' for y_1 are identical.

$S_4 = \{\langle i, j \rangle \,|\, \langle F \rangle_i \text{ does not contain } y_1 \text{ and } \langle F \rangle_j = \langle F \rangle_i(x/y_1)\}$. Thus $\langle H, H' \rangle$ exhibits S whenever $H = F^*(y_1/s, y_2/t)$ and $H' = F^*(y_1/f_x(s), y_2/t)$ for some terms s and t.

Now let T be the set of truth-assignments on the matrix F^M of F. Recall that a pair $\langle \mathfrak{T}, \mathfrak{T}' \rangle$ of members of T *fits* a set S of pairs of integers iff $\mathfrak{T}(\langle F \rangle_i)$

ISBN 0-201-02540-X

$= \mathfrak{T}(\langle F \rangle_j)$ for all pairs $\langle i, j \rangle$ in S; and that if $\mathfrak{A}$ is a truth-assignment on Herbrand instances, then $\mathfrak{A}$ on H *matches* $\mathfrak{T}$ iff $\mathfrak{A}(\langle H \rangle_i) = \mathfrak{T}(\langle F \rangle_i)$ for each i.

For each $\mathfrak{T}$ in **T** let $P_{\mathfrak{T}}$ be a monadic predicate letter. We shall so construct F_1 and F_2, using the predicate letters $P_{\mathfrak{T}}$, that $D(F_1 \wedge F_2, \omega)$ $= D(F, \omega)$. Given a truth-assignment $\mathfrak{A}$ on $E(F, \omega)$, each predicate letter $P_{\mathfrak{T}}$ is intended to be true of just those terms s in the domain such that $\mathfrak{T}$ is matched by $\mathfrak{A}$ on some Herbrand instance $F^*(y_1/s, y_2/t)$.

The schema F_1 is $\forall y (\bigwedge -P_{\mathfrak{T}} y)$, where the conjunction is over those members $\mathfrak{T}$ of **T** that *falsify* F^M. We let F_2 be the prenex schema whose prefix is $\forall y_1 \exists x$ and whose matrix is the conjunction of the following clauses G_1-G_4. G_1 is

$$(\bigvee P_{\mathfrak{T}} y_1) \wedge (\bigvee P_{\mathfrak{T}'} y_1),$$

where the first disjunction is over those $\mathfrak{T}$ such that $\langle \mathfrak{T}, \mathfrak{T} \rangle$ fits S_1, and the second disjunction is over those $\mathfrak{T}'$ such that $\langle \mathfrak{T}', \mathfrak{T}' \rangle$ fits S_2. G_2 is

$$\bigwedge -(P_{\mathfrak{T}} y_1 \wedge P_{\mathfrak{T}'} y_1),$$

where the conjunction is over those pairs $\langle \mathfrak{T}, \mathfrak{T}' \rangle$ that do *not* fit S_3. G_3 is the conjunction, for all $\mathfrak{T}$ in **T**, of formulas

$$P_{\mathfrak{T}} y_1 \Rightarrow \bigvee P_{\mathfrak{T}'} x,$$

where for each $\mathfrak{T}$ the disjunction is over those $\mathfrak{T}'$ such that $\langle \mathfrak{T}, \mathfrak{T}' \rangle$ fits S_4. Finally, G_4 is the conjunction, for all $\mathfrak{T}$ in **T**, of formulas

$$P_{\mathfrak{T}} x \Rightarrow \bigvee P_{\mathfrak{T}'} y_1,$$

where for each $\mathfrak{T}$ the disjunction is over those $\mathfrak{T}'$ such that $\langle \mathfrak{T}', \mathfrak{T} \rangle$ fits S_4.

Note that $F_1 \wedge F_2$ contains one x-variable x, and x is governed by one y-variable y_1. Hence $D(F_1 \wedge F_2, \omega) = D(F, \omega)$. We now prove properties (1) and (2).

Proof of (1). Let $\mathfrak{A}$ be any truth-assignment on $E(F, \omega)$. We define a truth-assignment $\mathfrak{B}$ on $E(F_1 \wedge F_2, \omega)$ by stipulating that $\mathfrak{B}$ verifies an atom $P_{\mathfrak{T}} s$ iff $\mathfrak{T}$ is matched by $\mathfrak{A}$ on some Herbrand instance $F^*(y_1/s, y_2/t)$. We show that $\mathfrak{B}$ verifies each instance of F_2^* over $D(F_1 \wedge F_2, \omega)$.

Let s be any term in $D(F_1 \wedge F_2, \omega)$. Then $\mathfrak{B}$ verifies $P_{\mathfrak{T}} s \wedge P_{\mathfrak{T}'} s$, where $\mathfrak{T}$ and $\mathfrak{T}'$ are the members of **T** matched by $\mathfrak{A}$ on $H = F^*(y_1/s, y_2/s)$ and on $H' = F^*(y_1/s, y_2/f_x(s))$, respectively. Since $\langle H, H \rangle$ exhibits S_1 and $\langle H', H' \rangle$ exhibits S_2, $\langle \mathfrak{T}, \mathfrak{T} \rangle$ fits S_1 and $\langle \mathfrak{T}', \mathfrak{T}' \rangle$ fits S_2. Hence $\mathfrak{B}$ verifies the instance $G_1(y_1/s)$. Now suppose $\mathfrak{B}$ verifies $P_{\mathfrak{T}} s$ and $P_{\mathfrak{T}'} s$ for some pair $\langle \mathfrak{T}, \mathfrak{T}' \rangle$. Then there are terms t and u such that $\mathfrak{A}$ on $H = F^*(y_1/s, y_2/t)$ matches $\mathfrak{T}$ and $\mathfrak{A}$ on $H' = F^*(y_1/s, y_2/u)$ matches $\mathfrak{T}'$. Since $\langle H, H' \rangle$

exhibits S_3, $\langle \mathfrak{T}, \mathfrak{T}' \rangle$ fits S_3. Hence $\mathfrak{B}$ verifies the instance $G_2(y_1/s)$. Suppose $\mathfrak{B}$ verifies $P_{\mathfrak{T}}s$: for some t, $\mathfrak{T}$ is matched by $\mathfrak{A}$ on $F^*(y_1/s, y_2/t)$. Let $\mathfrak{T}'$ be the member of **T** matched by $\mathfrak{A}$ on $F^*(y_1/f_x(s), y_2/t)$. Then $\langle \mathfrak{T}, \mathfrak{T}' \rangle$ fits S_4 and $\mathfrak{B}$ verifies $P_{\mathfrak{T}'}f_x(s)$. Hence $\mathfrak{B}$ verifies the instance $G_3(y_1/s, x/f_x(s))$. Finally, suppose $\mathfrak{B}$ verifies $P_{\mathfrak{T}}f_x(s)$: for some term t, $\mathfrak{T}$ is matched by $\mathfrak{A}$ on $F^*(y_1/f_x(s), y_2/t)$. Let $\mathfrak{T}'$ be the member of **T** matched by $\mathfrak{A}$ on $F^*(y_1/s, y_2/t)$. Then $\langle \mathfrak{T}', \mathfrak{T} \rangle$ fits S_4 and $\mathfrak{B}$ verifies $P_{\mathfrak{T}'}s$. Hence $\mathfrak{B}$ verifies the instance $G_4(y_1/s, x/f_x(s))$. Thus $\mathfrak{B}$ verifies every instance of $F_2{}^*$.

It remains to show that if $\mathfrak{A}$ verifies $E(F, \omega)$ then $\mathfrak{B}$ verifies every instance of $F_1{}^*$. Suppose then that $\mathfrak{B}$ verifies an atom $P_{\mathfrak{T}}s$. Thus $\mathfrak{A}$ on some Herbrand instance $F^*(y_1/s, y_2/t)$ matches $\mathfrak{T}$. Since $\mathfrak{A}$ verifies every Herbrand instance of F, it follows that $\mathfrak{T}$ verifies F^M. Hence if $\mathfrak{T}$ falsifies F^M, then $\mathfrak{B}$ verifies $-P_{\mathfrak{T}}s$. □

Proof of (2). Let $\mathfrak{B}$ be any truth-assignment on $E(F_1 \wedge F_2, \omega)$ that verifies every instance of $F_2{}^*$. We shall first construct a mapping Φ from pairs of terms to **T** that obeys the following conditions: for all terms s and t,

(a) $\mathfrak{B}$ verifies $P_{\Phi(s,t)}s$;

(b) $\langle \Phi(s, s), \Phi(s, s) \rangle$ fits S_1;

(c) $\langle \Phi(s, f_x(s)), \Phi(s, f_x(s)) \rangle$ fits S_2;

(d) $\langle \Phi(s, t), \Phi(f_x(s), t) \rangle$ fits S_4.

We define Φ by induction on the difference in height between s and t. First we so choose Φ on all pairs $\langle s, s \rangle$ and $\langle s, f_x(s) \rangle$ that (a), (b), and (c) are fulfilled. This may be done since $\mathfrak{B}$ verifies every instance $G_1(y_1/s)$ of G_1. Now suppose $h(t) > h(f_x(s))$ and $\Phi(f_x(s), t)$ is defined and fulfills (a). Then let $\Phi(s, t)$ be any member $\mathfrak{T}'$ of **T** such that $\langle \mathfrak{T}', \Phi(f_x(s), t) \rangle$ fits S_4 and $\mathfrak{B}$ verifies $P_{\mathfrak{T}'}s$. This may be done since $\mathfrak{B}$ verifies $G_4(y_1/s, x/f_x(s))$. Finally, suppose $h(t) < h(s)$ and $\Phi(\hat{y}_1s, t)$ is defined and fulfills (a). Let $\Phi(s, t)$ be any member $\mathfrak{T}'$ of **T** such that $\langle \Phi(\hat{y}_1s, t), \mathfrak{T}' \rangle$ fits S_4 and $\mathfrak{B}$ verifies $P_{\mathfrak{T}'}s$. This may be done since $\mathfrak{B}$ verifies $G_3(y_1/\hat{y}_1s, x/s)$. Moreover, our construction explicitly ensures condition (d) except when $t = f_x(s)$. To show that (d) holds in this case as well, we first note a consequence of condition (a):

(e)   for all terms s, t, and u, $\langle \Phi(s, t), \Phi(s, u) \rangle$ fits S_3.

This follows since $\mathfrak{B}$ verifies every instance of G_2. Now let $\langle i, j \rangle \in S_4$. We must show $\Phi(s, f_x(s))(\langle F \rangle_i) = \Phi(f_x(s), f_x(s))(\langle F \rangle_j)$. Suppose $\langle F \rangle_i$ and $\langle F \rangle_j$ contain the predicate letter P. By saturation there are components $\langle F \rangle_k = Pxx$ and $\langle F \rangle_l = Py_1y_1$. And then $\langle i, k \rangle \in S_2$, $\langle k, k \rangle \in S_3$, $\langle k, l \rangle \in S_4$, $\langle l, l \rangle \in S_3$, and $\langle l, j \rangle \in S_1$. By conditions (b), (c), (e), and condition (d) for $t = f$, we have $\Phi(s, f_x(s))(\langle F \rangle_i) = \Phi(s, f_x(s))(\langle F \rangle_k) =$

ISBN 0-201-02540-X

$\Phi(s, +)(\langle F \rangle_k)$ $=$ $\Phi(f_x(s), +)(\langle F \rangle_l)$ $=$ $\Phi(f_x(s), f_x(s))(\langle F \rangle_l)$ $=$ $\Phi(f_x(s), f_x(s))(\langle F \rangle_j)$. This proves the desired result.

We now define a truth-assignment $\mathfrak{A}$ on $E(F, \omega)$. For each predicate letter P in F and all terms s and t in $D(F, \omega)$, let

$$\mathfrak{A}(Ps\ t) = \Phi(s, t)(\langle F \rangle_i),$$

where $\langle F \rangle_i = Py_1y_2$. We show that for each Herbrand instance $H = F^*(y_1/s, y_2/t)$ of F, $\mathfrak{A}$ on H matches $\Phi(s, t)$. That is, $\mathfrak{A}(\langle H \rangle_i) = \Phi(s, t)(\langle F \rangle_i)$. This is immediate if $\langle F \rangle_i = Py_1y_2$. If $\langle F \rangle_i = Pxy_2$, so that $\langle H \rangle_i = Pf_x(s)\ t$, then let $\langle F \rangle_j = Py_1y_2$. Hence $\langle i, j \rangle \in S_4$ so that $\Phi(s, t)(\langle F \rangle_i) = \Phi(f_x(s), t)(\langle F \rangle_j) = \mathfrak{A}(Pf_x(s)\ t)$, as desired. If $\langle F \rangle_i = Py_1y_1$, so that $\langle H \rangle_i = Pss$, let $\langle F \rangle_j = Py_1y_2$. Then $\langle i, i \rangle \in S_3$ and $\langle i, j \rangle \in S_1$. Hence $\Phi(s, t)(\langle F \rangle_i) = \Phi(s, s)(\langle F \rangle_i) = \Phi(s, s)(\langle F \rangle_j) = \mathfrak{A}(Pss)$. If $\langle F \rangle_i = Pxy_1$, so $\langle H \rangle_i = Pf_x(s)\ s$, then let $\langle F \rangle_j = Pxy_2$. Hence $\langle i, i \rangle \in S_3$ and $\langle i, j \rangle \in S_1$. Hence again $\Phi(s, t)(\langle F \rangle_i) = \Phi(s, s)(\langle F \rangle_j)$. And by the second case treated in this paragraph, $\Phi(s, s)(\langle F \rangle_j) = \mathfrak{A}(Pf_x(s)\ s)$. If $\langle F \rangle_i = Py_1x$, so $\langle H \rangle_i = Ps\ f_x(s)$, let $\langle F \rangle_j = Py_1y_2$. Then $\langle i, i \rangle \in S_3$ and $\langle i, j \rangle \in S_2$. Hence $\Phi(s, t)(\langle F \rangle_i) = \Phi(s, f_x(s))(\langle F \rangle_j) = \mathfrak{A}(Ps\ f_x(s))$. If $\langle F \rangle_i = Pxx$ we proceed similarly, letting $\langle F \rangle_j = Pxy_2$. Finally, suppose $\langle F \rangle_i = Py_2y_2$, so that $\langle H \rangle_i = Ptt$. Then $\langle i, i \rangle \in S_4$. Hence, by condition (d), $\Phi(s, t)(\langle F \rangle_i) = \Phi(u, t)(\langle F \rangle_i)$ for all terms u. Now let $\langle F \rangle_j = Py_1y_2$; then $\langle i, j \rangle \in S_1$. Hence $\Phi(s, t)(\langle F \rangle_i) = \Phi(t, t)(\langle F \rangle_i) = \Phi(t, t)(\langle F \rangle_j) = \mathfrak{A}(Ptt)$.

We have shown that on each Herbrand instance H of F, the truth-assignment $\mathfrak{A}$ matches some member of the range of Φ. If $\mathfrak{B}$ verifies every instance of $F_1{}^*$, then $\mathfrak{B}$ verifies an atom $P\mathfrak{T}s$ only if $\mathfrak{T}$ verifies F^M. Hence, by condition (a), every member of the range of Φ verifies F^M, and so $\mathfrak{A}$ verifies every Herbrand instance of F. This completes the proof of property (2), and with it the proof that F has a model iff $F_1 \wedge F_2$ does. $\square$

REMARK.   Models for the monadic schema $F_1 \wedge F_2$ may be correlated with *collections* of subsets of $\mathbf{T}$. We may take any such correlation $\{\mathbf{T}_1,...,\mathbf{T}_n\}$ as the universe and interpret each predicate letter $P\mathfrak{T}$ over this universe as true of precisely those $\mathbf{T}_i$ that contain $\mathfrak{T}$. This interpretation yields a model for F_2 provided that $\{\mathbf{T}_1,...,\mathbf{T}_n\}$ fulfills the following three conditions:

(A)   each $\mathbf{T}_i$ contains a $\mathfrak{T}$ such that $\langle \mathfrak{T}, \mathfrak{T} \rangle$ fits S_1 and a $\mathfrak{T}'$ such that $\langle \mathfrak{T}', \mathfrak{T}' \rangle$ fits S_2;

(B)   no $\mathbf{T}_i$ contains truth-assignments $\mathfrak{T}$ and $\mathfrak{T}'$ such that $\langle \mathfrak{T}, \mathfrak{T}' \rangle$ fails to fit S_3;

(C)   there is a mapping $\sigma\colon \{1,...,n\} \to \{1,...,n\}$ such that, for $1 \le i \le n$,

   (i)   for each $\mathfrak{T}$ in $\mathbf{T}_i$ there is a $\mathfrak{T}'$ in $\mathbf{T}_{\sigma(i)}$ with $\langle \mathfrak{T}, \mathfrak{T}' \rangle$ fitting S_4;

   (ii)   for each $\mathfrak{T}$ in $\mathbf{T}_{\sigma(i)}$ there is a $\mathfrak{T}'$ in $\mathbf{T}_i$ with $\langle \mathfrak{T}', \mathfrak{T} \rangle$ fitting S_4.

ISBN 0-201-02540-X

(Conditions (A), (B), (Ci), and (Cii) correspond, of course, to the clauses G_1, G_2, G_3, and G_4 of F_2.) Moreover, our interpretation yields a model of F_1 if in addition

(D)   every truth-assignment in each of the $\mathbf{T}_i$ verifies F^M.

Hence our proofs of properties (1) and (2) amount to a proof that F is satisfiable iff there exists a collection $\{\mathbf{T}_1,...,\mathbf{T}_n\}$ of subsets of $\mathbf{T}$ that fulfills (A)–(D).   End of Remark.

Our reduction indicates the direction in which to pursue coinstantiation conditions for Class 5.1 analogous to those derived in §1 for the Gödel–Kalmár–Schütte Class. Here, instead of considering Herbrand instances singly we would consider classes of Herbrand instances, each class containing Herbrand instances that share the same substituent for y_1. We would then describe the coinstantiations occurring within classes, using the sets S_1, S_2, and S_3, and those occurring in members of "neighboring" classes, using the set S_4.

The reductions of this section show how a monadic schema may be used to encode facts about coinstantiations. It may well be that every amenable class can be reduced in this fashion to monadic, or alternatively, that there is for every amenable class a syntactic condition on the set of truth-assignments to the matrix of schemata in the class that is equivalent to satisfiability. In other words, those restrictions on coinstantiability necessary for a class to be amenable may in general permit a complete description in monadic language of the constraints on truth-assignments that coinstantiations produce.

Of course, every solvable class is trivially reducible to the Monadic Class, by the effective mapping that takes a schema in the class to $\exists x Px$ if the schema has a model and to $\exists x(Px \wedge -Px)$ if not. However, in contrast to this reduction "after the fact," those we have presented are in some intuitive sense natural. The monadic reductions above, which, as properties (1) and (2) show, use a monadic schema to encode the structure of expansions, might suggest a way of explicating this intuitive notion of natural reduction.

3. Schemata with Few Atomic Subformulas

We begin this section by giving a criterion for the simultaneous coinstantiability of atomic formulas. We then use this criterion to prove the solvability of the class of schemata that contain just two distinct atomic formulas. In Chapter 1 we showed that two atomic subformulas of a schema F are simultaneously coinstantiable if and only if they are coinstantiated in $\langle H, H \rangle$ for some H in $E(F, m + 1)$, where m is the number of x-variables of F. This provides an algorithm for deciding simultaneous coinstantiability.

ISBN 0-201-02540-X

But the criterion we give here yields a more direct algorithm, one that requires examination not of expansions but rather of the variables that occur in the atomic formulas.

Let A and B be atomic subformulas of a schema F. We assume A and B share the same predicate letter; otherwise they cannot be coinstantiable. Call two (not necessarily distinct) variables of F *linked* iff, for some integer k, one of the variables is the kth argument in A and the other variable is the kth argument in B. Hence A and B are coinstantiated in $\langle H, H \rangle$, where H is an Herbrand instance of F, iff the substituents in H for each pair of linked variables are identical. Let $\simeq$ be the equivalence relation on the set of variables of F that is the reflexive and transitive closure of the relation linked. Thus A and B are coinstantiated in $\langle H, H \rangle$ iff the substituents in H for all variables in each $\simeq$-equivalence class are identical. We claim that there is such an H iff

(1) no $\simeq$-equivalence class contains two distinct x-variables;

(2) no $\simeq$-equivalence class contains both an x-variable and a y-variable governing that x-variable.

Conditions (1) and (2) are necessary, since in no Herbrand instance are the substituents for distinct x-variables identical and in no Herbrand instance are the substituents for an x-variable and a y-variable governing that x-variable identical. The conditions are sufficient as well. For if they are fulfilled, then there is an Herbrand instance H in which the substituent for each y-variable y_i is $\mathbf{1}$ if y_i is $\simeq$-equivalent to no x-variable, and the substituent for y_i is identical to the substituent for the *unique* x-variable to which y_i is $\simeq$-equivalent otherwise. And if H is such an Herbrand instance, then A and B are coinstantiated in $\langle H, H \rangle$. (It is easily seen that this Herbrand instance H lies in $E(F, m + 1)$, where m is the number of x-variables of F. This is just the bound obtained in Chapter 1.)

REMARK. We may give a similar criterion for coinstantiability. Let A and B be two atomic subformulas of a schema F that share the same predicate letter. We shall construct an equivalence relation on the set Z of pairs $\langle v, i \rangle$, where v is a variable of F and i is 1 or 2. First we say two members $\langle v, 1 \rangle$ and $\langle w, 2 \rangle$ of Z are *latched* iff, for some k, the variable v is the kth argument in A and the variable w is the kth argument in B. Thus A and B are coinstantiated in $\langle H_1, H_2 \rangle$ iff the substituent in H_1 for v is identical with that in H_2 for w whenever $\langle v, 1 \rangle$ and $\langle w, 2 \rangle$ are latched. Note that neither v and w nor H_1 and H_2 need be distinct. Let $\simeq$ be the minimal equivalence relation on Z such that:‡

---

‡ That is, construed as a set of pairs of members of Z, $\simeq$ is the intersection of all equivalence relations possessing properties (a) and (b).

ISBN 0-201-02540-X

(a)   if $\langle v, i \rangle$ and $\langle w, j \rangle$ are latched, then $\langle v, i \rangle \simeq \langle w, j \rangle$;

(b)   if v is an x-variable and $\langle v, 1 \rangle \simeq \langle v, 2 \rangle$, then $\langle w, 1 \rangle \simeq \langle w, 2 \rangle$ for each variable w governing v. (The need for this clause arises from the fact that if the substituents in two Herbrand instances for an x-variable are identical then so are the substituents for each variable governing that x-variable.)

It quickly follows that A and B are coinstantiated in $\langle H_1, H_2 \rangle$ iff the substituent in H_i for v is identical with that in H_j for w whenever $\langle v, i \rangle \simeq \langle w, j \rangle$, $1 \leq i, j \leq 2$. We claim that there are such H_1 and H_2 iff

(1)   no $\simeq$-equivalence class contains pairs $\langle v, i \rangle$ and $\langle w, j \rangle$ for distinct x-variables v and w;

(2)   no $\simeq$-equivalence class contains pairs $\langle v, i \rangle$ and $\langle w, i \rangle$ for v an x-variable and w a y-variable governing v.

The necessity of (1) and (2) is evident. For sufficiency we need note merely the following: if (1) and (2) are fulfilled then there are Herbrand instances H_1 and H_2 such that the substituent in H_i for a y-variable v is 1 if $\langle v, i \rangle$ is $\simeq$-equivalent to no pair $\langle w, j \rangle$ when w is an x-variable; and the substituent in H_i for a y-variable v is identical with the substituent in H_j for w when w is an x-variable such that $\langle v, i \rangle \simeq \langle w, j \rangle$.   End of Remark.

We now turn to the class of schemata that contain exactly two distinct atomic subformulas. Without loss of generality we may restrict ourselves to prenex schemata: for, since all schemata are rectified, if a schema contains two atomic subformulas, then so do its prenex equivalents. With the help of the criterion for simultaneous coinstantiability given at the beginning of this section, we shall show the solvability and finite controllability of this class.

Class 7.1.   The class of prenex schemata that contain two distinct atomic subformulas is solvable and finitely controllable.

Proof.   Let F be in Class 7.1. We may assume that the matrix F^M of F is consistent, for otherwise F has no models. Hence we may take F^M to be a disjunction of one or more of the following conjunctions: $A \wedge B$, $-A \wedge -B$, $A \wedge -B$, $-A \wedge B$, where A and B are the two atomic subformulas.

(1)   If $A \wedge B$ is a disjunct in F^M, then F has models of every cardinality. Namely, over any universe we interpret the predicate letter(s) of A and B to be true everywhere.

(2)   If $-A \wedge -B$ is a disjunct in F^M, then again F has models of every cardinality.

(3)   If $F^M = A \wedge -B$ or $F^M = -A \wedge B$, then F falls under the Herbrand Class (1.2), and has models iff A and B are not coinstantiable. The finite controllability of the Herbrand Class was shown in Chapter 3, §2.

ISBN 0-201-02540-X

(4)   If $F^M = (A \wedge -B) \vee (-A \wedge B)$, there are several cases. If A and B are not coinstantiable, then by (3) above F has finite and infinite models. If A and B are simultaneously coinstantiable, then F has no models: indeed, if A and B are coinstantiated in $\langle H, H \rangle$ then H has the form $(C \wedge -C) \vee (-C \wedge C)$, and so H is inconsistent. Finally, if A and B are coinstantiable but not simultaneously coinstantiable, then we claim that F has models of every cardinality ≥ 2. The remainder of our proof is devoted to this claim.

Since A and B are coinstantiable they share the same predicate letter. Since they are not simultaneously coinstantiable, some $\approx$-equivalence class either contains two distinct x-variables or contains both an x-variable and a y-variable governing that x-variable, where $\approx$ is the equivalence relation specified as on page 191. Hence there is a sequence $v_1, \ldots, v_n$ of (not necessarily distinct) variables of F such that v_i and v_{i+1} are linked for each i, $1 \leq i < n$, and either

(1) v_1 and v_n are distinct x-variables, or
(2) v_1 is an x-variable and v_n is a y-variable governing v_1.

We pick such a sequence $v_1, \ldots, v_n$ minimal in length, and, for each $i < n$, we pick an integer k_i for which v_i is the k_ith argument in one of A or B while v_{i+1} is the k_ith argument in the other of A or B. Note that all the k_i are distinct; for were they not, the sequence could be shortened.

We define a function φ: $D(F, \omega) \to \{0, 1\}$; the definition depends on which of (1) or (2) above holds. If (1) holds, then $\varphi(t) = 0$ if $\chi(t) = v_1$ and $\varphi(t) = 1$ otherwise. If (2) holds, then $\varphi(t) = 0$ if $\chi(t) \neq v_1$ and $\varphi(t) = 1 - \varphi(\hat{v}_n t)$ if $\chi(t) = v_1$ (note that $\hat{v}_n t$ is defined in this case, since v_n governs v_1).

Using φ we define a truth-assignment $\mathfrak{A}$: $\mathfrak{A}$ verifies an atom $Pt_1 \cdots t_p$ iff $\Sigma_{i=1}^{n-1} \varphi(t_{k_i})$ is odd. We claim that $\mathfrak{A}$ verifies $E(F, \omega)$. This claim implies that F has a model. It implies moreover that F has a finite model of cardinality 2. For if we let δ be the term-mapping that takes each term t to the earliest term s for which $\varphi(t) = \varphi(s)$, then $\mathfrak{A}(P\delta t_1 \cdots \delta t_p) = \mathfrak{A}(Pt_1 \cdots t_p)$ for all terms $t_1, \ldots, t_p$. Thus $\mathfrak{A}$ verifies $\Delta[E(F, \omega)]$, where Δ is the δ-canonical formula-mapping. By the Finite Model Lemma, F has a model with universe $\delta[D(F, \omega)]$; and clearly $|\delta[D(F, \omega)]| = 2$.

Now let H be an Herbrand instance of F. Then $\mathfrak{A}$ verifies H provided that $\mathfrak{A}$ assigns opposite truth-values to $\langle H \rangle_1$ and $\langle H \rangle_2$ (note that $\langle H \rangle_1 = \langle H \rangle_3$ and $\langle H \rangle_2 = \langle H \rangle_4$). Let $\langle H \rangle_1 = Pt_1 \cdots t_p$ and $\langle H \rangle_2 = Ps_1 \cdots s_p$. Then $\mathfrak{A}$ assigns opposite truth-values to these atoms just in case $\eta = \Sigma_{i=1}^{n-1} (\varphi(t_{k_i}) + \varphi(s_{k_i}))$ is odd. For each variable v of F, let $\sigma(v)$ be the substituent in H for v. By the choice of the k_i, $\{t_{k_i}, s_{k_i}\} =$

ISBN 0-201-02540-X

$\{\sigma(v_i), \sigma(v_{i+1})\}$. Hence $\eta =$

$$\varphi(\sigma(v_1)) + \varphi(\sigma(v_2)) + \varphi(\sigma(v_2)) + \cdots + \varphi(\sigma(v_{n-1}))$$
$$+ \varphi(\sigma(v_{n-1})) + \varphi(\sigma(v_n)).$$

Consequently, η is odd iff $\varphi(\sigma(v_1)) + \varphi(\sigma(v_n))$ is odd. If condition (1) holds, then $\chi(\sigma(v_1)) = v_1$ and $\chi(\sigma(v_n)) = v_n$, since v_1 and v_n are x-variables. Hence $\varphi(\sigma(v_1)) + \varphi(\sigma(v_n)) = 0 + 1 = 1$. If condition (2) holds, then $\chi(\sigma(v_1)) = v_1$ and $\sigma(v_n) = \hat{v}_n\sigma(v_1)$, since v_1 is an x-variable and v_n is a y-variable governing v_1. Hence $\varphi(\sigma(v_1)) = 1 - \varphi(\sigma(v_n))$ and so, again, $\varphi(\sigma(v_1)) + \varphi(\sigma(v_n)) = 1$. In either case, then, η is odd as desired. $\square$

In contrast to Class 7.1, the class of prenex schemata that contain four distinct atomic subformulas is unsolvable (Goldfarb, 1975; Lewis, 1979). Indeed, even the following subclass is unsolvable: prenex schemata with prefixes $\forall y_1 \exists x \forall y_2 \cdots \forall y_n$ and matrices $(A \lor B) \land (-C \lor -D)$, where A, B, C, and D are atomic formulas sharing the same predicate letter. Alternatively we may restrict the matrix to the form $(A \land -B) \lor (C \land -D)$.

The status of the class of prenex schemata that contain three distinct atomic subformulas is open. We do know, however, that this class cannot be finitely controllable. For the schema

$$\forall y_1 \exists x \forall y_2 [(P y_1 y_2 \lor P y_2 y_1) \land (-P y_1 y_2 \lor -P y_2 x)]$$

has infinite models but no finite models.

If we look only at Skolem schemata, then we have the following result: the class of such schemata that contain five atomic subformulas is unsolvable, and remains so when the number of x-variables in the prefix is restricted to one (Goldfarb, 1975; see also Lewis, 1979). But three and four remain open; and we know of no schema in these classes that possesses only infinite models.

The unsolvable classes just specified all allow arbitrarily many y-variables. This is intrinsic: as we shall now show, every class of schemata restricted both in number of atomic subformulas and in number of y-variables is solvable. This result turns on a general lemma, a lemma that underscores the role of coinstantiations in determining the satisfiability of schemata.

Lemma.   Let F_1 and F_2 be schemata obeying the following conditions:

(1) $F_1{}^M$ and $F_2{}^M$ share the same truth-functional structure. That is, the truth-functional connectives occur homologously in $F_1{}^M$ and $F_2{}^M$.

(2)   There is a mapping $\Psi: E(F_1, \omega) \to E(F_2, \omega)$ such that, for all

ISBN 0-201-02540-X

Herbrand instances H and H' of F_1 and all integers i and j, if $\langle H \rangle_i = \langle H' \rangle_j$ then $\langle \Psi(H) \rangle_j = \langle \Psi(H') \rangle_j$.

Then F_1 has a model, provided that F_2 has a model.

Proof. Let $\mathfrak{A}$ verify $E(F_2, \omega)$. It is enough to show the existence of a truth-assignment $\mathfrak{B}$ such that $\mathfrak{B}(\langle H \rangle_i) = \mathfrak{A}(\langle \Psi(H) \rangle_i)$ for each Herbrand instance H of F_1 and each i. For by condition (1) any such truth-assignment $\mathfrak{B}$ verifies each Herbrand instance of F_1, and so F_1 has a model. Now such a $\mathfrak{B}$ fails to exist only if there are H and H' such that $\langle H \rangle_i = \langle H' \rangle_j$ yet $\mathfrak{A}(\langle \Psi(H) \rangle_i) \neq \mathfrak{A}(\langle \Psi(H') \rangle_j)$. But this is precluded by condition (2). $\square$

For all positive integers m and n, let $\mathfrak{S}(m, n)$ be the (infinite) class of schemata containing at most m distinct atomic subformulas and at most n y-variables. We shall show the solvability of each $\mathfrak{S}(m, n)$ by effectively reducing it to a finite class of schemata. We do not thereby exhibit decision procedures for the $\mathfrak{S}(m, n)$. Indeed, there cannot be any effective correlation of a decision procedure with each such class $\mathfrak{S}(m, n)$. For since the union of the $\mathfrak{S}(m, n)$ is the class of all schemata, any such correlation would yield a general decision procedure.

For all positive integers m and p, let $\mathfrak{R}(m, p)$ be the class of prenex schemata that contain at most m distinct atomic subformulas and in which each predicate letter of F is at most p-place. Each class $\mathfrak{R}(m, p)$ is solvable. For if a schema F is in $\mathfrak{R}(m, p)$, then F contains at most $m \cdot p$ variables. Hence for each m and p there is a *finite* class $\mathfrak{C}$ of schemata such that every F in $\mathfrak{R}(m, p)$ is either in $\mathfrak{C}$ or else is an alphabetic variant of some schema in $\mathfrak{C}$ (where here we include as alphabetic variants those schemata obtained by relettering predicate letters as well as variables).

Thus the following result suffices for the solvability of each class $\mathfrak{S}(m, n)$.

Class 7.2. For all positive integers m and n, the class $\mathfrak{S}(m, n)$ is effectively reducible to the class $\mathfrak{R}(m, (2n + 1)^m)$.

Proof. Let F be a schema in $\mathfrak{S}(m, n)$, and suppose F contains at least one p-place predicate letter P for $p > (2n + 1)^m$. We shall construct a schema G in $\mathfrak{S}(m, n)$ that has a model just when F has a model and that contains instead of P a number of $(p - 1)$-place predicate letters. By iterating this construction we eventually obtain a schema in $\mathfrak{R}(m, (2n + 1)^m)$ that has a model just when F does.

Suppose the distinct atomic subformulas of F that contain P are $Pv_1^1 \cdots v_p^1$, $Pv_1^2 \cdots v_p^2, \ldots, Pv_1^k \cdots v_p^k$. Then $k \leq m$, since F contains no more than m distinct atomic subformulas. Call two variables of F *equivalent* iff either they are identical or else they are distinct x-variables that are governed by the same y-variables of F. Since F contains at most n y-variables, there are at most $2n + 1$ equivalence classes of variables. Since $p > (2n + 1)^m$, by the pigeonhole argument there are at least two integers q and r, $1 \leq q$

ISBN 0-201-02540-X

$< r \le p$, such that $v_q{}^i$ and $v_r{}^i$ are equivalent variables for *each* i, $1 \le i \le k$.

By the choice of q and r, then, for all i and j, $1 \le i, j \le k$, two atomic formulas $Pv_1{}^i \cdots v_p{}^i$ and $Pv_1{}^j \cdots v_p{}^j$ are coinstantiable only if either

(a) $v_q{}^i = v_r{}^i$ and $v_q{}^j = v_r{}^j$; or

(b) $v_q{}^i$ and $v_r{}^i$ are distinct x-variables governed by the same y-variables, and $\langle v_q{}^i, v_r{}^i \rangle = \langle v_q{}^j, v_r{}^j \rangle$.

Let φ be the function on $\{1,\ldots,k\}$ that carries i to 0 if $v_q{}^i = v_r{}^i$ and carries i to $\langle v_q{}^i, v_r{}^i \rangle$ otherwise. For each object a in the range of φ, let P_a be a new $(p - 1)$-place predicate letter. Finally, let G be the schema obtained from F by replacing each atomic subformula $Pv_1{}^i \cdots v_p{}^i$ with $P_{\varphi(i)} v_1{}^i \cdots v_{r-1}^i v_{r+1}^i \cdots v_p{}^i$. We claim that F has a model iff G has a model.

Since F and G have the same prefix,‡ $D(F, \omega) = D(G, \omega)$. Hence there is a bijection $\Psi: E(F, \omega) \to E(G, \omega)$ defined by stipulating that the substituent in any Herbrand instance H for each variable is the same as the substituent in $\Psi(H)$ for that variable. It suffices to show that for all Herbrand instances H and H' of F and all integers i and j,

$$\langle H \rangle_i = \langle H' \rangle_j \quad \text{iff} \quad \langle \Psi(H) \rangle_i = \langle \Psi(H') \rangle_j.$$

For if this holds then two applications of the Lemma, one in each direction, show that F has a model just when G does.

Now if $\langle F \rangle_i$ and $\langle F \rangle_j$ do not contain P, then $\langle H \rangle_i = \langle \Psi(H) \rangle_i$ and $\langle H' \rangle_j = \langle \Psi(H') \rangle_j$. Suppose that $\langle F \rangle_i$ and $\langle F \rangle_j$ both contain P. If $\langle H \rangle_i = \langle H' \rangle_j$, then $\langle F \rangle_i$ and $\langle F \rangle_j$ are coinstantiable. Hence $\langle G \rangle_i$ and $\langle G \rangle_j$ share the same predicate letter P_a and the argument-sequences of both $\langle \Psi(H) \rangle_i$ and $\langle \Psi(H') \rangle_j$ are just the result of deleting the rth member from the argument-sequence of $\langle H \rangle_i$. Hence $\langle \Psi(H) \rangle_i = \langle \Psi(H') \rangle_j$. Conversely, suppose that $\langle \Psi(H) \rangle_i = \langle \Psi(H') \rangle_j = P_a s_1 \cdots s_{p-1}$. Then $\langle H \rangle_i = \langle H' \rangle_j = Pt_1 \cdots t_p$, where for $i < r$, $t_i = s_i$; for $i > r$, $t_i = s_{i-1}$; and t_r is determined thus:

(a) if $a = 0$, then $t_r = t_q = s_q$;

(b) if $a = \langle v, w \rangle$, where v and w are distinct x-variables, then $t_r = \hat{w} t_q = \hat{w} s_q$. (Note that $\hat{w} s_q$ is defined since in this case $\chi(x_q) = v$, and v and w are governed by the same y-variables.)

This concludes the proof. □

---

‡ There is a slight problem here if in the transformation of F into G all occurrences of an x-variable v disappear. For then G contains a vacuous quantifier, and so is not a schema. We handle this as follows. In the application of the Lemma we consider a schema G' like G but whose matrix includes an additional conjunct $Qv \lor -Qv$. Thus G' is in $\mathfrak{S}(m + 1, n)$. To obtain a schema in $\mathfrak{S}(m, n)$ as endproduct, the additional conjunct and the quantifier $\exists v$ in the prefix are deleted. Of course these deletions do not affect satisfiability.

ISBN 0-201-02540-X

We may easily specify exactly which classes $\mathfrak{S}(m, n)$ are finitely controllable. The finite controllability of the Initially-extended Ackermann Class (prenex schemata with prefixes $\exists z_1 \cdots \exists z_t \forall y \exists x_1 \cdots \exists x_n$) implies that of $\mathfrak{S}(m, 1)$ for every m. The finite controllability of Class 7.1 implies that of $\mathfrak{S}(2, n)$ for every n. And on page 194 we exhibited a schema in $\mathfrak{S}(3, 2)$ that has infinite models but no finite models.

ISBN 0-201-02540-X

Schemata with Identity

In this chapter we extend our quantificational language by adding a sign for the relation of identity. There is a simple reduction of formulas that contain the identity sign to formulas that do not. We shall see, however, that this reduction is of little use in obtaining solvable classes; more subtle methods of dealing with identity must be developed. We shall also see that many solvable classes of pure quantification theory become unsolvable when the identity sign is allowed.

1. Basic Facts

We shall use "$\equiv$" as the identity sign; this is to avoid confusion with our use of "$=$" in the metalanguage, when we talk of identity of terms, formulas, etc. (We shall follow our usual convention and let "$\equiv$" be a metalinguistic name for itself.) Our quantificational language is extended by adding to the stock of atomic formulas all strings of the form $t \equiv s$, where t and s are terms of the language, variable-free or not. Formulas of the extended language are constructed in the usual fashion from this enlarged stock of atomic formulas; all the syntactic notions introduced in Chapter 0 are then defined in the same way as for the unextended language. A schema of the extended language is satisfiable over a universe U iff it is true in U under some interpretation of its predicate letters and the following standard interpretation of the identity sign: $\equiv$ is assigned the set of all pairs $\langle u, u \rangle$ of elements u in U; that is, $y_1 \equiv y_2$ is true when and only when y_1 and y_2 have the same value in U. (For terminological convenience, we consider $\equiv$ to be a predicate letter. Thus, for example, in speaking of all atoms $Ps\ t$ we mean to include atoms $s \equiv t$. Moreover, we shall usually speak of a formula containing identity when we mean, of course, a formula containing the identity sign.) The classical elimination of identity is given by the following lemma.

Burton Dreben and Warren D. Goldfarb, The Decision Problem: Solvable Classes of Quantificational Formulas

ISBN 0-201-02540-X

Identity Elimination Lemma. For each schema F containing identity there is a schema G not containing identity such that F has a model iff G has a model and F has a finite model iff G has a finite model.

Proof. Let I be a dyadic predicate letter foreign to F and let F_I be the result of replacing each atomic subformula $v_1 \equiv v_2$ of F by the atomic formula Iv_1v_2. Suppose n is the largest number of argument places of any predicate letter of F_I (including I), and let J_F be the conjunction of all formulas

$$[Iy_1y_2 \Rightarrow (A \Leftrightarrow A(y_1/y_2))],$$

where A is any atomic formula constructed from either a predicate letter of F or the predicate letter I and variables among $y_1,\ldots,y_{n+1}$. Finally, let the schema G be a rectified equivalent of

$$F_I \wedge \forall y_1 \cdots \forall y_{n+1} J_F \wedge \forall y I y y.$$

Thus G does not contain identity. If F has a model with universe U, then so does G; we need only assign to the predicate letter I the set $\{\langle u, u \rangle | u \in U\}$. Conversely, suppose there are interpretations of the predicate letters of G over a universe U under which G is true. Since $\forall y_1 \cdots \forall y_{n+1} J_F \wedge \forall y I y y$ is true under these interpretations, the extension of I is an equivalence relation over U, and F is satisfiable over the universe U' whose members are just the I-equivalence classes of members of U. (The universe U' may be of lesser cardinality than U.) □

A schema without identity is either unsatisfiable, satisfiable over infinite universes only, or, for some $p > 0$, satisfiable over universes of every cardinality $\geq p$; whereas for a schema containing identity there are further possibilities.

Definition 8.1. The *spectrum* of a schema F is the set of positive integers p such that F is satisfiable over a (finite) universe of cardinality p.

Thus if F is satisfiable but does not contain identity, the spectrum of F is either empty or else a final segment of the integers (in the latter case F has both finite and infinite models). If, however, F does contain identity, then the spectrum of F may also be any finite set, any cofinite set (that is, any set whose complement is finite), or may be infinite but not cofinite. If the spectrum of F is infinite, then by the Compactness Theorem F is satisfiable over infinite universes; but if the spectrum is finite, F may or may not be so satisfiable.

We give some examples of spectra. Using no predicate letters aside from $\equiv$, we can construct for each $p > 0$ a schema F_p satisfiable only in universes of cardinality p: F_p is just

$$\exists x_1 \cdots \exists x_p [\wedge -x_i \equiv x_j \wedge \forall y(\vee y \equiv x_i)],$$

ISBN 0-201-02540-X

where the conjunction is over all i and j, $1 \leq i$, $j \leq p$, such that $i \neq j$ and the disjunction is over all i, $1 \leq i \leq p$. By taking disjunctions of various F_p we obtain for any finite set S of positive integers a schema F with spectrum S. By taking negations of these disjunctions we obtain for each cofinite set S a schema with spectrum S. No disjunction F of the F_p is satisfiable over infinite universes; but by disjoining a schema of pure quantification theory which is satisfiable only over infinite universe we obtain a schema with the same finite spectrum as F but satisfiable over infinite universes too. (The class of schemata containing no predicate letters but $\equiv$ is finitely controllable; no such schema has infinite models only.) More complex spectra may also be obtained: for example, by translating the axioms for a field into quantification theory with identity (that is, by replacing with predicate letters the function signs in the usual formulation of the axioms), we obtain a schema whose spectrum is the set of prime powers. Little is known about characterizing the sets of integers that are spectra. Clearly every such set is recursive; the converse is not true. A major open problem is whether for every schema F there is a schema G whose spectrum is the complement of that of F.

Thus we have a question additional to that of solvability when considering classes of schemata containing identity, namely, the characterization of the spectra of schemata in the class.

The extension of the Expansion Theorem needed for applications to schemata that contain identity is easily obtained from the Identity Elimination Lemma.

Definition 8.2.   Let F be a schema containing identity. A truth-assignment $\mathfrak{A}$ on $\tilde{L}(F, \omega)$ is *standard* iff

(1) $\mathfrak{A}$ verifies $s \equiv s$ for every term s in $D(F, \omega)$;

(2)   whenever $\mathfrak{A}$ verifies $s \equiv t$ for two terms s and t in $D(F, \omega)$, then $\mathfrak{A}$ assigns the same value to any two atoms that differ only in that one contains occurrences of s at some or all of those places where the other contains occurrences of t.

Extended Expansion Theorem.   A schema F containing identity is satisfiable iff, for each p, there is a standard truth-assignment verifying $E(F, p)$.

Proof.   Let G be the schema not containing identity constructed from F as in the Identity Elimination Lemma. Any standard truth-assignment $\mathfrak{A}$ verifying $E(F, p)$ can be transformed into a truth-assignment verifying $E(G, p)$ by stipulating that $\mathfrak{A}$ verifies an atom Ist iff $\mathfrak{A}$ verifies $s \equiv t$. Conversely, any truth-assignment $\mathfrak{A}$ verifying $E(G, \omega)$ can be transformed into a standard truth-assignment verifying $E(F, \omega)$ by stipulating that $\mathfrak{A}$ verifies $s \equiv t$ iff $\mathfrak{A}$ verifies Ist. The Extended Expansion Theorem then results from the Ex-

ISBN 0-201-02540-X

pansion Theorem for pure quantification theory and the Identity Elimination Lemma. $\square$

Hence the satisfiability of a schema F containing identity requires that the expansions of F have a property stronger than consistency, namely *standard consistency*. Obviously, if the truth-assignment $\mathfrak{A}$ on $\bar{L}(F, \omega)$ verifies an atom $s \equiv t$ iff $s = t$, then $\mathfrak{A}$ is automatically standard. The new problems encountered in solvability proofs arise from the fact that distinct terms may be "equated" by standard assignments $\mathfrak{A}$. When this occurs, the existence of a standard assignment verifying the expansion is not settled, we shall soon see, by coinstantiability considerations alone.

2. Solvability through Elimination of Identity

A natural way of treating decision problems for schemata with identity is simply to eliminate identity and then use solvability results for pure quantification theory. To do this, however, the Identity Elimination Lemma is of little help. In the proof of the Lemma we constructed a formula $F_I \wedge \forall y_1 \cdots \forall y_{n+1} J_F \wedge \forall y I y y$ for each schema F; but the formula J_F contains at least one dyadic letter I and is governed by at least three y-variables. Hence the reductions so obtained of most classes of schemata will not in general be subclasses of solvable classes of schemata without identity. Thus in many of those proofs below that proceed by elimination of $\equiv$, we use a more delicate procedure: we avoid adjoining the clause $\forall y_1 \cdots \forall y_{n+1} J_F$ by showing that the clause $\forall y I y y$ by itself suffices.

Negative Identity Reduction. Let F be a schema containing identity, and let F' be the result of replacing each *negative* occurrence of an atomic formula $v_1 \equiv v_2$ by $I v_1 v_2$, where I is a dyadic predicate letter foreign to F (a negative occurrence is one within the scopes of an odd number of negation sign occurrences). Let G be $\forall y I y y \wedge F'$. Then F and G are satisfiable over the same universes.

Proof. If F has a model with universe U, then by interpreting I as the identity over U, the schema G also has a model with universe U. Conversely, we first note that $\forall y I y y$ logically implies $\forall y_1 \forall y_2 (y_1 \equiv y_2 \Rightarrow I y_1 y_2)$.‡ Since F' differs from F in having I only where F has negative occurrences of $\equiv$, it follows that $\forall y_1 \forall y_2 (y_1 \equiv y_2 \Rightarrow I y_1 y_2) \wedge F'$ implies F. Thus G logically implies F. $\square$

The Negative Identity Reduction quickly yields a solvability proof for the Herbrand Class with identity.

Class 1.2 (Herbrand Class with Identity). Let F be a schema containing

---

‡ In this chapter, of course, we speak of logical implication in quantification theory with identity.

identity whose matrix is a conjunction of signed atomic formulas. Then either F has no model, or F is satisfiable only over universes of cardinality 1, or from F can be constructed in a uniform manner a schema G that is in Class 1.2, does not contain identity, and is satisfiable over the same universes as F.

Proof. If a schema F is in the Herbrand Class with identity, then so is the schema $G = \forall y I y y \wedge F'$ constructed from F by the Negative Identity Reduction. Hence it suffices to consider only those schemata F in the Herbrand Class with Identity that contain identity just in positively signed atomic formulas. We show how to eliminate these occurrences of identity as well.

(1) Suppose F contains a conjunct $v_1 \equiv v_2$, where v_1 and v_2 are distinct y-variables. Then F implies $\forall y_1 \forall y_2 (y_1 \equiv y_2)$; whence F either has no model or else is satisfiable only over universes of cardinality 1.

(2) Suppose F contains a conjunct $v_1 \equiv v_2$, where one of v_1, v_2 is an x-variable and the other is a y-variable governed by the x-variable. Then F implies $\exists x \forall y (x \equiv y)$; whence the conclusion of (1) holds.

(3) Suppose F contains a conjunct $v_1 \equiv v_2$, where one of v_1, v_2, say v_1, is an x-variable and the other governs it. Then F is equivalent to the schema obtained by deleting from F the quantifier binding v_1 and replacing all occurrences of v_1 in F by occurrences of v_2.

By iteration of (3), if necessary, we either reach the conclusion of (1) or else obtain a schema in which identity occurs only in conjuncts of the form $v \equiv v$. These conjuncts may be deleted without affecting satisfiability.

Since Class 1.2 without identity is finitely controllable, we obtain the following information about spectra: if F is a satisfiable schema in the Herbrand Class with identity, then the spectrum of F is either $\{1\}$ or else $\{p \mid p \geq n\}$ for some positive integer n uniformly calculable from F. □

The same argument works for Class 1.3 with identity: the matrix of a schema F in this class is a conjunction, each conjunct of which is either a positively signed atomic formula or a disjunction of negatively signed atomic formulas. The positive occurrences of identity may be eliminated by (1)–(3) above, and the negative ones by the Negative Identity Reduction. However, this argument does not work for the mirror-image of Class 1.3 with identity, the class of schemata whose matrices are conjunctions consisting of negatively signed atomic formulas and of disjunctions of positively signed atomic formulas. The positively signed atomic formulas containing $\equiv$ cannot be eliminated by (1)–(3), since they no longer occur as conjuncts in the matrix of F. Indeed, this mirror-image class is unsolvable, and hence stands in sharp contrast to what obtains in the absence of identity, where the same

ISBN 0-201-02540-X

elementary considerations yield the condensability and hence the solvability of both 1.3 and its mirror image.

This contrast merits a closer look. Let F be a schema in the mirror-image class without identity. Then F is unsatisfiable iff for some n there are Herbrand instances $H_1,...,H_n$, H of F such that, for some $k \geq 0$,

(a) $\langle H \rangle_{k+1} \vee \cdots \vee \langle H \rangle_{k+n}$ is a conjunct in H;

(b) for each i, $1 \leq i \leq n$, $-\langle H \rangle_{k+i}$ is a conjunct in H_i.

For a simple truth-functional argument shows that $E(F, \omega)$ is inconsistent just in case there are such Herbrand instances. This furnishes an effective criterion for satisfiability; for there is some inconsistent conjunction of Herbrand instances of F only if there is an inconsistent conjunction of $n + 1$ or fewer Herbrand instances of F, where n is the largest number of atomic formulas in a single conjunct in F. However, if F contains identity, then F may be unsatisfiable even though there is no finite set of Herbrand instances with properties (a) and (b); in that case $E(F, \omega)$ is consistent but not *standardly* consistent.

An example will make this clear. Let F be the schema

$$\forall y_1 \exists x_1 \exists x_2 \exists x_3 \forall y_2 [-Py_1x_1 \wedge -Py_1x_2 \wedge -Py_1x_3$$

$$\wedge (Py_2y_1 \vee y_1 \equiv y_2) \wedge -Rx_1x_2x_3 \wedge Ry_1y_1y_1].$$

Then F is unsatisfiable. For let t and s be any terms in $D(F, \omega)$; let $H = F^*(y_1/t, y_2/s)$ and, for $1 \leq j \leq 3$, $H_j = F^*(y_1/f_{x_j}(t), y_2/t)$. Then $H \wedge H_1 \wedge H_2 \wedge H_3$ truth-functionally implies the equations $f_{x_j}(t) \equiv t$, $1 \leq j \leq 3$. But H implies $-Rf_{x_1}(t) f_{x_2}(t) f_{x_3}(t) \wedge Rttt$; hence no standard truth-assignment verifies $H \wedge H_1 \wedge H_2 \wedge H_3$. However, there are nonstandard truth-assignments verifying $E(F, \omega)$; for example, that which verifies Pt_1t_2 iff $t_1 \neq \hat{y}_1t_2$, verifies $Rt_1t_2t_3$ iff $t_1 = t_2 = t_3$, and verifies $t_1 \equiv t_2$ for all terms t_1 and t_2.

In this class there are failures of standard consistency that result from even more complicated factors than those shown by our example. A conjunction of Herbrand instances may truth-functionally imply an equation $s_1 \equiv t_1$; another conjunction may imply, by truth-functional logic and substitutions of s_1 for t_1 and vice versa, another equation $s_2 \equiv t_2$; and so on. Then, finally, using the equations $s_i \equiv t_i$ so implied, a conjunction of Herbrand instances may become inconsistent under substitutions of s_i for t_i and vice versa. Thus, the lack of a standard truth-assignment verifying $E(F, \omega)$ may be due to the combined influence of an unbounded number of Herbrand instances—in contrast to this class without identity, where we need consider only a bounded number of Herbrand instances to decide satisfiability. This fact can be exploited to encode an undecidable word problem (namely, one concerning Thue systems) into this mirror-image class

ISBN 0-201-02540-X

with identity; we thereby obtain the unsolvability of this class (Harry Lewis, private communication, 1972).

The solvability of Class 1.4 with identity remains open.

We now turn to classes of $\forall\exists\forall$-schemata. For such schemata F we may deal in a convenient manner with those (troubling) truth-assignments $\mathfrak{A}$ that verify an equation $m \equiv n$, where m and n are distinct terms in $D(F, \omega)$. (We use the numerical notation for terms in $D(F, \omega)$, introduced in Chapter 5.) We consider the following three classes: Class 5.5, the $\forall\exists\forall$-Horn Class; Class 5.6, the $\forall\exists\forall$-Krom Class; and Class 5.1, in which every atomic subformula of F must be of one of the seven forms Py_1y_1, Pxx, Py_2y_2, Py_1x, Pxy_1, Py_1y_2, Pxy_2. The specifications of the extensions of these classes to schemata with identity are obvious. (Note, however, that for Class 5.1 with Identity we may actually allow any atomic subformula containing identity; for $v_1 \equiv v_2$ is logically equivalent to $v_2 \equiv v_1$, so that we may replace any identity containing y_2 with one in which y_2 is the right-hand argument.)

Classes 5.1, 5.5, and 5.6 with Identity.   There is a recursive function ψ such that, for every schema F in one of these three classes, F is satisfiable iff there is a standard truth-assignment verifying $E(F, \psi(F))$.

Proof.   The three classes without identity are all solvable. Hence there is a recursive function φ such that, for each schema G without identity in one of these classes, G is satisfiable iff $E(G, \varphi(G))$ is consistent. For each schema F with identity in one of these classes, we shall let $\psi(F) = \varphi(G)$ for a schema G not containing identity and constructed from F. The schema G will have prefix $\forall y_1 \exists x \forall y_2$, so that $D(F, \omega) = D(G, \omega)$, and will contain the new dyadic predicate letters I, P, and Q, whose intended interpretations over the domain $D(G, \omega)$ are these: Imn is to be true iff $m = n$, Pmn is to be true iff $m < n$, and Qmn is to be true iff $m > n$. Let J be

$$Iy_1y_1 \wedge Py_1x_1 \wedge Qxy_1 \wedge (Qy_1y_2 \Rightarrow Qxy_2)$$
$$\wedge (Pxy_2 \Rightarrow Py_1y_2) \wedge (Py_1y_2 \Rightarrow -Iy_1y_2) \wedge (Qy_1y_2 \Rightarrow -Iy_1y_2).$$

Note that J is equivalent to a Krom and Horn formula, and that each atomic subformula of J is of one of the seven forms allowed by the specification of Class 5.1. Now any truth-assignment that gives I, P, and Q their intended interpretations over $D(F, \omega)$ verifies all instances $J(y_1/m, x/m + 1, y_2/n)$ of J; and any truth-assignment that verifies all such instances of J gives I, P, and Q their intended interpretations. (Of course, we are really interested only in the interpretation of I; the letters P and Q are used solely to obtain a formula J that "enforces" this interpretation.)

Now let G be $\forall y_1 \exists x \forall y_2(J \wedge F_I{}^M)$, where $F_I{}^M$ is the result of replacing each atomic subformula $v_1 \equiv v_2$ in F^M by Iv_1v_2. Thus G is in one of the three $\forall\exists\forall$-classes without identity.

ISBN 0-201-02540-X

Suppose there is a standard truth-assignment $\mathfrak{A}$ on $\tilde{L}(F, \omega)$ that verifies $E(F, \psi(F)) = E(F, \varphi(G))$. We wish to show F satisfiable. There are two cases.

Case 1. $\mathfrak{A}$ verifies $m \equiv n$ for some distinct $m, n \leq \varphi(G)$. We show that in this case F has a finite model. Since m and n are distinct, we may assume $n > m$. In fact, let n be the least number $\leq \varphi(G)$ for which there exists a number $m < n$ such that $\mathfrak{A}$ verifies $m \equiv n$. But then, since $\mathfrak{A}$ is standard, $\mathfrak{A}$ provides a model for F^* over the universe $D(F, n) = \{0, 1, \ldots, n - 1\}$: the indicial function sign f_x of F^* is interpreted over this universe as carrying each $k < n - 1$ to $k + 1$ and carrying $n - 1$ to m.

Case 2. $\mathfrak{A}$ falsifies $m \equiv n$ for all distinct $m, n \leq \varphi(G)$. Extend $\mathfrak{A}$ to an assignment on $\tilde{L}(G, \omega)$ by giving values to atoms containing I, P, or Q in accord with the intended interpretations of these predicate letters. Thus, for $0 \leq m, n \leq \varphi(G)$, $\mathfrak{A}(Imn) = \mathfrak{A}(m \equiv n)$. But then the truth-assignment $\mathfrak{A}$ so extended verifies $E(G, \varphi(G))$. By the choice of φ it follows that G is satisfiable, that is, there is a truth-assignment $\mathfrak{B}$ on $\tilde{L}(G, \omega)$ verifying $E(G, \omega)$. As pointed out earlier, $\mathfrak{B}$ then verifies an atom Imn iff $m = n$. Extend $\mathfrak{B}$ to a truth-assignment on $\tilde{L}(F, \omega)$ by stipulating that $\mathfrak{B}$ verifies an equation $m \equiv n$ iff $m = n$. Thus $\mathfrak{B}$ is standard, and since the interpretation of $\equiv$ agrees with that of I, the truth-assignment $\mathfrak{B}$ so extended verifies $E(F, \omega)$. By the Extended Expansion Theorem, F is satisfiable.   □

Classes 5.1, 5.5, and 5.6 with Identity are not finitely controllable, since even without identity they are not. But they are docile. The proof of this parallels the docility proofs for the classes without identity; we elaborate in §4 of the Appendix. We do not have much information about the spectra of schemata in these classes. There are schemata in them whose spectra are neither finite nor cofinite. For example, let F be

$$\forall y_1 \exists x \forall y_2 [Rxy_1 \wedge Qy_1y_1 \wedge (Rxy_2 \Rightarrow y_1 \equiv y_2) \wedge (Qy_1y_2 \Rightarrow Pxy_2)$$

$$\wedge (Py_1y_2 \Rightarrow Qxy_2) \wedge (-Py_1y_2 \vee -Qy_1y_2)].$$

Then F is in each of the $\forall\exists\forall$-classes with identity. Moreover, F is satisfiable only over infinite universes and universes of even cardinality, that is, the spectrum of F is the set of even positive integers.

The foregoing solvability proof turns on there being exactly one term of height r in $D(F, \omega)$ for each r. In §5 of the Appendix we shall show the amenability of a class extending Class 5.1, a class containing schemata in which more than one x-variable may occur. (Hence the domains of the schemata contain many terms of height r for each $r > 1$.) This class becomes unsolvable once identity is allowed.

In Chapter 5 we also showed the amiability of three $\forall\exists\forall$-classes, Classes 5.2–5.4. Class 5.2 with Identity, however, is not finitely controllable: it contains the schema $\forall y_1 \exists x \forall y_2 [(y_1 \equiv y_2 \Rightarrow -Qy_1y_2) \wedge (y_1 \equiv y_2 \Rightarrow Qxy_2)$

$\wedge\,(Qy_1y_2 \Rightarrow Qxy_2)]$, which possesses infinite models only. Class 5.2 with Identity is solvable and docile, since it is subsumed by Class 5.1 with Identity. This is the only example we know of a natural class that remains solvable but ceases to be finitely controllable once identity is allowed. On the other hand, as we shall see in §4 of the Appendix, Classes 5.3 and 5.4 with Identity are finitely controllable.

Identity may also be eliminated from schemata in the Ackermann Class with Identity or in the Initially-extended Ackermann Class with Identity. These classes were first shown solvable by Ackermann (1954).

Class 2.2   (Ackermann Class with Identity).   Let F be a prenex schema containing identity with prefix $\forall y \exists x_1 \cdots \exists x_n$. Then we can construct in a uniform manner an Ackermann schema that does not contain identity and that is satisfiable over the same universes as F.

Proof.   Let $F = \forall y \exists x_1 \cdots \exists x_n F^M$. For each n-tuple $\bar{v} = \langle v_1, \ldots, v_n \rangle$ of variables among $y, x_1, \ldots, x_n$ let $F_{\bar{v}}^M = F^M(x_1/v_1, \ldots, x_n/v_n)$, and let $J_{\bar{v}}$ be

$$\bigwedge - y \equiv x_i \wedge \bigwedge - x_i \equiv x_j,$$

where the first conjunction is over those i such that x_i occurs in $\bar{v}$, and the second conjunction is over those distinct i and j such that x_i and x_j occur in $\bar{v}$. Let $G_{\bar{v}}$ result from $F_{\bar{v}}^M$ thus: first replace each atomic subformula $v \equiv v$ with a conventional sign for truth and each atomic subformula $v \equiv w$, where v and w are distinct variables, with a conventional sign for falsity; then eliminate these signs by the usual truth-functional rules. Finally, let G^M be the disjunction of all formulas $G_{\bar{v}} \wedge J_{\bar{v}}$, and let G be $\forall y \exists x_1 \cdots \exists x_n G^M$.

The schemata F and G are logically equivalent. Moreover, in G the identity sign occurs only negatively. Hence the Negative Identity Reduction is applicable: G is satisfiable over the same universes as the schema $\forall y \exists x_1 \cdots \exists x_n (Iyy \wedge G_I^M)$, where G_I^M results from G^M by replacing each atomic subformula $v \equiv w$ with Ivw. The schema $\forall y \exists x_1 \cdots \exists x_n (Iyy \wedge G_I^M)$ is an Ackermann schema not containing identity. $\square$

Since the Ackermann Class without identity is finitely controllable, we may infer that the spectrum of every satisfiable Ackermann schema with identity is a final segment of the integers.

Initially-extended Ackermann schemata with identity do not have quite such well-behaved spectra. For example, the schema $\exists z_1 \forall y \exists x_1 (y \equiv z_1 \wedge x_1 \equiv y)$ is satisfiable only over a universe of cardinality 1; and the schema $\exists z_1 \forall y \exists x_1 [y \equiv z_1 \vee (-x_1 \equiv z_1 \wedge (Py \Leftrightarrow -Px_1))]$ is satisfiable only over universes of cardinality 1 or cardinality ≥ 3. Thus there cannot be for each initially-extended Ackermann schema F with identity a schema without identity that is satisfiable over the same universes as F. Our procedure for eliminating identity from such schemata F will not preserve the size of finite models. Hence we shall treat the spectrum problem separately.

ISBN 0-201-02540-X

Let $F = \exists z_1\cdots\exists z_l\forall y\exists x_1\cdots\exists x_n F^M$ be an initially-extended Ackermann schema containing identity. We say F is in *special form* iff the matrix F^M is $J \wedge F_1$, where

(1) J is the conjunction of inequations $-z_i \equiv z_j$ for all distinct i, $j \le l$ and inequations $-z_i \equiv x_k$ for all $i \le l$ and $k \le n$; and

(2)   the only positive occurrences of identity in F_1, if any, are in atomic subformulas $z_i \equiv y$.

We show first that identity may be eliminated from schemata in special form, and second that the Initially-extended Ackermann Class with Identity is reducible to the subclass containing just the schemata in special form.

Initially-extended Ackermann Class with Identity, part 1. Let $F = \exists z_1\cdots\exists z_l\forall y\exists x_1\cdots\exists x_n(J \wedge F_1)$ be a schema in special form. Then we can construct in a uniform manner an initially-extended Ackermann schema G not containing identity such that F has a model iff G has a model, and F has a finite model iff G has a finite model.

Proof.   Let G_I result from $(J \wedge F_1)$ by replacing each atomic subformula $v \equiv w$ with Ivw, where I is a dyadic predicate letter foreign to F, and let $G = \exists z_1\cdots\exists z_l\forall y\exists x_1\cdots\exists x_n(Iyy \wedge G_I)$. Thus G is an initially-extended Ackermann schema without identity.

Any model for F becomes a model for G when I is interpreted as identity. Conversely, suppose G has a model $\mathfrak{M}$ with universe U. Thus there are $a_1,\ldots,a_l$ in U such that $\forall y\exists x_1\cdots\exists x_n(Iyy \wedge G_I)$ is true in $\mathfrak{M}$ when $z_1,\ldots,z_l$ take the values $a_1,\ldots,a_l$. Let $V = \{a_1,\ldots,a_l\} \cup \{b \mid b \in U$ and I is false in $\mathfrak{M}$ of $\langle a_i, b\rangle$ for each $i \le l\}$; and let $\mathfrak{N}$ be the submodel of $\mathfrak{M}$ with universe V. Since $-Iz_ix_k$ is a conjunct in G_I for each $i \le l$ and each $k \le n$, $\mathfrak{N}$ is also a model for G. Moreover, G logically implies $\forall y_1\forall y_2(y_1 \equiv y_2 \Rightarrow Iy_1y_2)$. Finally, since $-Iz_iz_j$ is a conjunct in G_I for all distinct i, $j \le l$, by the construction of V, $\forall y(Iz_iy \Rightarrow z_i \equiv y)$ is true in $\mathfrak{N}$ when z_i takes the value a_i, $1 \le i \le l$. By the definition of special form,

$$G \wedge \forall y_1\forall y_2(y_1 \equiv y_2 \Rightarrow Iy_1y_2) \wedge \bigwedge_{1 \le i \le l} \forall y(Iz_iy \Rightarrow z_i \equiv y)$$

logically implies $\forall y\exists x_1\cdots\exists x_n(J \wedge F_1)$. Hence F is true in $\mathfrak{N}$.

We have shown that F has a model iff G has a model, and F has a finite model iff G has a finite model. $\square$

Initially-extended Ackermann Class with Identity, part 2. Let $F = \exists z_1\cdots\exists z_l\forall y\exists x_1\cdots\exists x_n F^M$ be a schema containing identity. Then we can construct in a uniform manner a finite set $\{F_1,\ldots,F_m\}$ of initially-extended Ackermann schemata in special form such that any model for F of cardinality $\ge l + 1$ is a model for at least one of the F_i, and any model for one of the F_i is a model for F.

Proof. For each *l*-tuple $\vec{w} = \langle w_1,\ldots,w_l \rangle$ of variables among $z_1,\ldots,z_l$, let $F_{\vec{w}} = \exists z_1 \cdots \exists z_l \forall y \exists x_1 \cdots \exists x_n (J_1 \wedge C_{\vec{w}})$, where J_1 is the conjunction of inequations $-z_i \equiv z_j$ for all distinct i, $j \le l$, and $C_{\vec{w}} = F^M(z_1/w_1,\ldots,z_l/w_l)$. Every model for F of cardinality $\ge l$ is a model for at least one $F_{\vec{w}}$; conversely, each $F_{\vec{w}}$ logically implies F.

We now treat each $F_{\vec{w}}$ separately thus. Let J_2 be the conjunction of all inequations $-z_i \equiv x_k$, $1 \le i \le l$ and $1 \le k \le n$. For each *n*-tuple $\vec{v} = \langle v_1,\ldots,v_n \rangle$ of variables among $z_1,\ldots,z_l$, y, $x_1,\ldots,x_n$, let $J_{\vec{v}}$ be the conjunction of all inequations $-x_i \equiv x_j \wedge -x_i \equiv y$ for all distinct x_i and x_j in $\vec{v}$; and let $C_{\vec{w},\vec{v}} = C_{\vec{w}}(x_1/v_1,\ldots,x_n/v_n)$. Finally, let $G_{\vec{w}}$ be $\exists z_1 \cdots \exists z_l \forall y \exists x_1 \cdots \exists x_n (J_1 \wedge J_2 \wedge \bigvee (J_{\vec{v}} \wedge C_{\vec{w},v}))$, where the disjunction is over all *n*-tuples $\vec{v}$ of variables among $z_1,\ldots,z_l$, y, $x_1,\ldots,x_n$.

Any model for $F_{\vec{w}}$ of cardinality $\ge l+1$ is a model for $G_{\vec{w}}$. Conversely, $G_{\vec{w}}$ logically imples $F_{\vec{w}}$. We may eliminate all atomic subformulas $v \equiv w$ except those of the forms $z_i \equiv y$ and $y \equiv z_i$ from the subformulas $C_{\vec{w},\vec{v}}$ of $G_{\vec{w}}$ thus: first we replace each such atomic formula $v \equiv w$ by a symbol for truth if $v = w$ and by a symbol for falsity if $v \ne w$; these symbols are then eliminated in the usual way. The result is logically equivalent to $G_{\vec{w}}$. Moreover, all equations $y \equiv z_i$ may be replaced with equations $z_i \equiv y$. The result is an initially-extended Ackermann schema in special form. □

We conclude our treatment of the Initially-extended Ackermann Class with Identity by giving a characterization of spectra.

Initially-extended Ackermann Class with Identity, part 3. Let $F = \exists z_1 \cdots \exists z_l \forall y \exists x_1 \cdots \exists x_n F^M$ be a schema containing identity. If F has a model of cardinality $p \ge l + 1$, then F has models of every cardinality $\ge p$.

Proof. Let $\mathfrak{M}$ be a model for F with universe U. Then there are (not necessarily distinct) elements $a_1,\ldots,a_l$ of U such that $\forall y \exists x_1 \cdots \exists x_n F^M$ is true in $\mathfrak{M}$ when $z_1,\ldots,z_l$ take the values $a_1,\ldots,a_l$. If $|U| \ge l + 1$, then U also contains an element b distinct from $a_1,\ldots,a_l$. Let U' be any superset of U, and let $\gamma: U' \to U$ carry each element of U to itself and each element of $U' - U$ to b. Let $\mathfrak{N}$ have universe U', and in $\mathfrak{N}$ let each *k*-place predicate letter P of F, $k > 0$, be true of just those *k*-tuples $\langle b_1,\ldots,b_k \rangle$ such that P is true in $\mathfrak{M}$ of $\langle \gamma b_1,\ldots,\gamma b_k \rangle$. Note that $\mathfrak{M}$ is a submodel of $\mathfrak{N}$. We show that $\mathfrak{N}$ is a model for F. For this it suffices to show that, for each element c of U', the formula $\exists x_1 \cdots \exists x_n F^M$ is true in $\mathfrak{N}$ when the variables $z_1,\ldots,z_l$, y take the values $a_1,\ldots,a_l$, c.

Case 1. $c \in U$. Since $\exists x_1 \cdots \exists x_n F^M$ is true in $\mathfrak{M}$ when $z_1,\ldots,z_l$, y take the values $a_1,\ldots,a_l$, c, and sine $\mathfrak{M}$ is a submodel of $\mathfrak{N}$, it follows that $\exists x_1 \cdots \exists x_n F^M$ is true in $\mathfrak{N}$ for these values of the variables.

Case 2. $c \in U' - U$. There are elements $d_1,\ldots,d_n$ of U such that F^M is true in $\mathfrak{M}$ when $z_1,\ldots,z_l$, y, $x_1,\ldots x_n$ take the values $a_1,\ldots,a_l$, b, $d_1,\ldots,d_n$. For $1 \le i \le n$, let $e_i = d_i$ if $d_i \ne b$, and let $e_i = c$ if $d_i = b$. It

ISBN 0-201-02540-X

suffices to show that F^M is true in $\mathfrak{N}$ when $z_1,\ldots,z_l$, y, $x_1,\ldots,x_n$ take the values $a_1,\ldots,a_l$, c, $e_1,\ldots,e_n$. We show that for each atomic subformula A of F, A is true in $\mathfrak{N}$ when $z_1,\ldots,z_l$, y, $x_1,\ldots,x_n$ take the values $a_1,\ldots,a_l$, c, $e_1,\ldots,e_n$ iff A is true in $\mathfrak{M}$ when $z_1,\ldots,z_l$, y, $x_1,\ldots,x_n$ take the values $a_1,\ldots,a_l$, b, $d_1,\ldots,d_n$.

If A contains a predicate letter other than $\equiv$, then, since $\gamma a_i = a_i$, $\gamma c = b$, and $\gamma e_j = d_j$ for $1 \le i \le l$ and $1 \le j \le n$, the result is an immediate consequence of the specification of $\mathfrak{N}$. If A contains $\equiv$, then the result follows since γ is injective on $\{a_1,\ldots,a_l, c, e_1,\ldots,e_n\}$. $\square$

We have shown that the spectrum of a satisfiable schema $\exists z_1\cdots\exists x_l\forall y\exists x_1\cdots\exists x_n F^M$ is either a set of integers $\le l$, or a final segment of the integers, or else the union of a set of integers $\le l$ and a final segment of the integers.

3. Monadic and Bernays–Schönfinkel Classes with Identity

In this section we no longer seek to eliminate the identity sign from schemata. By treating $\equiv$ directly we show finite controllability and obtain spectrum-characterizations of two classes. Our argument for the Monadic Class with Identity is readily applicable to the Initially-extended Essentially Monadic Class with Identity. Hence, with this section we shall have exhausted the known positive results for schemata with identity (see §4).

Class 2.1 (Monadic Class with Identity). Let F be a schema containing identity and the monadic predicate letters $P_1,\ldots,P_p$. Let q be the maxium number of nested quantifiers in F. If F has a model of any cardinality $\ge 2^p \cdot q$, then F has models of every cardinality $\ge 2^p \cdot q$.

Proof. Let $E_1,\ldots,E_{2^p}$ be all the different conjunctions $\bigwedge_{1 \le k \le p} \pm P_k y$, where $\pm$ indicates that each $P_k y$ may occur either negated or not. A *structure* $\mathfrak{M}$ (for F) is a nonempty set M of objects together with interpretations of the predicate letters P_k over M. If $\mathfrak{M}$ is a structure with universe M, then for each i, $1 \le i \le 2^p$, let $M_i = \{a \in M | E_i$ is true in $\mathfrak{M}$ when the variable y is assigned the value $a\}$. Two structures $\mathfrak{M}$ and $\mathfrak{N}$ with universes M and N are *similar* iff, for each $i \le 2^p$, either $M_i = N_i$ or $M_i \cap N_i$ has cardinality $\ge q$, and for all $j \ne i$, $M_i \cap N_j$ is empty.

Lemma. Let $\mathfrak{M}$ and $\mathfrak{N}$ be similar structures with universes M and N. For all (not necessarily proper) subformulas G of F and each assignment of values in $M \cap N$ to the free variables (if any) of G, G is true in $\mathfrak{M}$ under the assignment iff G is true in $\mathfrak{N}$ under the assignment.

Proof. By induction on the complexity of G. If G is atomic, then G is $P_k v$ or $v \equiv w$, and the result is obvious. If G is $H_1 \vee H_2$, $H_1 \wedge H_2$, or $-H_1$, then the result follows immediately if it holds for H_1 and H_2.

Let G be $\exists vH$, and suppose an assignment of values in $M \cap N$ to the free variables of G is fixed. Let X be the set of values so assigned. Since G is a subformula of F, it contains at most $q - 1$ free variables; hence $|X| < q$. Suppose G is true in $\mathfrak{M}$ under the assignment. Hence there is an a in M such that H is true in $\mathfrak{M}$ when, additionally, the variable v is assigned the value a. If $a \in N$, then by the induction hypothesis, H is true in $\mathfrak{N}$ under the augmented assignment, whence $\exists vH$ is true in $\mathfrak{N}$ under the original assignment. Suppose $a \notin N$, and let $a \in M_i$, where $1 \leq i \leq 2^p$. Thus $M_i \neq N_i$, so that $M_i \cap N_i$ has cardinality $\geq q$. But then there is an element b of $(M_i \cap N_i) - X$. Let $\gamma: M \to M$ carry a to b, b to a, and every other member of M to itself. Then γ is an automorphism of the structure $\mathfrak{M}$, and γ is the identity on X. Hence H is true in $\mathfrak{M}$ under the original assignment augmented by assigning b to v. And then, by the induction hypothesis, H is true in $\mathfrak{N}$ under this augmented assignment, so $\exists vH$ is true in $\mathfrak{N}$ under the original assignment. We have shown that if G is true in $\mathfrak{M}$, then it is true in $\mathfrak{N}$. The converse is symmetrical.

The case $G = \forall vH$ remains. Here G is logically equivalent to $-\exists v-H$, so the preceding arguments suffice. $\square$

The lemma implies that if $\mathfrak{M}$ and $\mathfrak{N}$ are similar structures, then $\mathfrak{M}$ is a model for F just in case $\mathfrak{N}$ is a model for F. Suppose $\mathfrak{M}$ is a model for F with universe M of cardinality $\geq 2^p \cdot q$. Then $|M_i| \geq q$ for at least one $i \leq 2^p$. And then for each $m \geq 2^p \cdot q$ there is a structure $\mathfrak{N}$ similar to $\mathfrak{M}$ whose universe has cardinality m. Hence there are models for F for all cardinalities $\geq 2^p \cdot q$. $\square$

A refinement of the argument above works for the Initially-extended Essentially Monadic Class with Identity, in which each atomic subformula containing a predicate letter other than $\equiv$ contains at most one variable aside from initial x-variables (see Chapter 2, §5). Indeed, suppose $F = \exists z_1 \cdots \exists z_l F_1$ is in this class, where F_1 contains at most q nested quantifiers. A structure $\mathfrak{M}$ (for F_1) is a universe M together with interpretations over M of the predicate letters of F and elements $a_1^{\mathfrak{M}},\ldots,a_l^{\mathfrak{M}}$ of M, which are taken as the fixed values for $z_1,\ldots,z_l$. Let $E_1,\ldots,E_r$ be the maximal consistent conjunctions of signed atomic formulas each of which contains no variable aside from $y, z_1,\ldots,z_l$; thus $r = d(F, l + 1)$. If $\mathfrak{M}$ is a structure with universe M, then for each i, $1 \leq i \leq r$, let $M_i = \{a \in M \,|\, E_i$ is true in $\mathfrak{M}$ when $z_1,\ldots,z_l$ are assigned the values $a_1^{\mathfrak{M}},\ldots,a_l^{\mathfrak{M}}$ and y is assigned the value $a\}$. Similarity of structures is defined as before. Note that since $M_i = \{a_k^{\mathfrak{M}}\}$ whenever $y \equiv z_k$ is a conjunct in E_i, $a_k^{\mathfrak{M}} = a_k^{\mathfrak{N}}$ for $1 \leq k \leq l$ whenever $\mathfrak{M}$ and $\mathfrak{N}$ are similar. A Lemma like that in the argument for Class 2.1 with Identity is then forthcoming, although the proof is somewhat subtler. We may conclude that if F has a model of cardinality $\geq l + q \cdot d(F, l + 1)$, then F has models of all cardinalities $\geq l + q \cdot d(F, l + 1)$.

ISBN 0-201-02540-X

Class 1.1 consists of schemata F whose functional forms F^* contain no k-place function sign for $k > 0$. Its finite controllability—even when identity is present—is trivial. For a schema F containing identity in this class has a model iff there is a standard truth-assignment verifying $E(F, \omega)$; since $E(F, \omega) = E(F, 1)$, any such truth-assignment provides a finite model for F. However, to characterize spectra of schemata in Class 1.1 with Identity, we need a rather elaborate proof. We shall limit our attention to Bernays–Schönfinkel schemata: prenex schemata F with prefixes $\exists z_1 \cdots \exists z_l \forall y_1 \cdots \forall y_m$. Every schema in Class 1.1 can be transformed into a Bernays–Schönfinkel schema by the usual prenexing rules.

Let F be a Bernays–Schönfinkel schema, and let $C = \{c_{z_1}, \ldots, c_{z_l}\}$ be the set of indicial constants in F^*. Then F has a model of cardinality p iff there is a set D of terms and a standard truth-assignment $\mathfrak{A}$ such that

(1) $\mathfrak{A}$ verifies every instance of F^* over $C \cup D$;
(2)   there are exactly p $\mathfrak{A}$-distinct terms in $C \cup D$.

(Two terms s and t are $\mathfrak{A}$-*distinct,* naturally, iff $\mathfrak{A}$ falsifies $s \equiv t$.) Indeed, such a truth-assignment $\mathfrak{A}$ provides a model of cardinality p for F whose universe contains, for each s in $C \cup D$, exactly one representative of the equivalence class $\{t \mid t \in C \cup D$ and $\mathfrak{A}$ verifies $s \equiv t\}$. Conversely, given a model $\mathfrak{M}$ for F of cardinality p, we may let the constants in C denote the (not necessarily distinct) elements of the model that are appropriate values of the initial x-variables of F, and pick any D large enough so that every other element of the model can be denoted by a term in D. The truth-assignment $\mathfrak{A}$ may then be defined in accord with the interpretations of the predicate letters of F provided by $\mathfrak{M}$. Note that we may so choose D that the terms in D are pairwise $\mathfrak{A}$-distinct and are $\mathfrak{A}$-distinct from each term in C. (Here p may be finite or infinite.)

We shall show that the spectrum of a Bernays–Schönfinkel schema with identity is either finite or cofinite. The proof requires a combinatorial result first shown by Ramsey (1928) for precisely this purpose. If S is a set and $m > 0$, we let $S^{(m)}$ be the class of m-membered subsets of S.

Ramsey's Theorem B.   For every m, n, and $p > 0$ there is a $q > 0$ with the following property: if S is any set of cardinality $\geq q$ and f is any function on $S^{(m)}$ with range of cardinality $\leq n$, then there is an $R \subseteq S$ of cardinality p such that $f[R^{(m)}]$ contains only one member.

For each m, n, and p we let $\zeta(m, n, p)$ be the least q with the property above. A primitive recursive bound on ζ can be found in Ramsey's paper, and much work since Ramsey's has been done to improve this bound. (Incidentally, Ramsey's Theorem is more familiar and easier to prove in its infinitary form (Ramsey's Theorem A): for every infinite set S, every

ISBN 0-201-02540-X

$m > 0$, and every function f on $S^{(m)}$ with finite range, there is an infinite R $\subseteq S$ such that $f[R^{(m)}]$ has cardinality 1. In this form the theorem has many applications to set theory and model theory. The theorem is often used to yield sets of "order-indiscernibles". Such applications are indeed closely analogous to the one we make below of the finitary Ramsey's Theorem.)

Bernays–Schönfinkel Class with Identity.   Let $F = \exists z_1 \cdots \exists z_l \forall y_1 \cdots \forall y_m F^M$ be a schema containing identity.

(a)   If F has a model $\mathfrak{M}$ of cardinality $\geq l$, then F has models of every cardinality $\geq l$ and $\leq |\mathfrak{M}|$.

(b)   If F has a model of cardinality $\geq \zeta(m, d(F, m + l), m + 1) + l$, then F has models of every cardinality $\geq l$.

Proof.   (a)   Suppose $\mathfrak{M}$ is a (finite or infinite) model for F. Thus there are (not necessarily distinct) elements $a_1,\ldots,a_l$ of this model such that $\forall y_1 \cdots \forall y_m F^M$ is true in $\mathfrak{M}$ when $z_1,\ldots,z_l$ take the values $a_1,\ldots,a_l$. But then any submodel of $\mathfrak{M}$ that contains $a_1,\ldots,a_l$ is also a model for F.

(b)   Suppose that F has a model of cardinality $\geq \zeta(m, d(F, m + l), m + 1) + l$. Then there is a set D of terms and a standard truth-assignment $\mathfrak{A}$ such that

(1) $\mathfrak{A}$ verifies every instance of F^* over $C \cup D$;
(2)   the terms in D are pairwise $\mathfrak{A}$-distinct, and $\mathfrak{A}$-distinct from the terms in C;
(3) $|D| \geq \zeta(m, d(F, m + l), m + 1)$.

As always we assume a fixed well-ordering of D. For any terms $s_1,\ldots,s_k$ in D, we write $\mathfrak{p}_\mathfrak{A}^e(s_1,\ldots,s_k)$ for the profile $\mathfrak{p}_\mathfrak{A}(c_{z_1},\ldots,c_{z_l}, s_1,\ldots,s_k)$. Note that on m-tuples of terms the function $\mathfrak{p}_\mathfrak{A}^e$ takes on at most $d(F, m + l)$ distinct values. Hence, by Ramsey's Theorem B, there is an $R \subseteq D$ of cardinality $m + 1$ such that

$$\mathfrak{p}_\mathfrak{A}^e(s_1,\ldots,s_m) = \mathfrak{p}_\mathfrak{A}^e(t_1,\ldots,t_m)$$

whenever $\langle s_1,\ldots,s_m \rangle$ and $\langle t_1,\ldots,t_m \rangle$ are arrangements—in order—of two m-membered subsets of R. We now construct an infinite model for F; together with (a) this yields (b).

Let T be an infinite set of terms disjoint from C. If $S \subseteq T$ has cardinality $\leq m + 1$, then the *S-good* term-mapping is the term-mapping that carries every term not in S to itself and carries the earliest member of S to the earliest member of R, the second earliest member of S to the second earliest member of R, and so on. Our application of Ramsey's Theorem was designed to yield the following crucial consequence.

Lemma.   Let $t_1,\ldots,t_{k+1}$ be any distinct terms in T, $k \leq m$, let γ' be the

ISBN 0-201-02540-X

$\{t_1,\ldots,t_k\}$-good term-mapping, and let γ be the $\{t_1,\ldots,t_{k+1}\}$-good term-mapping. Then $\mathfrak{p}_{\mathfrak{A}}{}^e(\gamma't_1,\ldots,\gamma't_k) = \mathfrak{p}_{\mathfrak{A}}{}^e(\gamma t_1,\ldots,\gamma t_k)$.

Proof. If t_{k+1} is later than all of $t_1,\ldots,t_k$, then $\langle\gamma't_1,\ldots,\gamma't_k\rangle = \langle\gamma t_1,\ldots,\gamma t_k\rangle$. Otherwise, let $\langle u_1,\ldots,u_{m+1}\rangle$ be the arrangement of R in order, and suppose t_{k+1} is the jth earliest member of $\{t_1,\ldots,t_{k+1}\}$. Then $\langle\gamma't_1,\ldots,\gamma't_k\rangle$ is some permutation of $\langle u_1,\ldots,u_k\rangle$ whereas $\langle\gamma t_1,\ldots,\gamma t_k\rangle$ is the same permutation of $\langle u_1,\ldots,u_{j-1}, u_{j+1},\ldots,u_{k+1}\rangle$. But by the choice of R, $\quad\mathfrak{p}_{\mathfrak{A}}{}^e(u_1,\ldots,u_m) \quad = \quad \mathfrak{p}_{\mathfrak{A}}{}^e(u_1,\ldots,u_{j-1}, u_{j+1},\ldots,u_{m+1})$. Hence $\mathfrak{p}_{\mathfrak{A}}{}^e(\gamma't_1,\ldots,\gamma't_k) = \mathfrak{p}_{\mathfrak{A}}{}^e(\gamma t_1,\ldots,\gamma t_k)$. $\square$

For terminological convenience we extend the definition of good term-mapping thus: if S is any subset of $C \cup T$ that contains at most $m + 1$ terms from T, then the S-good term-mapping is the $(S \cap T)$-good term-mapping. Since every good term-mapping is the identity on C, the basic properties of the profile and iterated applications of the lemma yield a

Corollary. Let S be a subset of $C \cup T$ that contains at most $m + 1$ terms from T, and let $\{t_1,\ldots,t_k\}$ be any subset of S. Then $\mathfrak{p}_{\mathfrak{A}}{}^e(\gamma't_1,\ldots,\gamma't_k) = \mathfrak{p}_{\mathfrak{A}}{}^e(\gamma t_1,\ldots,\gamma t_k)$, where γ' is the $\{t_1,\ldots,t_k\}$-good term-mapping and γ is the S-good term-mapping. $\square$

We shall define a standard truth-assignment $\mathfrak{B}$ that verifies each instance of F^* over $C \cup T$. We define $\mathfrak{B}$ only on atoms containing at most $m + 1$ terms in T; this suffices since each instance of F^* over $C \cup T$ contains at most m terms in T. If A is any such atom, let S be the set of terms occurring in A, let γ be the S-good term-mapping, and Γ be the γ-canonical formula-mapping. We then let

$$\mathfrak{B}(A) = \mathfrak{A}(\Gamma(A)).$$

We show first that $\mathfrak{B}$ is standard. Note that the terms in T are pairwise $\mathfrak{B}$-distinct. For if t_1 and t_2 are distinct terms in T, then the $\{t_1, t_2\}$-good term-mapping carries t_1 and t_2 to distinct terms in R, and by the choice of $\mathfrak{A}$ these terms in R are $\mathfrak{A}$-distinct. Similarly, each term in T is $\mathfrak{B}$-distinct from each term in C. Finally, since every good term-mapping is the identity on C, the only nontrivial equations verified by $\mathfrak{B}$ are those equations, if any, between members of C that are verified by $\mathfrak{A}$. If $c_{z_i} \equiv c_{z_j}$ is such an equation, and if an atom A differs from an atom B only in that A has occurrences of c_{z_i} where B has occurrences of c_{z_j}, then the same good term-mapping is used to define $\mathfrak{B}$ on A and $\mathfrak{B}$ on B; hence $\Gamma(A)$ differs from $\Gamma(B)$ only in having occurrences of c_{z_i} rather than of c_{z_j}. Since $\mathfrak{A}$ is standard, $\mathfrak{A}(\Gamma(A)) = \mathfrak{A}(\Gamma(B))$; whence $\mathfrak{B}(A) = \mathfrak{B}(B)$. Hence $\mathfrak{B}$ is standard.

It remains to show that $\mathfrak{B}$ verifies each instance of F^* over $C \cup T$. Let $H = F^*(y_1/t_1,\ldots,y_m/t_m)$ be such an instance, let γ be the $\{t_1,\ldots,t_m\}$-good term-mapping, and let $\Psi(H) = F^*(y_1/\gamma t_1,\ldots,y_m/\gamma t_m)$. Then $\Psi(H)$ is an

ISBN 0-201-02540-X

instance of F^* over $C \cup D$, and is verified by $\mathfrak{A}$. Thus it suffices to show that, for each i,

$$\mathfrak{B}(\langle H \rangle_i) = \mathfrak{A}(\langle \Psi(H) \rangle_i).$$

Suppose $\langle H \rangle_i = Ps_1 \cdots s_p$. Then $\mathfrak{B}(\langle H \rangle_i) = \mathfrak{A}(P\gamma' s_1 \cdots \gamma' s_p)$, where γ' is the $\{s_1,\ldots,s_p\}$-good term-mapping. But $\{s_1,\ldots,s_p\}$ is a subset of $\{t_1,\ldots,t_m\} \cup C$. Hence, by the corollary, $\mathfrak{A}(P\gamma' s_1 \cdots \gamma' s_p) = \mathfrak{A}(P\gamma s_1 \cdots \gamma s_p)$; and $P\gamma s_1 \cdots \gamma s_p = \langle \Psi(H) \rangle_i$, since γ carries every member of C to itself.

This concludes the proof, for we have shown that $\mathfrak{B}$ is standard, that the terms in T are pairwise $\mathfrak{B}$-distinct, and that $\mathfrak{B}$ verifies every instance of F^* over $C \cup T$. Thus $\mathfrak{B}$ provides an infinite model for F. $\square$

4. The Gödel–Kalmár–Schütte Class with Identity

A number of the classes shown solvable in the preceding chapters become unsolvable once identity is allowed. Indeed, a simple reduction establishes the unsolvability of Classes 1.5, 1.6, and 2.4 with Identity. Let F be an arbitrary schema $\forall y_1 \forall y_2 \forall y_3 \exists x F^M$; the class of all such schemata is, as we have often noted, unsolvable. Let G be

$$\forall y_1 \forall y_2 \forall y_3 \exists x \exists x_1 \exists x_2 \exists x_3 [y_1 \equiv x_1 \wedge y_2 \equiv x_2 \wedge y_3 \equiv x_3$$

$$\wedge F^M(y_1/x_1, y_2/x_2, y_3/x_3)].$$

Then G is logically equivalent to F. Moreover, no atomic subformula of G aside from $y_1 \equiv x_1$, $y_2 \equiv x_2$, and $y_3 \equiv x_3$ contains y-variables; in particular, no distinct atomic subformulas of G are coinstantiable. Consequently, G is in Classes 1.5, 1.6, and 2.4 with Identity, whence these classes are unsolvable.

We note in addition that Class 4.1, the Maslov Class, becomes unsolvable once identity is allowed (Chang and Keisler, 1962).

We have not yet mentioned Classes 2.3 and 2.5. Class 2.5, the Gödel–Kalmár–Schütte Class, contains all prenex schemata with prefixes $\forall y_1 \forall y_2 \exists x_1 \cdots \exists x_n$; Class 2.3, the Minimal Gödel–Kalmár–Schütte Class, contains all prenex schemata with prefixes $\forall y_1 \forall y_2 \exists x$ in which only dyadic predicate letters occur. The status of Class 2.5 with Identity is open; indeed, it is not known even whether the minimal subclass 2.3 with Identity is solvable. We take the status of these classes to be the most important open problem in the theory of solvable classes, and shall devote the rest of this chapter to a discussion of it.

At the end of his 1933 paper Gödel claimed, "theorem I [the finite controllability of Class 2.5 without identity] also holds for formulas that contain the $=$ sign, which can be proved by the same procedure." However, the considerations adduced below show that Gödel's proof does not gen-

ISBN 0-201-02540-X

eralize to Class 2.5 with Identity, nor do the finite controllability proofs of Schütte (1934a) and Ackermann (1954). Moreover, the known solvability proofs for Class 2.5 without identity are also inapplicable to Class 2.5 with Identity. Indeed, our considerations suggest that Class 2.5 with Identity might be unsolvable.

We start by investigating the possibility of eliminating identity in favor of an ordinary dyadic predicate letter I. The Identity Elimination Lemma affords no help here. For suppose $F = \forall y_1 \forall y_2 \exists x F^M$ is a schema in Class 2.3 with Identity. The lemma applied to F yields the formula $\forall y I y y \wedge \forall y_1 \forall y_2 \forall y_3 J_F \wedge F_I$. This formula is equivalent to the schema

$$\forall y_1 \forall y_2 \forall y_3 \exists x [I y_1 y_1 \wedge J_F \wedge F_I{}^M],$$

where $F_I{}^M$ comes from the matrix F^M of F by replacing $\equiv$ with the predicate letter I; alternatively, the quantifier $\forall y_3$ can appear in the prefix after the quantifier $\exists x$. In either case, however, the class of schemata obtained from schemata F in Class 2.3 with Identity by this reduction is not a subclass of any known solvable class without identity: there are three y-variables and a noninitial x-variable, and all combinations of y_1, y_2, y_3 taken two at a time occur as arguments of atomic subformulas.

In §2 of this chapter we gained solvability results by eliminating identity in favor of a dyadic predicate letter I and adjoining clauses logically weaker than those used in the Identity Elimination Lemma. Here too we might try to devise a reduction with such weaker clauses. To avoid introducing any new y-variables, we might try the following clause J': the conjunction of all formulas

$$I w_1 w_2 \Rightarrow (P v_1 v_2 \Leftrightarrow P v_1 v_2 (w_1/w_2)),$$

where P is either a predicate letter of F or I itself, and w_1, w_2, v_1, v_2 are any variables among y_1, y_2, x. Then let G' be

$$\forall y_1 \forall y_2 \exists x (I y_1 y_1 \wedge J' \wedge F_I{}^M).$$

Thus G' is in Class 2.3 without identity. However, the clauses we have added will not do: they do not sufficiently constrain the interpretation of I. That is, there are unsatisfiable schemata F containing identity such that the schemata G' obtained by this procedure are satisfiable.

Our example of such a schema comes from Stål Aanderaa. Let F be $\forall y_1 \forall y_2 \exists x F^M$, where F^M is the conjunction of (1)–(4):

(1) $(Q y_1 \wedge Q y_2) \Rightarrow y_1 \equiv y_2$;

(2) $(-Q y_1 \wedge -Q y_2) \Leftrightarrow Q x$;

(3) $(Q x \wedge -y_1 \equiv y_2) \Rightarrow (P x y_1 \Leftrightarrow -P x y_2)$;

(4) $Q y_1 \Rightarrow (P y_1 x \wedge -x \equiv y_2)$.

ISBN 0-201-02540-X

This schema is unsatisfiable. For suppose U is a universe over which F is satisfiable. Under the (assumed) interpretations of the predicate letters P and Q, from (1) and (2) there is exactly one element a in U of which Q is true. By (2) and (4) there are at least two elements b and c of U such that Q is false of b and of c, and P is true of $\langle a, b \rangle$ and $\langle a, c \rangle$. But this conflicts with (3).

On the other hand, the schema G' not containing identity that is obtained from F by the procedure suggested above is satisfiable. For it is easy to check that the following truth-assignment $\mathfrak{A}$ verifies every Herbrand instance of G': for all terms s and t in $D(G', \omega)$, $\mathfrak{A}$ verifies Qs iff $s \neq 1$ and $\mathfrak{A}$ verifies $-Q\hat{y}_1 s \wedge -Q\hat{y}_2 s$; $\mathfrak{A}$ verifies Pst iff $\mathfrak{A}$ verifies $Qs \wedge -Qt$ and either $s = \hat{y}_1 t$ or $s = \hat{y}_2 t$ or $t = \hat{y}_1 s$; and $\mathfrak{A}$ verifies Ist iff either $\mathfrak{A}$ verifies $Qs \wedge Qt$ or else $s = t$. Thus our attempted reduction fails to preserve unsatisfiability.

Were we to define $\mathfrak{A}$ as above except for reading $\equiv$ instead of I, we would obtain a truth-assignment verifying the expansion $E(F, \omega)$. But of course this truth-assignment is not standard. It is merely *locally standard* in the following sense: for all s, t, u in $D(F, \omega)$,

(a) $\mathfrak{A}$ verifies $s \equiv s$;

(b) if $\mathfrak{A}$ verifies $s \equiv t$ and if s, t, and u all occur together in some Herbrand instance of F, then $\mathfrak{p}_{\mathfrak{A}}(s, u) = \mathfrak{p}_{\mathfrak{A}}(t, u)$.

For standardness we need the stronger condition that if $\mathfrak{A}$ verifies $s \equiv t$, then $\mathfrak{p}_{\mathfrak{A}}(s, u) = \mathfrak{p}_{\mathfrak{A}}(t, u)$ for *every* term u in $D(F, \omega)$. Thus Aanderaa's example shows that the existence of a standard truth-assignment verifying the expansion cannot be inferred from the existence of a locally standard one. The suggested procedure fails because the clauses $Iy_1 y_1$ and J' can do nothing more than require local standardness.

The example also shows that the solvability proofs of Gödel (1932), Kalmár (1933), and Schütte (1934) for Class 2.5 do not generalize to Class 2.5 with Identity. Each of these proofs proceeds by deriving a syntactic criterion necessary and sufficient for the satisfiability of a schema in Class 2.5. This criterion is essentially the GKS Condition given in Chapter 7, particularly the reformulation of the Condition that appears in the Remark on page 179. Now if F contains identity, then clearly the GKS Condition takes no account of the fact that $\equiv$ cannot be treated like a predicate letter. That is, such a schema F fulfills the Condition just in case the schema F_I obtained by replacing $\equiv$ with a dyadic predicate letter I fulfills the Condition. Thus the Condition remains necessary but is no longer sufficient for satisfiability; for the truth-assignment $\mathfrak{A}$ constructed in the sufficiency proof will not in general be standard. Or even locally standard: for example, nothing in the Condition bars truth-assignments $\mathfrak{T}$ from $\mathbf{T}$ even though $\mathfrak{T}$

ISBN 0-201-02540-X

may falsify some atomic subformula $v \equiv v$ of F. The Condition may be strengthened to require that each member $\mathfrak{T}$ of **T** treat $\equiv$ appropriately, that is, each $\mathfrak{T}$ in **T** verifies $v \equiv v$ for every variable v, and if $\mathfrak{T}$ verifies $v_1 \equiv v_2$, then $\mathfrak{T}(A) = \mathfrak{T}(A(v_1/v_2))$ for all atomic subformulas A of F. But all this does is to make the Condition sufficient for the existence of a *locally* standard truth-assignment $\mathfrak{A}$ verifying $E(F, \omega)$, and as Aanderaa's example showed, this is not enough for the satisfiability of F.

The finite controllability proofs of Gödel and Schütte are thereby also shown to fail for Class 2.5 with Identity. Both of these proofs are based on the following property of schemata F in Class 2.5 without identity: if F meets the GKS Condition, then F has a finite model. But as we just pointed out, there are unsatisfiable schemata F in Class 2.5 with Identity that meet the GKS Condition (or any straightforward strengthening of the GKS Condition). Hence the central assertion of these finite controllability proofs is false for Class 2.5 with Identity.

The source of the difficulty will become clearer through an examination of where the amenability arguments used in Chapter 2 break down when applied to Class 2.5 with Identity. We consider the simple case of a schema $F = \forall y_1 \forall y_2 \exists x(-x \equiv y_1 \wedge -x \equiv y_2 \wedge M)$, where M is quantifier-free and contains only dyadic predicate letters (including identity). Suppose that $\mathfrak{A}$ is a truth-assignment that verifies the expansion $E(F, p)$ for some suitably large p. We wish to construct a term-mapping γ whose range is included in $D(F, p + 1)$ and a formula-mapping Γ with the following property (cf. page 44);

($\ddagger$)   For each predicate letter P of F, each term t of height >1, and all variables v and w of F, $\mathfrak{A}(\Gamma(P\hat{v}t \ \hat{w}t)) = \mathfrak{A}(P\hat{v}\gamma t \ \hat{w}\gamma t)$.

Moreover, since identity occurs in F, we must require that the induced truth-assignment $\mathfrak{A}(\Gamma)$ be standard whenever the given truth-assignment $\mathfrak{A}$ is standard.

In the amenability proof for the Minimal Gödel–Kalmár–Schütte Class, the formula-mapping Γ is constructed from γ thus:

(I)   if $s = \hat{y}_i t$, then $\Gamma(Pst) = P\hat{y}_i\gamma t \ \gamma t$ and $\Gamma(Pts) = P\gamma t \ \hat{y}_i\gamma t$, $i = 1$, 2;

(II)   if $\Gamma(Pst)$ is not determined by (I), then $\Gamma(Pst) = P\gamma s \ \gamma t$.

This construction of Γ, however, does not suffice when F contains identity; $\mathfrak{A}(\Gamma)$ may fail to be standard even though $\mathfrak{A}$ is standard. There may be terms s and t such that $s \neq t$ but $s \equiv t$ falls under (II) and $\mathfrak{A}$ verifies $\gamma s \equiv \gamma t$; hence $\mathfrak{A}(\Gamma)$ verifies $s \equiv t$. But then, for $\mathfrak{A}(\Gamma)$ to be standard we must have $\mathfrak{A}(\Gamma(P\hat{y}_i \ t)) = \mathfrak{A}(\Gamma(P\hat{y}_i \ s))$, that is, $\mathfrak{A}(P\hat{y}_i\gamma t \ \gamma t) = \mathfrak{A}(\Gamma(P\hat{y}_i \ s))$. Since $\mathfrak{A}$ is assumed standard, we do have $\mathfrak{A}(P\hat{y}_i\gamma t \ \gamma t) = \mathfrak{A}(P\hat{y}_i\gamma t \ \gamma s)$; but

ISBN 0-201-02540-X

since we cannot be assured that $\Gamma(P\hat{y}_i t\ s) = P\hat{y}_i \gamma t\ \gamma s$, the equivalence required for $\mathfrak{A}(\Gamma)$ to be standard is not assured.

Our difficulties would be over if we could so change the construction of Γ that $\mathfrak{A}(\Gamma)$ verifies $s \equiv t$ only if $s = t$. Since $-x \equiv y_1$ and $-x \equiv y_2$ are conjuncts in the matrix of F, $\mathfrak{A}$ falsifies $t \equiv \hat{y}_i t$ whenever $1 < h(t) \leq p + 1$, $i = 1, 2$. Hence clause (I) of the definition of Γ can remain as is; it contributes no verified equations $s \equiv t$. We may also retain clause (II) when either $s = t$ or $\mathfrak{A}$ falsifies $\gamma s \equiv \gamma t$. But if $s \neq t$ and $\mathfrak{A}$ verifies $\gamma s \equiv \gamma t$, then we need $\mathfrak{A}$-*distinct* terms s' and t' that permit us to define $\Gamma(Pst) = Ps't'$. In fact, Γ should be so defined that $\mathfrak{p}_{\mathfrak{A}(\Gamma)}(s, t) = \mathfrak{p}_{\mathfrak{A}}(s', t')$. (This will have repercussions on the definition of γ, but these need not detain us here.) This constrains the choice of s' and t'; in particular it requires that $\mathfrak{p}_{\mathfrak{A}(\Gamma)}(s) = \mathfrak{p}_{\mathfrak{A}}(s')$ and $\mathfrak{p}_{\mathfrak{A}(\Gamma)}(t) = \mathfrak{p}_{\mathfrak{A}}(t')$. We have already determined these profiles by the retained part of clause (II); $\mathfrak{p}_{\mathfrak{A}(\Gamma)}(s) = \mathfrak{p}_{\mathfrak{A}}(\gamma s)$ and $\mathfrak{p}_{\mathfrak{A}(\Gamma)}(t) = \mathfrak{p}_{\mathfrak{A}}(\gamma t)$. Hence suitable s' and t' can be found if for each t_1 there is a t_2 such that $\mathfrak{A}$ falsifies $t_1 \equiv t_2$ and $\mathfrak{p}_{\mathfrak{A}}(t_1) = \mathfrak{p}_{\mathfrak{A}}(t_2)$.

A basic problem for this proposed construction of Γ arises, however, if there is a term t_1 such that, for all terms t_2, if $\mathfrak{p}_{\mathfrak{A}}(t_1) = \mathfrak{p}_{\mathfrak{A}}(t_2)$, then $\mathfrak{A}$ verifies $t_1 \equiv t_2$. In other words, a problem arises if there is a "distinguished profile", a profile such that all terms with this profile under $\mathfrak{A}$ are identified by $\mathfrak{A}$. (Indeed, this is the crux of Aanderaa's example F above. The unsatisfiability of F is due precisely to the fact that all elements of which Q is true have to be identical.)

We have encountered a special property of schemata in Class 2.5 with Identity. Suppose we have a model, with universe U, for a schema F that contains only dyadic predicate letters; and let us call two elements a and b of U *equivalent* iff P is true of $\langle a, a \rangle$ just in case P is true of $\langle b, b \rangle$ for all predicate letters P of F. (This corresponds, in terms of expansions and truth-assignments, to the condition $\mathfrak{p}_{\mathfrak{A}}(a) = \mathfrak{p}_{\mathfrak{A}}(b)$.) Then there are satisfiable F in Class 2.5 with Identity every model of which contains a one-membered equivalence class. This is not true if F is in Class 2.5 without identity; indeed, it is not true for any schema of quantification theory without identity. We can always add to a model for a schema F without identity a new element, and extend the interpretations of the predicate letters so that this new element is indiscernible by means of the predicate letters of F from an element in the original model.

We return to the amenability proof and the case in which there is a distinguished profile: there is a term t such that for each term s if $\mathfrak{p}_{\mathfrak{A}}(s) = \mathfrak{p}_{\mathfrak{A}}(t)$ then $\mathfrak{A}$ verifies $s \equiv t$. Let u be the earliest term with this profile under $\mathfrak{A}$. Now we can again try to find a formula-mapping Γ that yields a standard induced truth-assignment $\mathfrak{A}(\Gamma)$, provided $\mathfrak{A}$ is standard. We alter clause (II) of the construction of Γ thus: $\Gamma(Pst) = P\gamma s\ \gamma t$ if either $s = t$ or $\mathfrak{A}$ falsifies

ISBN 0-201-02540-X

$\gamma s \equiv \gamma t$ or $\mathfrak{p}_\mathfrak{A}(\gamma s) = \mathfrak{p}_\mathfrak{A}(\gamma t) = \mathfrak{p}_\mathfrak{A}(u)$; otherwise $\Gamma(Pst) = Ps't'$ for the earliest $\mathfrak{A}$-distinct s' and t' with $\mathfrak{p}_\mathfrak{A}(s') = \mathfrak{p}_\mathfrak{A}(\gamma s)$ and $\mathfrak{p}_\mathfrak{A}(t') = \mathfrak{p}_\mathfrak{A}(\gamma t)$. Hence the only equations $s \equiv t$ verified by $\mathfrak{A}(\Gamma)$ are those for which $s = t$ and those for which $\mathfrak{p}_\mathfrak{A}(\gamma s) = \mathfrak{p}_\mathfrak{A}(\gamma t) = \mathfrak{p}_\mathfrak{A}(u)$. Unfortunately, $\mathfrak{A}(\Gamma)$ still need not be standard. For example, we can have the following: $\mathfrak{A}(\Gamma)$ verifies $t_1 \equiv t_2$ because (by hypothesis) $\mathfrak{p}_\mathfrak{A}(\gamma t_1) = \mathfrak{p}_\mathfrak{A}(\gamma t_2) = \mathfrak{p}_\mathfrak{A}(u)$; $\Gamma(P\hat{y}_1 t_1 \ t_1)$ $= P\hat{y}_1 t_1 \ \gamma t_1$ (by clause (I)) and $\Gamma(P\hat{y}_1 t_1 \ t_2) = P\gamma \hat{y}_1 t_1 \ \gamma t_2$ (because we assume $\mathfrak{A}$ falsifies $\gamma \hat{y}_1 t \equiv \gamma t_2$). For $\mathfrak{A}(\Gamma)$ to be standard we must have $\mathfrak{A}(P\hat{y}_1 \gamma t_1 \ \gamma t_1) = \mathfrak{A}(P\gamma \hat{y}_1 t_1 \ \gamma t_2)$ or, since $\mathfrak{A}$ is itself standard and verifies $\gamma t_1 \equiv \gamma t_2 \wedge \gamma t_1 \equiv u$, we must have $\mathfrak{A}(P\hat{y}_1 \gamma t_1 \ u) = \mathfrak{A}(P\gamma \hat{y}_1 t_1 \ u)$. Thus, in general, to assure that $\mathfrak{A}(\Gamma)$ is standard we want γ to have the property $\mathfrak{p}_\mathfrak{A}(u, \hat{y}_i \gamma t) = \mathfrak{p}_\mathfrak{A}(u, \gamma \hat{y}_i \ t)$ for each t. However, to secure this while preserving property ($\ddagger$) we cannot pick $\Gamma(Pst) = Ps't'$ for s' and t' as above. Rather, s' and t' must be $\mathfrak{A}$-distinct terms with the stronger property that $\mathfrak{p}_\mathfrak{A}(u, s') = \mathfrak{p}_\mathfrak{A}(u, \gamma s)$ and $\mathfrak{p}_\mathfrak{A}(u, t') = \mathfrak{p}_\mathfrak{A}(u, \gamma t)$. This can be obtained if for each s_1 that is $\mathfrak{A}$-distinct from u there is an s_2 such that $\mathfrak{A}$ falsifies $s_1 \equiv s_2$ and $\mathfrak{p}_\mathfrak{A}(u, s_1) = \mathfrak{p}_\mathfrak{A}(u, s_2)$. But if this requirement is not met, then we cannot restrict $\mathfrak{A}(\Gamma)$ to verify only equations $s \equiv t$ such that either $s = t$ or else γs and γt have the distinguished profile. And this requirement may not be met: even though for each s_1 that is $\mathfrak{A}$-distinct from u there is an s_2 such that $\mathfrak{A}$ falsifies $s_1 \equiv s_2$ and $\mathfrak{p}_\mathfrak{A}(s_1) = \mathfrak{p}_\mathfrak{A}(s_2)$, that is, even though $\mathfrak{p}_\mathfrak{A}(s_1)$ is not a distinguished profile for any s_1 that is $\mathfrak{A}$-distinct form u, there may be a term s_1 $\mathfrak{A}$-distinct from u such that for every term s_2 if $\mathfrak{p}_\mathfrak{A}(u, s_1) = \mathfrak{p}_\mathfrak{A}(u, s_2)$, then $\mathfrak{A}$ verifies $s_1 \equiv s_2$. Thus we have to consider equations that must be verified because of relations to an element determined by a distinguished profile, that is, there may be "second level" distinguished profiles. This phenomenon can occur repeatedly: at each level we may be forced to equate all terms with certain relations to terms previously ascertained to have a uniqueness property. Herein lies the heart of the difficulty with Class 2.5 with Identity.

We now introduce some terminology that enables us to encapsulate the general problem. Let F be $\forall y_1 \forall y_2 \exists x_1 \cdots \exists x_n F^M$, and suppose $\mathfrak{A}$ is a standard truth-assignment verifying $E(F, \omega)$.

A *special term of level 1* (under $\mathfrak{A}$) is a term t such that, for each s, if $\mathfrak{p}_\mathfrak{A}(s) = \mathfrak{p}_\mathfrak{A}(t)$, then $\mathfrak{A}$ verifies $s \equiv t$. The special terms of level 1 divide into equivalence classes of $\mathfrak{A}$-identical terms. The earliest term in each equivalence class is a *distinguished term of level 1* (under $\mathfrak{A}$). Since each equivalence class corresponds to what we called above a distinguished profile, there are at most $d(F, 1)$ distinguished terms of level 1. (If there are exactly $d(F, 1)$ such terms, then $\mathfrak{A}$ provides a model for F of cardinality $d(F, 1)$.) Now suppose we have defined the special and distinguished terms of all levels $\leq k$, and let the distinguished terms of levels $\leq k$ be $t_1, \ldots, t_m$. A *special term of level $k + 1$* (under $\mathfrak{A}$) is a term t that is not a special term of level $\leq k$ and such that, for all terms s, if $\mathfrak{p}_{\mathfrak{A}}(t_1,\ldots,t_m, s) = \mathfrak{p}_{\mathfrak{A}}(t_1,\ldots,t_m, t)$, then $\mathfrak{A}$ verifies $s \equiv t$. The special terms of level $k + 1$ divide into at most $d(F, m + 1)$ equivalence classes of $\mathfrak{A}$-identical terms; the earliest member of each such class is a *distinguished term of level $k + 1$* (under $\mathfrak{A}$). There may or may not be distinguished terms of level $k + 1$; if there are, then there must also be distinguished terms of each level $\leq k$. Finally, say F is *k-strongly satisfiable* iff there is a standard truth-assignment $\mathfrak{A}$ verifying $E(F, \omega)$ under which there are no distinguished terms of level $k + 1$.

By the methods of Chapter 3 we may show that if F is k-strongly satisfiable for some k, then F has a finite model (see the Appendix, §6). Indeed, there is a primitive recursive function ψ such that if F is k-strongly satisfiable, then F has a model of cardinality $\leq \psi(F, k)$. Conversely, if F has a finite model, then F is k-strongly satisfiable for some k: for if F has a finite model, then there is a standard truth-assignment $\mathfrak{A}$ verifying $E(F, \omega)$ under which there are only finitely many $\mathfrak{A}$-distinct terms.

In short, the following proposition (P) is equivalent to the finite controllability of Class 2.5 with Identity:

(P) Each satisfiable schema F in Class 2.5 with Identity is k-strongly satisfiable for some k.

We have no idea whether (P) is true. The evidence is conflicting. We shall soon show that there is no primitive recursive function that *finitely controls* Class 2.5 with Identity: there is no primitive recursive function φ such that if F is a satisfiable schema in Class 2.5 with Identity then F has a model of cardinality $\leq \varphi(F)$. This would support the supposition that (P) is false. On the other hand, we have not been able to find a schema in Class 2.5 with Identity that has only infinite models.

If (P) is false, it is of course possible that Class 2.5 with Identity be solvable, even though not finitely controllable. However, the example that establishes that Class 2.5 with Identity is not primitive recursively finitely controlled can be used to show that there is no primitive recursive decision procedure for the class. And we might expect that any example that defeats (P) would similarly lead to an encoding of some unsolvable problem.

Before presenting the crucial example, we give a simpler one, which exhibits more clearly how distinguished elements may be generated. For each $k > 0$ we construct a schema G_k in Class 2.5 with Identity that is k-strongly satisfiable but not $(k - 1)$-strongly satisfiable. Each G_k contains only one x-variable, and the dyadic predicate letters P, $Q_1,\ldots,Q_{k-1}$, and $\equiv$. We let G_k be $\forall y_1 \forall y_2 \exists x(-x \equiv y_1 \wedge -x \equiv y_2 \wedge M)$, where M is the

ISBN 0-201-02540-X

conjunction of the following clauses:

(1) $Q_1 y_1 y_2 \Rightarrow (-P y_1 y_1 \wedge P y_2 y_2)$;

(2) $(P y_1 y_1 \wedge P y_2 y_2) \Rightarrow (y_1 \equiv y_2 \wedge P x y_2 \wedge Q_1 x y_2)$;

(3) $\bigwedge_{2 \le i < k} [Q_i y_1 y_2 \Rightarrow (P y_2 x \wedge -P y_1 x \wedge Q_{i-1} y_1 x)]$

(4) $\bigwedge_{1 \le i \le k-2} [(Q_i y_2 y_1 \wedge P y_2 y_1) \Rightarrow (P x y_2 \wedge Q_i x y_1 \wedge Q_{i+1} x y_2)]$

(5) $(Q_{k-1} y_2 y_1 \wedge P y_2 y_1) \Rightarrow (P x y_2 \wedge Q_{k-1} x y_1)$;

(6) $[\bigwedge_{1 \le i \le k-1} (-Q_i y_1 y_2 \wedge -Q_i y_2 y_1)] \Rightarrow [(Q_{k-1} y_1 x \wedge Q_{k-1} y_2 x)$
$\vee y_1 \equiv y_2]$.

The schema G_k is satisfiable over universes of cardinality $\ge k$. Consider the universe $\{1,\ldots,m\}$, $m \ge k$. We define interpretations of the predicate letters of G_k over this universe thus: P is true of $\langle 1, 1 \rangle$ and of $\langle p + 1, p \rangle$ for $1 \le p \le k - 1$, but false for all other pairs; for each i, $1 \le i < k$, Q_i is true of $\langle p, i \rangle$ for $i < p \le m$, but false for all other pairs. These interpretations provide a model for G_k.

Conversely, suppose we have interpretations over a universe U that provide a model for G_k. Then there is exactly one element a_1 in U such that P is true of $\langle a_1, a_1 \rangle$; and for each i, $1 \le i \le k - 1$, there is exactly one element a_{i+1} such that P and Q_i are both true of $\langle a_{i+1}, a_i \rangle$ and Q_1 is true of $\langle a_{i+1}, a_1 \rangle$. These elements $a_1,\ldots,a_k$ correspond then to distinguished terms of levels $1,\ldots,k$.

We turn finally to the proof that there is no primitive recursive function that finitely controls Class 2.5 with Identity, first announced by Goldfarb (1979a). Define a function ψ thus: $\psi(0, i) = i + 1$; $\psi(m + 1, 0) = \psi(m, 1)$; $\psi(m + 1, i + 1) = \psi(m, \psi(m + 1, i))$. Then the function that carries each n to $\psi(n, 0)$ is not primitive recursive. Indeed, if g is any one-place primitive recursive function, then $g(n) < \psi(n, 0)$ for all but a finite number of integers n. (The function ψ is discussed at length by Péter (1967, §9).)

For each n we shall construct a satisfiable schema $F_n = \forall y_1 \forall y_2 \exists x_1 \cdots \exists x_{4n+4} F_n^M$ that contains identity, two monadic predicate letters P and Q, and $2n + 2$ dyadic predicate letters $J_0,\ldots,J_n$, $K_0,\ldots,K_n$ such that the smallest model for F_n has cardinality $> \psi(n, 0)$.

Before constructing F_n, we give the number-theoretic facts that underlie the intended interpretations of its predicate letters. For all $m, k \ge 0$ let $\varphi_m(k)$ be the least integer i such that $k \le \psi(m, i)$. Since, for each m, $\psi(m, i)$ is a strictly increasing function of i, if $\varphi_m(k) = i > 0$, then $\psi(m, i - 1) < k \le \psi(m, i)$.

Lemma 1. (a) $\varphi_{m+1}(k) = 0$ iff $\varphi_m(k) \le 1$.

(b) If $\varphi_{m+1}(k) > 0$, then $\varphi_{m+1}(k) = \varphi_{m+1}(\varphi_m(k)) + 1$.

ISBN 0-201-02540-X

Proof. (a) $\varphi_{m+1}(k) = 0$ iff $k \leq \psi(m + 1, 0)$; $\psi(m + 1, 0) = \psi(m, 1)$; and $\varphi_m(k) \leq 1$ iff $k \leq \psi(m, 1)$.

(b) Let $\varphi_{m+1}(k) = i > 0$. Thus $k \leq \psi(m + 1, i) = \psi(m, \psi(m + 1, i - 1))$. Hence $\varphi_m(k) \leq \psi(m + 1, i - 1)$; whence $\varphi_{m+1}(\varphi_m(k)) \leq i - 1$. This suffices if $i = 1$. If $i > 1$, then $k > \psi(m + 1, i - 1) = \psi(m, \psi(m + 1, i - 2))$. Hence $\varphi_m(k) > \psi(m + 1, i - 2)$; whence $\varphi_{m+1}(\varphi_m(k)) > i - 2$. We may conclude that $\varphi_{m+1}(\varphi_m(k)) = i - 1$. $\square$

The schema F_n has the following intended infinite model: the universe is the nonnegative integers; Pp is true iff $p = 0$; Qp is true iff $p = 1$; for each $m \leq n$, $J_m p\, q$ is true iff $p > 0$ and $q = \varphi_m(p)$, and $K_m p\, q$ is true iff $p > 0$ and $q < \varphi_m(p)$.

Let F_n be a prenex equivalent of $\forall y_1 \forall y_2 M$, where M is the conjunction of the following eight clauses:

(1) $(Py_1 \wedge Py_2) \Rightarrow y_1 \equiv y_2$

(2) $(Qy_1 \wedge Qy_2) \Rightarrow (y_1 \equiv y_2 \wedge -Py_1)$

(3) $\exists x(J_0 x y_1)$

(4) $\bigwedge_{0 \leq m < n} [(J_m y_1 y_2 \wedge -Py_2 \wedge -Qy_2)$
$\Rightarrow \exists x_1 \exists x_2 (J_{m+1} y_1 x_1 \wedge J_{m+1} y_2 x_2 \wedge J_0 x_1 x_2)]$

(5) $\bigwedge_{0 \leq m < n} [(J_{m+1} y_1 y_2 \wedge -Py_2)$
$\Rightarrow \exists x_1 \exists x_2 (J_0 y_2 x_1 \wedge J_{m+1} x_2 x_1 \wedge J_m y_1 x_2)]$

(6) $\bigwedge_{0 \leq m \leq n} (K_m y_1 y_2 \Rightarrow -J_m y_1 y_2)$

(7) $-Py_1 \Rightarrow \exists x[J_0 y_1 x \wedge (-y_1 \equiv y_2 \Rightarrow -J_0 y_2 x)$
$\wedge \bigwedge_{0 \leq m \leq n} ((K_m y_2 y_1 \vee J_m y_2 y_1) \Rightarrow K_m y_2 x)]$

(8) $\exists x_1 \exists x_2 [Px_1 \wedge Qx_2 \wedge \bigwedge_{0 \leq m < n} ((J_m y_1 x_1 \vee J_m y_1 x_2) \Leftrightarrow J_{m+1} y_1 x_1)$
$\wedge \bigwedge_{0 \leq m \leq n} (-J_m x_1 y_1 \wedge (J_m x_2 y_1 \Rightarrow y_1 \equiv x_1))$
$\wedge ((J_n y_1 y_2 \wedge J_n y_1 x_1) \Rightarrow y_2 \equiv x_1)].$

It is a routine matter to check that the intended infinite model is indeed a model for F_n; thus F_n is satisfiable. There also exist finite models for F_n. We shall not describe these models; to provide a finite model, the interpretations of the predicate letters must depart in various intricate ways from the intended ones.

Let $\mathfrak{M}$ be any model for F_n, and let M be the universe of $\mathfrak{M}$. To show that M contains more than $\psi(n, 0)$ elements, we shall choose by induction distinct elements $\bar{0}, \bar{1}, \ldots, \overline{\psi(n, 0)}$ of M. Our induction proceeds along the following lines. Suppose we have so chosen $\bar{0}, \ldots, \overline{k - 1}$, where $k \leq \psi(n, 0)$, that $\bar{0}, \ldots, \overline{k - 1}$ behave in $\mathfrak{M}$ just as the integers $0, \ldots, k - 1$ behave in the intended infinite model. We choose an element $\bar{k}$ of M that

ISBN 0-201-02540-X

bears the relation J_0 to $\overline{\varphi_0(k)}$ $(=\overline{k-1})$; $\bar{k}$ is then distinct from $\bar{0},\dots,\overline{k-1}$. To continue the induction, we must show that $\bar{k}$ behaves in $\mathfrak{M}$ just as k does in the intended model. Using clause (4), we shall show, for each $m \le n$, first that $\bar{k}$ bears J_m to $\overline{\varphi_m(k)}$ and second that $\bar{k}$ bears J_m to nothing but $\overline{\varphi_m(k)}$. This second fact is critical, and is proved for m given the hypothesis that it holds for $m+1$. But then, to start off, we must verify this fact for $m=n$: we must show that $\bar{k}$ bears J_n to nothing but $\overline{\varphi_n(k)}$. Here we use the last conjunct in clause (8), which assures us that for any b in M, if b bears J_n to $\bar{0}$, then b bears J_n to nothing but $\bar{0}$. Hence the desired result is forthcoming if $\varphi_n(k) = 0$. For this reason our argument works only if $k \le \psi(n,0)$. If $k > \psi(n,0)$, then $\varphi_n(k) > 0$, and we shall not be able to prove what we need. (It is precisely this breakdown that allows us to construct finite models for F_n.)

More precisely, by induction on k we find distinct members $\bar{0},\dots,\overline{k-1}$ of M for $k \le \psi(n,0) + 1$ so that the following hypotheses are fulfilled: for all $m \le n$, all c in M, and all $p < k$,

(A) Pc iff $c = \bar{0}$, and Qc iff $c = \bar{1}$;
(B) if $J_m\bar{p}\ c$, then $p > 0$ and $c = \overline{\varphi_m(p)}$;
(C) if $\varphi_m(k) > p$ and $J_m c\ \bar{p}$, then $c = \bar{q}$ for some $q < k$.

(We write an expression like "J_mbc", where b and c are members of M, as short for "J_m is true in $\mathfrak{M}$ of the pair $\langle b, c\rangle$.") Since $\varphi_0(q) = q - 1$ for each $q > 0$, (B) and (C) imply, for each $p < k$ and each c in M,

(D) if $J_0\bar{p}\ c$, then $p > 0$ and $c = \overline{p-1}$; if $p < k - 1$ and $J_0c\ \bar{p}$, then $c = \overline{p+1}$.

We begin by so choosing $\bar{0}$ and $\bar{1}$ that (A)–(C) are fulfilled for $k = 2$. By clauses (1), (2), and (8) there is a unique element $\bar{0}$ of M such that $P\bar{0}$ and a unique element $\bar{1}$ of M such that $Q\bar{1}$. By clause (2) $-P\bar{1}$, so $\bar{0}$ and $\bar{1}$ are distinct. The x-variables x_1 and x_2 of clause (8) must then take the values $\bar{0}$ and $\bar{1}$. Hence, for each c in M and each $m \le n$, $-J_m\bar{0}\ c$, and $J_m\bar{1}\ c$ only if $c = \bar{0}$. Thus (A) and (B) hold for $k = 2$. To verify (C) we need consider only the case $m = 0$, since $\varphi_m(2) = 0$ when $m > 0$. Suppose c is any member of M distinct from $\bar{1}$; we show that $-J_0c\ \bar{0}$. Since $-P\bar{1} \wedge -\bar{1} \equiv c$, by clause (7) there exists d in M such that $J_0\bar{1}\ d \wedge -J_0c\ d$. Since $J_0\bar{1}\ d$, we have $d = \bar{0}$, and (C) holds for $k = 2$.

Now suppose we have selected distinct $\bar{0},\dots,\overline{k-1}$ in M, where $2 \le k \le \psi(n,0)$, so that (A)–(C), and hence also (D), hold for k. We seek an element $\bar{k}$ of M such that (A)–(C) hold for $k + 1$. Let $X = \{b \in M| J_0b\ \overline{\varphi_0(k)}\}$. By clause (3), X is nonempty. By hypothesis (B), X is disjoint from $\{\bar{0},\dots,\overline{k-1}\}$. Let b be any member of X.

Lemma 2. For each $m \le n$, $J_mb\ \overline{\varphi_m(k)}$.

Proof. For $m = 0$ the lemma is immediate. Let $0 \le m < n$, and suppose

ISBN 0-201-02540-X

that $J_m b \overline{\varphi_m(k)}$. If $\varphi_m(k)$ is 0 or 1, then $\varphi_{m+1}(k) = 0$ by Lemma 1(a). Moreover, by clause (8), $(J_m b \, \bar{1} \lor J_m b \, \bar{0}) \Rightarrow J_{m+1} b \, \bar{0}$. Hence $J_{m+1} b \overline{\varphi_{m+1}(k)}$. If $\varphi_m(k) > 1$, then by Lemma 1(b) $\varphi_{m+1}(k) = \varphi_{m+1}(\varphi_m(k)) + 1$. By clause (4) there exist c and d in M such that

$$J_{m+1} b \, c \land J_{m+1} \overline{\varphi_m(k)} \, d \land J_0 c \, d.$$

By (B), $d = \overline{\varphi_{m+1}(\varphi_m(k))}$. Since $\varphi_{m+1}(\varphi_m(k)) < k - 1$, by (D) $c = \overline{\varphi_{m+1}(\varphi_m(k)) + 1} = \overline{\varphi_{m+1}(k)}$. Thus $J_{m+1} b \overline{\varphi_{m+1}(k)}$. □

Lemma 3. For each $m \le n$ and each $q < k$, if $J_m b \, \bar{q}$ then $q = \varphi_m(k)$.

Proof. Let $m \le n$. By (A), (D), and clause (7), if $0 < p < k$ then $[(J_m b \, \bar{p} \lor K_m b \, \bar{p}) \Rightarrow K_m b \overline{p - 1}]$. Consequently, if $0 \le r < p < k$ and $J_m b \, \bar{p}$, then $K_m b \, \bar{r}$, whence $-J_m b \, \bar{r}$ by clause (6). In short, if p and r are distinct integers $<k$, then $-(J_m b \, \bar{p} \land J_m b \, \bar{r})$. And then Lemma 2 yields the desired result. □

Lemma 4. For each $m \le n$ and each c in M, if $J_m b \, c$, then $c = \overline{\varphi_m(k)}$.

Proof. Suppose $m = n$ and $J_n b \, c$. Since $k \le \psi(n, 0)$, $\varphi_n(k) = 0$. By Lemma 2, $J_n b \, \bar{0}$. By clause (8), $(J_n b \, c \land J_n b \, 0) \Rightarrow c \equiv 0$. Hence $c = \bar{0} = \overline{\varphi_n(k)}$.

Suppose the lemma true for $m + 1$, where $0 \le m < n$, and let $J_m b \, c$. By Lemma 3, if c is $\bar{0}$ or $\bar{1}$, then $c = \overline{\varphi_m(k)}$. Assume that c is neither $\bar{0}$ nor $\bar{1}$. By clause (4) there are d and e in M such that

$$J_{m+1} b \, d \land J_{m+1} c \, e \land J_0 d \, e.$$

By the induction hypothesis, $d = \overline{\varphi_{m+1}(k)}$. By (D), $e = \overline{\varphi_{m+1}(k) - 1}$. Hence, by (C), $c = \bar{q}$ for some $q < k$. The result then follows by Lemma 3. □

Lemma 5. $|X| = 1$.

Proof. Suppose b_1 and b_2 were distinct members of X. Since $-Pb_1$, by clause (7) there exists c in M such that $J_0 b_1 c \land -J_0 b_2 c$. By Lemma 4, $c = \overline{\varphi_0(k)}$. But then $b_2 \notin X$, a contradiction. □

Now let $\bar{k}$ be the unique member of X. We have so far verified (A) and (B) for $k + 1$. We turn now to (C).

Lemma 6. For each $m \le n$, each $p < k$, and each c in M, if $\varphi_m(k + 1) > p$ and $J_m c \, \bar{p}$, then $c = \bar{q}$ for some $q < k + 1$.

Proof. Let $m = 0$, $p < \varphi_0(k + 1) = k$, and $J_0 c \, \bar{p}$. If $p < \varphi_0(k)$, then the result follows from hypothesis (C) for k. If $p = \varphi_0(k)$, then the result follows from Lemma 5.

Let $0 \le m < n$ and suppose the lemma proved for m. Let $p < \varphi_{m+1}(k + 1)$ and $J_{m+1} c \, \bar{p}$. If $p = 0$, then by clause (8), $[J_{m+1} c \, \bar{0} \Rightarrow (J_m c \, \bar{0} \lor J_m c \, \bar{1})]$. By Lemma 1(a), $0 < \varphi_{m+1}(k + 1)$ implies $1 < \varphi_m(k + 1)$.

ISBN 0-201-02540-X

Hence, since the lemma holds for m, $c = \bar{q}$ for some $q < k + 1$. If $p > 0$ then by clause (5) there are d and e in M such that

$$J_0 \bar{p}\, d \wedge J_{m+1} e\, d \wedge J_m c\, e.$$

By (D), $d = \overline{p - 1}$. Since $p < \varphi_{m+1}(k + 1)$, $p - 1 < \varphi_{m+1}(k)$. By hypothesis (C) for k, $e = \bar{q}$ for some $q < k$. But then $q < \varphi_m(k + 1)$. For if $q \geq \varphi_m(k + 1)$, then $\varphi_{m+1}(q) \geq \varphi_{m+1}(\varphi_m(k + 1)) = \varphi_{m+1}(k + 1) - 1 > p - 1$; whence by (B) $-J_{m+1}\bar{q}\ \overline{p - 1}$, that is, $-J_{m+1}e\, d$. Since then $q < \varphi_m(k + 1)$, and the Lemma holds for m, $c = \bar{r}$ for some $r < k + 1$. $\square$

We have verified (A)–(C) for $k + 1$, and so the induction may continue. Hence any model for F contains more than $\psi(n, 0)$ distinct elements.

As we have mentioned, if g is a one-place primitive recursive function, then $g(n) < \psi(n, 0)$ for all but a finite number of integers n. Since the construction of the schemata F_n is clearly primitive recursive, we have shown that there is no primitive recursive function χ such that, for each F in Class 2.5 with Identity that has a finite model, F has a model of cardinality $\leq \chi(F)$.

Our construction may be exploited to show that there is no primitive recursive decision procedure for Class 2.5 with Identity. Using the clauses of F_n, we may (primitive recursively) construct, for each Turing machine T and each n, a schema $F_{T,n}$ in Class 2.5 with Identity that is satisfiable iff the machine T halts in $\leq \psi(n, 0)$ steps. But there is no primitive recursive procedure for deciding, given any Turing machine T and any n, whether T halts in $\leq \psi(n, 0)$ steps. Thus there is no primitive recursive procedure for deciding, given any F in Class 2.5 with Identity, whether F is satisfiable. (Full details of this argument will appear in Goldfarb, 1980.) We know of no natural class of schemata that is solvable but not primitive recursively solvable.

ISBN 0-201-02540-X

Appendix

1. Remarks on Condensability

We start by specifying two classes similar to Class 1.4 that contain only condensable schemata. The notion of Horn formula is defined on page 126. A *positive occurrence* of a predicate letter in a Horn formula is an occurrence of that predicate letter in a positively signed atomic subformula; similarly for a negative occurrence of a predicate letter.

Class A.1. Let F be a schema with Horn matrix in which the number of negative occurrences of each predicate letter is at most one. Then F is n-condensable, where n is the number of conjuncts in the matrix of F.

Class A.2. Let F be a schema with Horn matrix in which the number of positive occurrences of each predicate letter is at most one. Then F is q^n-condensable, where n is the number of conjuncts in the matrix of F and q is the largest number of negative occurrences of a predicate letter in F.

The proofs are due to Harry Lewis, and closely parallel that for Class 1.4.

Proof for A.1. Call a (finite) set S of disjunctions of signed atomic formulas *sparse*$_1$ iff the conjunction of members of S is Horn and contains no more than one negative occurrence of each predicate letter. It suffices to show

(#) Any inconsistent conjunction of instances of members of a sparse$_1$ set S contains an inconsistent subconjunction consisting of at most one instance of each member of S.

We prove (#) by induction on the cardinality of S. If $|S| = 1$, then (#) holds vacuously. Suppose $|S| = n + 1$ and that (#) holds for all sparse$_1$ sets of cardinality $\leq n$. Let K be an inconsistent conjunction of members of S; we may assume that K contains at least one instance of each member of S.

Let T be the set of members of S that contain at least one negatively signed atomic formula, and let m be the number of predicate letters of S.

Burton Dreben and Warren D. Goldfarb, The Decision Problem: Solvable Classes of Quantificational Formulas

ISBN 0-201-02540-X

Then $|T| \leq m$, since no predicate letter possesses more than one negative occurrence. Moreover, at least one member of T must contain only negatively signed atomic formulas: otherwise every conjunct in K would contain a positively signed atomic formula, and then K would be consistent. Consequently, since every member of T contains at most one positively signed atomic formula, some predicate letter has no positive occurrences in T. That is, all the positive occurrences of some predicate letter P in S are in members $D_1,\ldots,D_k$ of S each of which is just a single positive signed atomic formula.

If P does not occur in T, then the subconjunction of K obtained by deleting all instances of $D_1,\ldots,D_k$ is inconsistent, since it does not contain P at all. In this case (#) follows from the induction hypothesis. Hence we may assume that P occurs (negatively) in T, and so there is a unique disjunction D in T that contains P.

Suppose D contains only one disjunct, that is, D is a negatively signed atomic formula. If there is an atom A containing P such that A and $-A$ are conjuncts in K, then $A \wedge -A$ is the desired subconjunction of K. And otherwise the subconjunction of K obtained by deleting all instances of $D_1,\ldots,D_k$ and D is inconsistent, so (#) follows from the induction hypothesis.

Now suppose D contains more than one disjunct. Let D' be obtained by D by deleting the (negatively signed) disjunct that contains P, and let $S' = S \cup \{D'\} - \{D_1,\ldots,D_k, D\}$. Then S' is a sparse$_1$ set of cardinality $\leq n$. Let J be the conjunction formed thus:

(1) J contains every instance of a member of $S - \{D_1,\ldots,D_k, D\}$ that is a conjunct in K;

(2) J contains every E such that, for some atom A, A is an instance of one of $D_1,\ldots,D_k$ that is a conjunct in K and $-A \vee E$ is (up to permutation of disjuncts) an instance of D that is a conjunct in K.

Note that J is a conjunction of instances of members of S', and the predicate letter P does not occur in J.

J is inconsistent. For were there a truth-assignment $\mathfrak{A}$ verifying J, then we could so extend $\mathfrak{A}$ as to verify each atom containing P iff that atom is an instance of one of $D_1,\ldots,D_k$ that is a conjunct in K; and this extension of $\mathfrak{A}$ would verify K.

By the induction hypothesis it follows that there is an inconsistent subconjunction J' of J that contains at most one instance of each member of S'. If J' contains no instance of D', then J' is the desired subconjunction of K. Otherwise $J' = J'' \wedge E$, where J'' is a conjunction of at most one instance of each member of $S - \{D_1,\ldots,D_k, D\}$ and where, for some atom A, A is an instance of one of $D_1,\ldots,D_k$ that is a conjunct in K and

ISBN 0-201-02540-X

$-A \vee E$ is an instance of D that is a conjunct in K. And then $J'' \wedge (-A \vee E) \wedge A$ is the desired subconjunction of K. □

Proof for A.2. Call a (finite) set S of disjunctions of signed atomic formulas *sparse*$_2$ iff the conjunction of members of S is Horn and contains no more than one positive occurrence of each predicate letter. We show

(#)   Any inconsistent conjunction of instances of members of a sparse$_2$ set S contains an inconsistent subconjunction consisting of at most q^n instances of each member of S, where $n = |S|$ and q is the largest number of negative occurrences of a predicate letter in S.

As before, if $|S| = 1$, then (#) holds vacuously. Suppose $|S| = n + 1$ and that (#) holds for all sparse$_2$ sets of cardinality $\leq n$. Let K be an inconsistent conjunction of instances of members of S.

S contains a member D that is a single positively signed atomic formula. For otherwise every conjunct in K would contain at least one negatively signed atomic formula, and so K would be consistent. Let P be the predicate letter occurring in D, and let $D_1,...,D_k$ be the other members of S in which P occurs. Then $k \leq q$.

Suppose some D_i is $-B_1 \vee \cdots \vee -B_j$, where $B_1,...,B_j$ are all atomic formulas containing P. If there are atoms $A_1,...,A_j$ such that $-A_1 \vee \cdots \vee -A_j$ is an instance of D_i that is a conjunct in K and $A_1,...,A_j$ are instances of D that are conjuncts in K, then $A_1 \wedge \cdots \wedge A_j \wedge (-A_1 \vee \cdots \vee -A_j)$ is the desired subconjunction of K. Otherwise, the conjunction obtained from K by deleting all instances of D_i is inconsistent, and (#) follows from the induction hypothesis.

Suppose each D_i contains at least one atomic formula in which P does not occur. Let $D_1',...,D_k'$ result from $D_1,...,D_k$ by deleting all disjuncts containing P; and let $S' = S - \{D, D_1,...,D_k\} \cup \{D_1',...,D_k'\}$. Then S' is a sparse$_2$ set of cardinality n. Moreover, from K we may construct an inconsistent conjunction J of instances of members of S'. By the induction hypothesis, there is an inconsistent subconjunction J' of J that contains at most q^n instances of each member of S'. If J' contains no instances of $D_1',...,D_k'$, then J' is the desired subconjunction of K. Otherwise, the desired subconjunction of K results by reintroducing negatively signed atoms containing P into each conjunct in J' that is an instance of one of $D_1',...,D_k'$, and then conjoining the result with positively signed atoms containing P. Since the total number of negative occurrences of P in $D_1,...,D_k$ is at most q, and since J' contains at most q^n instances of each D_i, at most $q^n \cdot q = q^{n+1}$ atoms need be introduced. Thus one needs to conjoin at most q^{n+1} instances of D. □

An unsolvable class that contrasts with Classes A.1 and A.2 is the following: schemata F with Horn matrices such that, for each predicate letter

ISBN 0-201-02540-X

P in F, either F contains at most one positive occurrence of P or else F contains at most one negative occurrence of P.

By exploiting the notion of coinstantiability, we may obtain solvable extensions of Classes 1.4, A.1, and A.2. These extensions contain schemata that are not themselves condensable, but that closely resemble schemata that are.

Call a schema G a *special variant* of a schema F iff G results from F by relettering predicate letter occurrences, subject to the following restrictions: no occurrences of distinct predicate letters are relettered the same; and if $\langle F \rangle_i$ and $\langle F \rangle_j$ are coinstantiable, then $\langle G \rangle_i$ and $\langle G \rangle_j$ share the same predicate letter. Thus, to obtain G not all occurrences of a predicate letter P need be relettered as the same predicate letter; but any two occurrences of P in coinstantiable atomic subformulas of F must be so relettered. A schema F may have condensable special variants, even though it is not itself condensable. For example, if F is $\exists x_1 \exists x_2 \exists x_3 \exists x_4 [Px_1 \wedge (Px_2 \vee -Px_3) \wedge -Px_4]$, then F is not condensable, but its special variant $\exists x_1 \exists x_2 \exists x_3 \exists x_4 [Px_1 \wedge (Qx_2 \vee -Rx_3) \wedge -Sx_4]$ is, trivially, 1-condensable.

If G is a special variant of F, then F and G are alike but for predicate letters. Hence $D(F, \omega) = D(G, \omega)$, and there is a natural bijection $\theta \colon E(F, \omega) \to E(G, \omega)$ such that, for every Herbrand instance H of F, the substituents for each variable in H and in $\theta(H)$ are identical. Moreover, from the restriction on relettering it follows, for all Herbrand instances H and H' of F and all i and j, that $\langle H \rangle_i = \langle H' \rangle_j$ iff $\langle \theta(H) \rangle_i = \langle \theta(H') \rangle_j$. Hence, by the Lemma of Chapter 7, §3, $E(F, \omega)$ is consistent iff $E(G, \omega)$ is consistent. This immediately yields the solvability of the class of schemata that possess special variants in Class 1.4, Class A.1, or Class A.2.

A schema F possesses a special variant in Class 1.4 just in case the matrix of F is in conjunctive normal form and, for each $i \leq c(F)$, there is at most one $j \leq c(F)$ distinct from i such that $\langle F \rangle_i$ and $\langle F \rangle_j$ are coinstantiable. Similar specifications of the extensions of Classes A.1 and A.2 may easily be formulated.

In addition, these extended classes are finitely controllable. Suppose F has a special variant G that is n-condensable, and that F has a model. Then G has a model and, as we showed in Chapter 3, §2, there is a term-mapping δ with finite range such that $\Delta[E(G, \omega)]$ is consistent, where Δ is the δ-canonical formula-mapping (on $\bar{L}(G, \omega)$). Since $D(F, \omega) = D(G, \omega)$, we may also use Δ for the δ-canonical formula-mapping on $\bar{L}(F, \omega)$. To show that F has a finite model it suffices to show that $\Delta[E(F, \omega)]$ is consistent. But by an argument just like that for the Lemma of Chapter 7, §3, this will follow from the consistency of $\Delta[E(G, \omega)]$ provided that, for all Herbrand instances H and H' of F and all i and j, if $\Delta(\langle H \rangle_i) = \Delta(\langle H' \rangle_j)$, then $\Delta(\langle \theta(H) \rangle_i) = \Delta(\langle \theta(H') \rangle_j)$. Given the definition of special variant, for this it

ISBN 0-201-02540-X

suffices to show that if $\Delta(\langle H\rangle_i) = \Delta(\langle H'\rangle_j)$, then $\langle G\rangle_i$ and $\langle G\rangle_j$ share the same predicate letter. And this follows from the construction of the term-mapping δ in the Chapter 3 proof. For that construction assures (as long as $n \geq 2$, which we may assume without loss of generality) that there is a formula-mapping Γ that carries $\Delta(H)$ and $\Delta(H')$ to Herbrand instances of F. And then, since $\Gamma(\Delta(\langle H\rangle_i)) = \Gamma(\Delta(\langle H'\rangle_j))$, $\langle F\rangle_i$ and $\langle F\rangle_j$ are coinstantiable, in which case $\langle G\rangle_i$ and $\langle G\rangle_j$ must share the same predicate letter.

We may extend Classes A.1 and A.2 yet further, by paying attention to positive and negative occurrences of atomic formulas. Let F be a schema whose matrix is in conjunctive normal form; let $\text{pos}(F)$ be the set of numbers $i \leq c(F)$ such that the ith atomic formula occurrence in F is positively signed, and let $\text{neg}(F)$ be the set of numbers $i \leq c(F)$ such that the ith atomic formula occurrence in F is negatively signed. A schema G is a *very special variant* of F iff G results from F by relettering predicate letter occurrences, subject to the following restrictions: no occurrences of distinct predicate letters are relettered the same; and if $\langle F\rangle_i$ and $\langle F\rangle_j$ are coinstantiable, $i \in \text{pos}(F)$, and $j \in \text{neg}(F)$, then $\langle G\rangle_i$ and $\langle G\rangle_j$ share the same predicate letter.

If G is a very special variant of F, then again $D(F, \omega) = D(G, \omega)$ and there is a natural bijection $\theta: E(F, \omega) \rightarrow E(G, \omega)$. Moreover, for all Herbrand instances H and H' of G and all i and j, $\langle H\rangle_i = \langle H'\rangle_j$ only if $\langle \theta^{-1}(H)\rangle_i = \langle \theta^{-1}(H')\rangle_j$. Hence, by the Lemma of Chapter 7, §3, if $E(F,\omega)$ is consistent then $E(G, \omega)$ is consistent. Conversely, for all Herbrand instances H and H' of F, if $\langle H\rangle_i = \langle H'\rangle_j$, $i \in \text{pos}(F)$, and $j \in \text{neg}(F)$, then $\langle \theta(H)\rangle_i = \langle \theta(H')\rangle_j$. This is enough to ensure that the consistency of $E(G, \omega)$ implies that of $E(F, \omega)$. For suppose $\mathfrak{A}$ verifies $E(G, \omega)$. Define a truth-assignment $\mathfrak{B}$ on $\tilde{L}(F, \omega)$ thus: $\mathfrak{B}$ verifies an atom A iff, for some Herbrand instance H of F and some i in $\text{pos}(F)$, $A = \langle H\rangle_i$ and $\mathfrak{A}$ verifies $\langle \theta(H)\rangle_i$. We then have, for each Herbrand instance H of F and each j,

   (i) if $j \in \text{pos}(F)$ and $\mathfrak{A}$ verifies $\langle \theta(H)\rangle_j$, then $\mathfrak{B}$ verifies $\langle H\rangle_j$;
   (ii) if $j \in \text{neg}(F)$ and $\mathfrak{A}$ falsifies $\langle \theta(H)\rangle_j$, then $\mathfrak{B}$ falsifies $\langle H\rangle_j$.

Property (i) is immediate from the definition of $\mathfrak{B}$. Property (ii) follows, since $\mathfrak{B}$ verifies $\langle H\rangle_j$ iff there is an Herbrand instance H' of F and a number i in $\text{pos}(F)$ such that $\langle H\rangle_j = \langle H'\rangle_i$ and $\mathfrak{A}$ verifies $\langle \theta(H')\rangle_i$. But in this case $\langle \theta(H)\rangle_j = \langle \theta(H')\rangle_i$, so that $\mathfrak{A}$ verifies $\langle \theta(H)\rangle_j$.

Since $\mathfrak{A}$ verifies $\theta(H)$ for every Herbrand instance H of F, properties (i) and (ii) suffice to show that $\mathfrak{B}$ verifies every H.

Thus the class of schemata that possess very special variants in Class A.1 or Class A.2 is solvable. A schema F possesses a very special variant in Class A.1 iff the matrix of F is Horn and, for each i in $\text{pos}(F)$, there is at

ISBN 0-201-02540-X

most one j in neg(F) such that $\langle F \rangle_i$ and $\langle F \rangle_j$ are coinstantiable. The description of schemata possessing very special variants in Class A.2 is similar.

By arguments like those used before, we may infer the finite controllability of these extended classes from the finite controllability proof for condensable schemata.

2. Combinatorics

In the finite controllability proofs of Chapter 3, §§5–6, and Chapter 4, §2, we used several combinatorial lemmas. Here we prove these lemmas; in each case we first prove a sublemma, and then show that the desired results follow by applying the sublemma to domains $D(F, \omega)$.

Sublemma 1. For each $m > 0$ there is a finite set S and a function φ: $S^m \to S$ such that, for all $a_1, \ldots, a_m$, $b_1, \ldots, b_m$ in S, if $\varphi(a_1, \ldots, a_m) = b_j$, then $\varphi(b_1, \ldots, b_m) \neq a_i$, $1 \leq i, j \leq m$.

Proof. Let S be the set of $(m + 1)$-tuples of positive integers less than or equal to $m^2 + m + 1$. Then let $\varphi(a_1, \ldots, a_m) = \langle k_1, \ldots, k_{m+1} \rangle$, where

(1)   for each i, $1 \leq i \leq m$, k_i is the last member of the $(m + 1)$-tuple a_i;
(2) k_{m+1} is the least positive integer not occurring in any of the $(m + 1)$-tuples $a_1, \ldots, a_m$.

Suppose that $\varphi(a_1, \ldots, a_m) = b_j$ and $\varphi(b_1, \ldots, b_m) = a_i$. We derive a contradiction. Since $\varphi(a_1, \ldots, a_m) = b_j$, the last member of a_i occurs in b_j, namely, as the ith coordinate. Since $\varphi(b_1, \ldots, b_m) = a_i$, however, the last member of a_i does not occur in any of $b_1, \ldots, b_m$. This is a contradiction.   □

REMARK. For each $m > 0$ let κ_m be the cardinality of the smallest set S for which there exists a function φ with the property stated in the sublemma. Our proof yields $(m^2 + m + 1)^{m+1}$ as an upper bound for κ_m. Undoubtedly κ_m is in general much smaller. For example, $\kappa_2 = 7$ and $\kappa_3 = 19$; we also know that $\kappa_4 < 1000$. (Bounds smaller than $(m^2 + m + 1)^{m+1}$ can be obtained from the more complex proofs of Sublemma 1 given by Erdös (1963) and by Andrew Gleason (1956, unpublished).)   End of Remark.

The integers κ_m and κ_{m+n-1} referred to in the following lemmas are the ones specified at the beginning of the foregoing remark.

Combinatorial Lemma 1. Let $F = \forall y_1 \cdots \forall y_m \exists x_1 \cdots \exists x_n F^M$ be a Skolem schema. Then there is a function α: $D(F, \omega) \to \{1, \ldots, \kappa_m\}$ such that, for all terms s and t, if $\alpha(s) = \alpha(\hat{y}_i t)$, then $\alpha(t) \neq \alpha(\hat{y}_j s)$, $1 \leq i, j \leq m$, and $\alpha(s) = \alpha(\hat{x}_k s)$, $1 \leq k \leq n$.

Proof. Let $S = \{1, \ldots, \kappa_m\}$, and let φ: $S^m \to S$ have the property of

ISBN 0-201-02540-X

Sublemma 1. Define α on $D(F, \omega)$ inductively thus:

$$\alpha(\bot) = 1;$$

$$\alpha(f_{x_k}(t_1\cdots t_m)) = \varphi(\alpha(t_1),\ldots,\alpha(t_m)).$$

The required properties of α are immediate from this definition and the choice of φ. $\square$

Generalized Combinatorial Lemma 1. Let F be a normal schema containing m x-variables and n y-variables. Then there is a function $\alpha: D(F, \omega) \to \{1,\ldots,\kappa_{m+n-1}\}$ such that, for all terms s and t and all variables v governing $\chi(t)$ and w governing $\chi(s)$, if $\alpha(s) = \alpha(\hat{v}t)$, then $\alpha(t) \neq \alpha(\hat{w}s)$.

Proof. Let $S = \{1,\ldots,\kappa_{m+n-1}\}$ and let $\varphi: S^{m+n-1} \to S$ be a function with the property given in Sublemma 1. We extend φ to a function on S^k for each k, $0 \leq k \leq m + n - 1$, by stipulating that $\varphi(i_1,\ldots,i_k) = \varphi(i_1,\ldots,i_k, i_k,\ldots,i_k)$ if $0 < k < m+n-1$, and that φ carry the empty sequence to 1. Define α on $D(F, \omega)$ inductively thus: $\alpha(\bot) = 1$; if $s \neq \bot$, then $\alpha(s) = \varphi(\alpha(\hat{v}_1 s),\ldots,\alpha(\hat{v}_k s))$, where $v_1,\ldots,v_k$ are the variables governing $\chi(s)$. Note that $k \leq m + n - 1$, since there are only $m + n$ variables in F.

Again, the required properties of α are immediate from this definition and the choice of φ. $\square$

Sublemma 2. Let p and m be positive integers, and let S be a set of cardinality $\leq 2^p$. There is a function ψ on S^m whose range has cardinality $\leq(2p +1)^{m^2}$ such that, for all $a_1,\ldots,a_m$, $b_1,\ldots,b_m$ in S and all i and j, $1 \leq i, j \leq m$, if $\psi(a_1,\ldots,a_m) = \psi(b_1,\ldots,b_m)$ and $a_i = b_j$, then $a_i = a_j = b_i = b_j$.

Proof. We may assume without loss of generality that S is a set of p-tuples of zeroes and ones. If $a \in S$ and $1 \leq k \leq p$, we write $(a)_k$ for the kth member of a. For all distinct a and b in S let $\theta(a, b)$ be the pair $\langle k, (a)_k \rangle$, where k is the least integer such that $(a)_k \neq (b)_k$. For each a in S let $\theta(a, a) = 0$. Hence $|\theta[S^2]| \leq 2p + 1$. Finally, for all $a_1,\ldots,a_m$ in S let $\psi(a_1,\ldots,a_m)$ be the set

$$\{\langle i, j, \theta(a_i, a_j)\rangle \mid 1 \leq i, j \leq m\}.$$

Hence $|\psi[S^m]| \leq (2p + 1)^{m^2}$.

Suppose $\psi(a_1,\ldots,a_m) = \psi(b_1,\ldots,b_m)$ and $a_i = b_j$. Then $\theta(a_i, a_j) = \theta(b_i, b_j)$. If $\theta(a_i, a_j) = \langle k, q\rangle$, then $(a_i)_k = q$; and since $\theta(b_i, b_j) = \langle k, q\rangle$, $(b_j)_k \neq q$. But this contradicts $a_i = b_j$. Hence $\theta(a_i, a_j) = \theta(b_i, b_j) = 0$; whence $a_i = a_j$ and $b_i = b_j$. $\square$

Combinatorial Lemma 2. Let $F = \forall y_1\cdots\forall y_m \exists x_1\cdots\exists x_n FM$ be a Skolem schema, and let E be an equivalence relation of order $\leq r$ on $D(F, \omega)$. Then there is an equivalence relation $\simeq$ of order $\leq r\cdot(2rm^4 + 1)^{m^2}$ on $D(F, \omega)$

ISBN 0-201-02540-X

such that for all terms s and t and all i and j, $1 \leq i, j \leq m$,

(1) if $s \simeq t$, then sEt;
(2) if $s \simeq t$ and $\hat{y}_i s \simeq \hat{y}_j t$, then $\hat{y}_i s \simeq \hat{y}_j s \simeq \hat{y}_j t \simeq \hat{y}_i t$.

Proof. Let η be the term-mapping that carries each term s to the earliest term such that $\eta s E s$. Thus $|\eta[D(F, \omega)]| \leq r$. Let $R = \{1,\ldots,(2rm^4 + 1)^{m^2}\}$, and let $S = \eta[D(F, \omega)] \times R$. Thus $|S| \leq r \cdot (2rm^4 + 1)^{m^2}$. Since we may assume that $r, m > 2$, $|S| \leq 2^{rm^4}$. Hence there is a function $\psi: S^m \to R$ with the property given in Sublemma 2. We define a function $\zeta: D(F, \omega) \to S$ inductively thus:

$$\zeta(\textpm) = \langle \eta\textpm, 1 \rangle;$$

$$\zeta(s) = \langle \eta s, \psi(\zeta(\hat{y}_1 s),\ldots,\zeta(\hat{y}_m s)) \rangle \quad \text{if} \quad h(s) > 1.$$

Finally, we let $s \simeq t$ iff $\zeta(s) = \zeta(t)$. Clearly $\simeq$ has order at most $|S|$, and $s \simeq t$ only if sEt. Suppose now that $s \simeq t$ and $\hat{y}_i s \simeq \hat{y}_j t$. Then $\psi(\zeta(\hat{y}_1 s),\ldots,\zeta(\hat{y}_m s)) = \psi(\zeta(\hat{y}_1 t),\ldots,\zeta(\hat{y}_m t))$, and $\zeta(\hat{y}_i s) = \zeta(\hat{y}_j t)$. By the choice of ψ, $\zeta(\hat{y}_i s) = \zeta(\hat{y}_j s) = \zeta(\hat{y}_i t) = \zeta(\hat{y}_j t)$; that is, $\hat{y}_i s \simeq \hat{y}_j s \simeq \hat{y}_i t \simeq \hat{y}_j t$. $\square$

REMARK. The construction of the equivalence relation $\simeq$ from the relation E can easily be seen to have the property required by the Amplified form of Combinatorial Lemma 2, used in Chapter 4, §2. That is, let E' be an equivalence relation of order $\leq r$ that agrees with E on $D(F, q)$. Then the mapping η' that carries each s to the earliest term with $\eta' s E' s$ will agree with η on $D(F, q)$; and then the function ζ' will agree with ζ on $D(F, q)$. From E', then, we construct an equivalence relation $\simeq'$ that agrees with $\simeq$ on $D(F, q)$.   End of Remark.

Generalized Combinatorial Lemma 2. Let F be a normal schema containing m y-variables and n x-variables, and let E be a faithful equivalence of order $\leq r$ on $D(F, \omega)$. Then there is an equivalence relation $\simeq$ of order $\leq r \cdot (2r \cdot (m + n - 1)^4 + 1)^{(m+n-1)^2}$ on $D(F, \omega)$ such that

(1) $s \simeq t$ only if sEt;
(2) if $s \simeq t$ and $\hat{v}s \simeq \hat{w}t$, for some variables v and w governing $\chi(s)$, then $\hat{v}s \simeq \hat{v}t \simeq \hat{w}s \simeq \hat{w}t$.

Proof. Let η carry each term s to the earliest term such that $\eta s E s$. Let $R = \{1,\ldots,(2r(m + n - 1)^4 + 1)^{(m+n-1)^2}\}$ and let $S = \eta[D(F, \omega)] \times R$. Then $|S| \leq 2^{r(m+n-1)^4}$, and so there is a function $\psi: S^{m+n-1} \to S$ with the property given in Sublemma 2. Define a function $\zeta: D(F, \omega) \to S$ inductively thus:

$$\zeta(\textpm) = \langle \eta\textpm, 1 \rangle;$$

$$\zeta(s) = \langle \eta s, \psi(\zeta(\hat{v}_1 s),\ldots,\zeta(\hat{v}_k s), \zeta(\hat{v}_k s),\ldots,\zeta(\hat{v}_k s)) \rangle$$

ISBN 0-201-02540-X

if $h(s) > 1$, where $v_1,\ldots,v_k$ are the variables governing $\chi(s)$, in left-to-right order of their occurrence in F, and $\langle v_1,\ldots,v_k, v_k,\ldots,v_k \rangle$ indicates the $(m + n - 1)$-tuple obtained by repeating v_k $(m + n - 1) - k$ times. Finally, let $s \simeq t$ iff $\zeta(s) = \zeta(t)$.

Clearly the order of $\simeq$ is at most $|S|$, and $s \simeq t$ only if sEt. Suppose $s \simeq t$ and $\hat{v}s \simeq \hat{w}t$. Since E is faithful, $\chi(s) = \chi(t)$. Hence $\psi(\zeta(\hat{v}_1 s),\ldots,\zeta(\hat{v}_k s), \zeta(\hat{v}_k s),\ldots,\zeta(\hat{v}_k s)) = \psi(\zeta(\hat{v}_1 t),\ldots,\zeta(\hat{v}_k t), \zeta(\hat{v}_k t),\ldots,\zeta(\hat{v}_k t))$, and v and w are among $v_1,\ldots,v_k$. And then, by the choice of ψ, $\zeta(\hat{v}s) = \zeta(\hat{w}s) = \zeta(\hat{v}t) = \zeta(\hat{w}t)$. That is, $\hat{v}s \simeq \hat{w}s \simeq \hat{v}t \simeq \hat{w}t$. $\square$

3. The ∀∃∀-Krom Class

We show here the solvability and docility of Class 5.6, the class of ∀∃∀-schemata whose matrices are Krom and contain dyadic predicate letters only. (The reduction in Chapter 6, §5 shows that the limitation to dyadic predicate letters entails no loss of generality.) Our solvability proof closely follows that of Aanderaa and Lewis (1973), and our docility proof rests on an extension of their method.

We shall exploit a truth-functional property of Krom formulas, given in the next paragraph. Let A be a signed atomic formula; then $\bar{A}$ is the signed atomic formula equivalent to $-A$. That is, $\bar{A}$ is $-A$ if A is atomic, and $\bar{A}$ is B if A is $-B$. Let A and B be signed atomic formulas; then a *chain from A to B* is a conjunction $K = (\bar{A}_1 \vee A_2) \wedge (\bar{A}_2 \vee A_3) \wedge \cdots \wedge (\bar{A}_{n-1} \vee A_n)$, where $A_1,\ldots,A_n$ are signed atomic formulas, $A = \bar{A}_1$, and $B = A_n$. A Krom formula M *contains* K iff for each i, $1 \le i < n$, either $(\bar{A}_i \vee A_{i+1})$ or $(A_{i+1} \vee \bar{A}_i)$ is a conjunct in M. A set of Krom formulas *contains* K iff some conjunction of members of the set contains K. Note that any chain from A to B truth-functionally implies $(A \vee B)$, and that a Krom formula contains a chain from A to B iff it contains a chain from B to A.

Chain Criterion. A Krom formula M is inconsistent iff M contains chains from A to A and from $\bar{A}$ to $\bar{A}$ for some atomic formula A.

Proof. If M contains such chains, then M truth-functionally implies $A \wedge \bar{A}$, so that M is inconsistent. The converse is proved by induction on the number of conjuncts in M. We leave this to the reader (alternatively, proofs may be found in Krom (1967a) and in Aanderaa and Lewis (1973)). $\square$

Chains from A to A and from $\bar{A}$ to $\bar{A}$ can be rearranged and combined in many ways.

Lemma 1. Suppose a Krom formula M contains chains J from A to A and K from $\bar{A}$ to $\bar{A}$, and let B be any signed atomic formula occurring in either J or K. Then M contains chains from B to B and from $\bar{B}$ to $\bar{B}$.

Proof. We suppose that B occurs in J; the other case is symmetrical.

ISBN 0-201-02540-X

We suppose moreover that B occurs as the right-hand disjunct in some conjunct in J. For if B occurs as the left-hand disjunct, then $\bar{B}$ occurs as the right-hand disjunct in the previous conjunct in J, and we may carry out the proof for $\bar{B}$. Given these assumptions, $J = J_1 \wedge J_2$, where J_1 is a chain from A to B and J_2 is a chain from $\bar{B}$ to A. Hence M contains chains from B to A, from $\bar{A}$ to $\bar{A}$, and from A to B. The conjunction of these chains is a chain from B to B. And M contains chains from $\bar{B}$ to A, from $\bar{A}$ to $\bar{A}$, and from A to $\bar{B}$. The conjunction of these is a chain from $\bar{B}$ to $\bar{B}$. $\square$

A *signed predicate letter* is a predicate letter either alone or preceded by a negation sign. The first part of our solvability proof is a reduction of Class 5.6 to a more tractable subclass: the class of $\forall\exists\forall$-Krom schemata $\forall y_1 \exists x \forall y_2 G^M$ such that each conjunct in G^M has one of the five forms $\varphi y_1 y_2$ $\vee\ \psi y_1 y_2$, $\varphi y_2 y_1 \vee\ \psi y_2 y_1$, $\varphi y_1 y_2 \vee\ \psi x y_2$, $\varphi y_2 y_1 \vee\ \psi y_2 x$, $\varphi y_1 y_1 \vee\ \psi y_1 y_1$, where φ and ψ are signed predicate letters. Binary disjunctions of one of the first four forms are called *elementary*; those of the fifth form are called *monadic*.

Let G_0 be any member of Class 5.6. We first reduce G_0 to an $\forall\exists\forall$-Krom schema G_1 that contains no atomic subformulas $Py_1 x$ and Pxy_1, and that contains atomic subformulas $Py_2 x$ and Pxy_2 only in elementary conjuncts. For each predicate letter P of G_0 let P_1 and P_2 be new dyadic predicate letters. Let M_1 result from the matrix $G_0{}^M$ of G_0 by replacing all atomic subformulas Pxy_k and $Py_k x$, $k = 1, 2$, with $P_1 y_1 y_k$ and $P_2 y_k y_1$, respectively; and let M_2 be the conjunction of all biconditionals $(P_1 y_1 y_2 \Leftrightarrow Pxy_2)$ and $(P_2 y_2 y_1 \Leftrightarrow Py_2 x)$, for P a predicate letter of G_0. (We may take $(A \Leftrightarrow B)$ to be just $(-A \vee B) \wedge (A \vee -B)$. Hence each of these biconditionals is the conjunction of two elementary conjuncts.) Finally, let $G_1 = \forall y_1 \exists x \forall y_2 (M_1 \wedge M_2)$. Any truth-assignment $\mathfrak{A}$ verifying $E(G_0, \omega)$ can be extended to one verifying $E(G_1, \omega)$ by letting $\mathfrak{A}(P_1 m\ n) = \mathfrak{A}(Pm + 1\ n)$ and $\mathfrak{A}(P_2 m\ n) = \mathfrak{A}(Pm\ n + 1)$ for all predicate letters P of G_0 and all m and n. (We use the numerical notation for terms, introduced in Chapter 5.) Conversely, $E(G_1, \omega)$ truth-functionally implies $(P_1 m\ n \Leftrightarrow Pm + 1\ n)$ and $(P_2 m\ n \Leftrightarrow Pm\ n + 1)$ for all $m, n \geq 0$; whence $E(G_1, \omega)$ truth-functionally implies $E(G_0, \omega)$.

Next we eliminate from G_1 all atomic subformulas Pxx and $Py_2 y_2$, and restrict atomic subformulas $Py_1 y_1$ to monadic conjuncts. For each predicate letter P of G_1 let P' and P'' be new dyadic predicate letters. Let N_1 result from the matrix $(M_1 \wedge M_2)$ of G_1 by replacing atomic subformulas Pxx with $P''y_1 y_2$, $Py_2 y_2$ with $P'y_2 y_1$, and $Py_1 y_1$ with $P'y_1 y_2$; let N_2 be the conjunction, for each predicate letter P of G_1, of all biconditionals $(P'y_1 y_1 \Leftrightarrow Py_1 y_1)$ and $(P'y_2 y_1 \Leftrightarrow P'y_2 x)$ and $(P''y_1 y_2 \Leftrightarrow P'xy_2)$. Finally, let $G_2 = \forall y_1 \exists x \forall y_2 (N_1 \wedge N_2)$. Any truth-assignment $\mathfrak{A}$ verifying $E(G_1, \omega)$ can

ISBN 0-201-02540-X

be extended to one verifying $E(G_2, \omega)$ by letting $\mathfrak{A}(P'm\ n) = \mathfrak{A}(Pm\ m)$ and $\mathfrak{A}(P''m\ n) = \mathfrak{A}(Pm + 1\ m + 1)$. Conversely, since $E(G_2, \omega)$ truth-functionally implies $(P'm\ n \Leftrightarrow P'm\ q) \wedge (P''m\ n \Leftrightarrow P'm + 1\ n)$ for all new predicate letters P' and P'' and all m, n, and q, it follows that $E(G_2, \omega)$ truth-functionally implies $E(G_1, \omega)$.

We are left, then, with an $\forall\exists\forall$-Krom Schema G_2 in whose matrix each conjunct is either elementary, monadic, or of one of the forms $\varphi y_1 y_2 \vee \psi y_2 y_1$ and $\varphi y_2 y_1 \vee \psi y_1 y_2$. The final stage of our reduction aims at eliminating the conjuncts of the last two forms.

To do this we require several further definitions, which will also be used hereafter. Let F be an arbitrary $\forall\exists\forall$-Krom schema with matrix F^M. An *elementary instance* (for F) is a binary disjunction $(C_1 \vee C_2)(y_1/m, x/m + 1, y_2/n)$ such that either $(C_1 \vee C_2)$ or $(C_2 \vee C_1)$ is an elementary conjunct in F^M. Similary, a *monadic instance* (for F) is a binary disjunction $(C_1 \vee C_2)(y_1/m)$ such that either $(C_1 \vee C_2)$ or $(C_2 \vee C_1)$ is a monadic conjunction in F^M. An *elementary chain* (for F) is a chain in which each conjunct is an elementary instance for F. Note that if a formula-mapping Γ carries each atom Pij to $Pi + p\ j + q$ for some p, $q \geq 0$, then $\Gamma(D)$ is an elementary instance whenever D is an elementary instance; hence $\Gamma(K)$ is an elementary chain whenever K is an elementary chain. But if D is a monadic instance, then we must have $p = q$ for $\Gamma(D)$ to be a monadic instance. Hence we must have $p = q$ if Γ is to carry every chain contained in $E(F, \omega)$ to a chain contained in $E(F, \omega)$.

We now return to the schema G_2, from which we wish to eliminate conjuncts $\varphi y_1 y_2 \vee \psi y_2 y_1$ and $\varphi y_2 y_1 \vee \psi y_1 y_2$. For each predicate letter P of G_2 let $\check{P}$ be a new dyadic predicate letter. For each predicate letter P of G_2 and all terms or variables u_1 and u_2 let $\Pi(Pu_1u_2) = \check{P}u_2u_1$ and $\Pi(\check{P}u_1u_2) = Pu_2u_1$; and for each atomic formula A let $\Pi(-A) = -\Pi(A)$. Then let $G_3 = \forall y_1 \exists x \forall y_2(C_1 \wedge C_2 \wedge C_3)$, where

(1) C_1 is the conjunction of all formulas $(A \vee B) \wedge (\Pi(A) \vee \Pi(B))$ for $(A \vee B)$ an elementary or monadic conjunct in the matrix of G_2;

(2) C_2 is the conjunction of all formulas $(A \vee \Pi(B)) \wedge (\Pi(A) \vee B)$ for $(A \vee B)$ a conjunct in the matrix of G_2 that is neither elementary nor monadic;

(3) C_3 is the conjunction of all biconditionals $(Py_1 y_1 \Leftrightarrow \check{P}y_1 y_1)$ for P a predicate letter of G_2.

Clearly every conjunct in the matrix of G_3 is either elementary or monadic. Moreover, any model for G_2 can be transformed into a model for G_3 by interpreting each new predicate letter $\check{P}$ as the converse relation of P. It remains to show that if $E(G_2, \omega)$ is inconsistent, then so is $E(G_3, \omega)$.

Suppose $E(G_2, \omega)$ inconsistent. Then some conjunction of Herbrand

ISBN 0-201-02540-X

instances of G_2 is inconsistent, and so, by the Chain Criterion, $E(G_2, \omega)$ contains a chain J from A to A and a chain K from $\bar{A}$ to $\bar{A}$, for some A. Moreover, by Lemma 1 we may assume that either A is essentially monadic (that is, has the form φii) or else *no* signed atom occurring in J or in K is essentially monadic.

Note that if $(B_1 \vee B_2)$ is a conjunct in an Herbrand instance of G_2, then either (a) both $(B_1 \vee B_2)$ and $(\Pi(B_1) \vee \Pi(B_2))$ are conjuncts in Herbrand instances of G_3, or else (b) both $(B_1 \vee \Pi(B_2))$ and $(\Pi(B_1) \vee B_2)$ are conjuncts in Herbrand instances of G_3. Hence, for all signed atoms B_1 and B_2, if $E(G_2, \omega)$ contains a chain from B_1 to B_2, then $E(G_3, \omega)$ contains either a chain from B_1 to B_2 or else a chain from B_1 to $\Pi(B_2)$. Consequently, $E(G_3, \omega)$ contains a chain J' either from A to A or from A to $\Pi(A)$, and contains a chain K' either from $\bar{A}$ to $\bar{A}$ or from $\bar{A}$ to $\Pi(\bar{A})$.

If A is essentially monadic then $(A \vee \Pi(\bar{A}))$ and $(\bar{A} \vee \Pi(A))$ are conjuncts in Herbrand instances of G_3, by dint of the biconditionals included in the subconjunction C_3 of the matrix of G_3. If necessary, these conjuncts may be conjoined to J' and to K' to obtain chains from A to A and from $\bar{A}$ to $\bar{A}$. Thus $E(G_3, \omega)$ is inconsistent.

If J and K contain no essentially monadic signed atom, then neither do J' and K'. And since G_3 contains only monadic and elementary conjuncts, J' and K' are then elementary chains (for G_3). Suppose $A = \varphi mn$; let $p = \max(m, n) - m$ and $q = \max(m, n) - n$; and let Γ carry each atom Pij to $Pi + p\ j + q$. Then $\Gamma(J')$ and $\Gamma(K')$ are elementary chains (for G_3): the former is from $\Gamma(A)$ either to $\Gamma(A)$ or to $\Gamma(\Pi(A))$, and the latter is from $\Gamma(\bar{A})$ to either $\Gamma(\bar{A})$ or $\Gamma(\Pi(\bar{A}))$. But $\Gamma(A) = \varphi rr$ for $r = \max(m, n)$; that is, $\Gamma(A)$ is essentially monadic. Hence, reasoning as before, $E(G_3, \omega)$ contains chains from $\Gamma(A)$ to $\Gamma(A)$ and from $\Gamma(\bar{A})$ to $\Gamma(\bar{A})$; whence $E(G_3, \omega)$ is once again inconsistent.

This completes the reduction: we now restrict our attention to $\forall\exists\forall$-Krom schemata whose matrices are composed entirely of elementary and monadic conjuncts. For the remainder of the solvability proof we let F be such a schema. Moreover, let s be twice the number of predicate letters of F; thus there are s signed predicate letters constructed from the predicate letters of F. The method used in the last step of the reduction also yields

Lemma 2. F has no model iff $E(F, \omega)$ contains chains from A to A and from $\bar{A}$ to $\bar{A}$ for some essentially monadic atom A.   □

Our task is now to analyze the chains contained in $E(F, \omega)$. For all signed predicate letters φ and ψ, let $\Xi(\varphi, \psi)$ be the set of pairs of integers $\langle p, q \rangle$ such that, for some m and n, there is an elementary chain from φmn to $\psi m + p\ n + q$; and let $\Theta(\varphi, \psi)$ be the set of pairs $\langle p, p \rangle$ such that, for some m, there is a chain contained in $E(F, \omega)$ from φmm to $\psi m + p\ m + p$. (Here $p, q \in \mathbb{Z}$; that is, p and q are integers, positive, negative, or zero.)

ISBN 0-201-02540-X

We seek to characterize the sets $\Xi(\varphi, \psi)$ and $\Theta(\varphi, \psi)$ in a uniform and effective manner.

If $\mathbf{p}, \mathbf{p}_1,...,\mathbf{p}_m$ are pairs in $\mathbb{Z}^2$, then let $L(\mathbf{p}, \{\mathbf{p}_1,...,\mathbf{p}_m\})$ be the set $\{\mathbf{p} + n_1\mathbf{p}_1 + \cdots + n_m\mathbf{p}_m | n_1,...,n_m \geq 0\}$. (Addition of pairs and multiplication of pairs by integers is componentwise.) Thus $L(\mathbf{p}, \{\mathbf{p}_1,...,\mathbf{p}_m\})$ is the smallest set that contains $\mathbf{p}$ and that contains $\mathbf{q} + \mathbf{p}_1,...,\mathbf{q} + \mathbf{p}_m$ whenever it contains $\mathbf{q}$. A set $L(\mathbf{p}, \{\mathbf{p}_1,...,\mathbf{p}_m\})$ is called *linear*; and $\langle \mathbf{p}, \{\mathbf{p}_1,...,\mathbf{p}_m\}\rangle$ is a *presentation* of it. (Presentations are not unique). A *semilinear* set is a finite union of linear sets; a presentation of a semilinear set is a finite set of presentations of its component linear sets. We seek, then, to show the semilinearity of the sets $\Xi(\varphi, \psi)$ and $\Theta(\varphi, \psi)$, and to show how presentations of these sets may be constructed.

We shall often treat sequences of signed predicate letters as proxies for chains. If λ is a signed predicate letter, then $\bar{\lambda}$ is the signed predicate letter containing the same predicate letter as does λ but with opposite sign. A $\langle \varphi, \psi \rangle$-*sequence* is a sequence $\langle \varphi_1,...,\varphi_k \rangle$ of signed predicate letters, $k > 1$, such that $\varphi = \bar{\varphi}$ and $\psi = \varphi_k$. Suppose that, for all signed predicate letters φ and ψ, $U(\varphi, \psi)$ is a subset of $\mathbb{Z}^2$. Then for each sequence $\Phi = \langle \varphi_1,...,\varphi_k \rangle$ of signed predicate letters we let

$$\Sigma_U(\Phi) = \{\mathbf{p}_1 + \cdots + \mathbf{p}_{k-1} | \mathbf{p}_i \in U(\bar{\varphi}_i, \varphi_{i+1}), 1 \leq i < k\}.$$

(If $U(\bar{\varphi}_i, \varphi_{i+1})$ is empty for some i, then $\Sigma_U(\Phi)$ is empty.) The principal combinatorial fact that we use concerning sets $\Sigma_U(\Phi)$ is

Lemma 3. For all signed predicate letters φ and ψ and each $\mathbf{p} \in \mathbb{Z}^2$, there is a $\langle \varphi, \psi \rangle$-sequence Φ such that $\mathbf{p} \in \Sigma_U(\Phi)$ iff there are sequences $\Psi, \Lambda_1,...,\Lambda_m, m \geq 0$, each of which has length $\leq s^2 + 1$, such that

(1) Ψ is a $\langle \varphi, \psi \rangle$-sequence;

(2)   for each $i \leq m$ there is a member λ of Ψ such that Λ_i is a $\langle \bar{\lambda}, \lambda \rangle$-sequence;

(3) $\mathbf{p} = \mathbf{p}_0 + \mathbf{p}_1 + \cdots + \mathbf{p}_m$, where $\mathbf{p}_0 \in \Sigma_U(\Psi)$ and $\mathbf{p}_i \in \Sigma_U(\Lambda_i), 1 \leq i \leq m$.

Proof. If $\Psi = \langle \varphi_1,...,\varphi_k \rangle$ is any $\langle \varphi, \psi \rangle$-sequence and $\Lambda = \langle \varphi_i, \psi_1,...,\psi_j, \varphi_i \rangle$ is any $\langle \bar{\varphi}_i, \varphi_i \rangle$-sequence for some $i \leq k$, then clearly $\Phi = \langle \varphi_1,...,\varphi_i, \psi_1,...,\psi_j, \varphi_i, \varphi_{i+1},...,\varphi_k \rangle$ is a $\langle \varphi, \psi \rangle$-sequence, and $\Sigma_U(\Phi) = \{\mathbf{q}_1 + \mathbf{q}_2 | \mathbf{q}_1 \in \Sigma_U(\Psi)$ and $\mathbf{q}_2 \in \Sigma_U(\Lambda)\}$. Iteration of this procedure shows that if there exist $\Psi, \Lambda_1,...,\Lambda_m$ fulfilling (1)–(3), then $\mathbf{p} \in \Sigma_U(\Phi)$.

Conversely, suppose $\Phi = \langle \varphi_1,...,\varphi_k \rangle$ is a $\langle \varphi, \psi \rangle$-sequence of length $> s^2 + 1$. For each $i \leq k$ let $\xi(i) = \langle \varphi_i, \{\varphi_1,...,\varphi_i\}\rangle$; since there are s signed predicate letters, by the pigeonhole argument there are i and j, $i + j \leq k$ and $1 \leq j \leq s^2$, such that $\xi(i) = \xi(i + j)$. Let $\Psi = \langle \varphi_1,...,\varphi_i, \varphi_{i+j+1},...,\varphi_k \rangle$ and let $\Lambda = \langle \varphi_i,...,\varphi_{i+j} \rangle$. Then Ψ is a $\langle \varphi, \psi \rangle$-sequence, Λ is a $\langle \bar{\varphi}_i, \varphi_i \rangle$-

ISBN 0-201-02540-X

sequence of length $\leq s^2 + 1$, the length of Ψ is less than that of Φ, every member of Φ is a member of Ψ, and $\Sigma_U(\Phi) = \{\mathbf{q}_1 + \mathbf{q}_2 | \mathbf{q}_1 \in \Sigma_U(\Phi)$ and $\mathbf{q}_2 \in \Sigma_U(\Lambda)\}$. Iteration of this procedure shows that if $\mathbf{p} \in \Sigma_U(\Phi)$ for any $\langle \varphi, \psi \rangle$-sequence Φ then there are Ψ, $\Lambda_1,...,\Lambda_m$ fulfilling (a)–(c). $\quad\square$

Our first application of Lemma 3 is to elementary chains. For all signed predicate letters φ and ψ let $W(\varphi, \psi) \subseteq Z^2$ be defined thus: $\langle p, q \rangle \in W(\varphi, \psi)$ iff $\varphi mn \vee \psi m + p \ n + q$ is an elementary instance for all m, $n \geq 1$. Hence $W(\varphi, \psi) \subseteq \{\langle 0, 0 \rangle, \langle 0, 1 \rangle, \langle 1, 0 \rangle, \langle 0, -1 \rangle, \langle -1, 0 \rangle\}$. Moroever, each set $W(\varphi, \psi)$ can be obtained uniformly and effectively from F^M. For example, $\langle 0, -1 \rangle \in W(\varphi, \psi)$ iff either $(\varphi y_2 x \vee \psi y_2 y_1)$ or $(\psi y_2 y_1 \vee \varphi y_2 x)$ is a conjunct in F^M. From the definition of $W(\varphi, \psi)$ it follows quickly by induction on k that if there is a $\langle \varphi, \psi \rangle$-sequence Φ of length k such that $\langle p, q \rangle \in \Sigma_W(\Phi)$, then there are elementary chains from φmn to $\psi m + p \ n + q$ for all m, $n \geq k - 1$. And if for some m and n there is an elementary chain K from φmn to $\psi m + p \ n + q$, then there is a $\langle \varphi, \psi \rangle$-sequence Φ such that $\langle p, q \rangle \in \Sigma_W(\Phi)$: indeed, if K contains k conjuncts, we may take $\Phi = \langle \bar{\varphi}, \psi_1,...,\psi_k \rangle$, where each ψ_i is the signed predicate letter in the right-hand disjunct in the ith conjunct in K (thus $\psi_k = \psi$). Hence $\Xi(\varphi, \psi)$ is just the union of all sets $\Sigma_W(\Phi)$ for Φ a $\langle \varphi, \psi \rangle$-sequence. By Lemma 3 we then have

Lemma 4. For all signed predicate letters φ and ψ, $\Xi(\varphi, \psi)$ is the union of linear sets

$$L(\mathbf{p}, \Sigma_W(\Lambda_1) \cup \cdots \cup \Sigma_W(\Lambda_k)),$$

where for some $\langle \varphi, \psi \rangle$-sequence Ψ of length $\leq s^2 + 1$, $\mathbf{p} \in \Sigma_W(\Psi)$ and $\Lambda_1,...,\Lambda_k$ are all the $\langle \bar{\lambda}, \lambda \rangle$-sequences of length $\leq s^2 + 1$ for some member λ of Ψ. $\quad\square$

REMARK. Since each set $W(\varphi, \psi)$ contains only pairs $\langle p, q \rangle$ for $-1 \leq p, q \leq 1$, each set $\Sigma_W(\Psi)$ for Ψ of length $\leq s^2 + 1$ contains only pairs $\langle p, q \rangle$ for $-s^2 \leq p, q \leq s^2$. That is, $\Xi(\varphi, \psi)$ is the union of linear sets $L(\mathbf{p}, S)$ where every integer occurring in $\mathbf{p}$ or in a member of S has absolute value $\leq s^2$. End of Remark.

For all signed predicate letters φ and ψ, let $X(\varphi, \psi)$ be $\{\langle p, p \rangle | \langle p, p \rangle \in \Xi(\varphi, \psi)\}$. Thus if for some m there is an elementary chain from φmm to $\psi m + p \ m + p$, then $\langle p, p \rangle \in X(\varphi, \psi)$; and if $\langle p, p \rangle \in X(\varphi, \psi)$, then there are elementary chains from φmm to $\psi m + p \ m + p$ for all sufficiently large m.

Lemma 5. For all φ and ψ, $X(\varphi, \psi)$ is a semilinear set a presentation of which may be effectively obtained from a presentation of $\Xi(\varphi, \psi)$.

Proof. (See also Aanderaa and Lewis, 1973.) The lemma is a quick consequence of the following fact, whose proof we omit: let $L =$

$L(\langle p_0, q_0 \rangle, \{\langle p_1, q_1 \rangle, ..., \langle p_m, q_m \rangle\})$ be a linear set, and let k be the maximum of the absolute values of the p_i and q_i. Then the set $\{\langle p, p \rangle \mid \langle p, p \rangle \in L\}$ is the union of all linear sets $L(\langle r, r \rangle, S)$, where

(a) r has absolute value $\leq 2(m + 1)k^2$ and $\langle r, r \rangle \in L$;

(b) $S = \{\langle q, q \rangle \mid q = p_i q_j - p_j q_i, p_j - q_j \leq 0, p_i - q_i \geq 0, 1 \leq i, j \leq m\} \cup \{\langle p_i, p_i \rangle \mid p_i = q_i, 1 \leq i \leq m\}$. $\square$

For all signed predicate letters φ and ψ, let $Y(\varphi, \psi)$ be $X(\varphi, \psi) \cup \{\langle 0, 0 \rangle\}$ if $\varphi mm \vee \psi mm$ is a monadic instance (for all m), and let $Y(\varphi, \psi)$ be $X(\varphi, \psi)$ otherwise. Each set $Y(\varphi, \psi)$ is semilinear, and a presentation of it can be effectively obtained from one of $X(\varphi, \psi)$. If $\langle p, p \rangle \in Y(\varphi, \psi)$, then for all sufficiently large m there is a chain K from φmm to $\psi m + p\ m + p$ contained in $E(F, \omega)$; either K is elementary or else K consists of just one monadic instance. Hence, if Φ is a $\langle \varphi, \psi \rangle$-sequence and $\langle p, p \rangle \in \Sigma_Y(\Phi)$, then for all sufficiently large m there are chains from φmm to $\psi m + p\ m + p$ contained in $E(F, \omega)$; that is, $\langle p, p \rangle \in \Theta(\varphi, \psi)$. Conversely, every chain from φmm to $\psi m + p\ m + p$ that is contained in $E(F, \omega)$ is a conjunction $K_1 \wedge \cdots \wedge K_k$ for some $k \geq 1$, where each K_i is either a monadic instance or an elementary chain from and to essentially monadic signed atoms. Hence if $\langle p, p \rangle \in \Theta(\varphi, \psi)$, then there is a $\langle \varphi, \psi \rangle$-sequence Φ such that $\langle p, p \rangle \in \Sigma_Y(\Phi)$. We have shown that $\Theta(\varphi, \psi)$ is the union of all sets $\Sigma_Y(\Phi)$ for Φ a $\langle \varphi, \psi \rangle$-sequence. Once again we apply Lemma 3.

Lemma 6. For all φ and ψ the set $\Theta(\varphi, \psi)$ is semilinear, and a presentation of it may be uniformly and effectively constructed from presentations of the sets $Y(\lambda, \mu)$.

Proof. Note first that for each sequence $\Phi = \langle \varphi_1, ..., \varphi_k \rangle$ of signed predicate letters the set $\Sigma_Y(\Phi)$ is semilinear. Indeed, $\Sigma_Y(\Phi)$ is the union of all sets $L(\mathbf{p}_1 + \cdots + \mathbf{p}_{k-1}, S_1 \cup \cdots \cup S_{k-1})$ such that, for each $i < k$, $\langle \mathbf{p}_i, S_i \rangle$ is in the given presentation of $Y(\bar{\varphi}_i, \varphi_{i+1})$. Thus we can construct from presentations of the sets $Y(\varphi, \psi)$ a presentation of each set $\Sigma_Y(\Phi)$ for Φ of length $\leq s^2 + 1$.

For each sequence Ψ of length $\leq s^2 + 1$ let $\mathfrak{S}_\Psi$ be the set $\{\mathbf{p} + \mathbf{q}_1 + \cdots + \mathbf{q}_m \mid \mathbf{p} \in \Sigma_Y(\Psi), m \geq 0$, and each $\mathbf{q}_i$ is in $\Sigma_Y(\Lambda)$ for some $\langle \bar{\lambda}, \lambda \rangle$-sequence Λ of length $\leq s^2 + 1$ and some member λ of $\Psi\}$. Then $\mathfrak{S}_\Psi$ is semilinear: it is the union of all sets $L(\mathbf{r}_0 + \mathbf{r}_1 + \cdots + \mathbf{r}_l, S_0 \cup S_1 \cup \cdots \cup S_l \cup \{\mathbf{r}_1, ..., \mathbf{r}_l\})$ such that $\langle \mathbf{r}_0, S_0 \rangle$ is a member of the given presentation of $\Sigma_Y(\Phi)$, $\langle \mathbf{r}_1, S_1 \rangle, ..., \langle \mathbf{r}_l, S_l \rangle$ are distinct, and each $\langle \mathbf{r}_i, S_i \rangle$ for $i \geq 1$ is a member of the given presentation of some $\Sigma_Y(\Lambda)$ for Λ a $\langle \bar{\lambda}, \lambda \rangle$-sequence of length $\leq s^2 + 1$ and λ a member of Ψ.

Moreover, by Lemma 3, $\Theta(\varphi, \psi)$ is the union of the sets $\mathfrak{S}_\Psi$ for each $\langle \varphi, \psi \rangle$-sequence Ψ of length $\leq s^2 + 1$. Hence $\Theta(\varphi, \psi)$ is semilinear, and we have shown how to construct a presentation of it. $\square$

Now the schema F has no model iff there is a signed predicate letter φ

ISBN 0-201-02540-X

such that $\langle 0, 0 \rangle \in \Theta(\varphi, \varphi)$ and $\langle 0, 0 \rangle \in \Theta(\bar{\varphi}, \bar{\varphi})$. For if $\langle 0, 0 \rangle$ is in both $\Theta(\varphi, \varphi)$ and $\Theta(\bar{\varphi}, \bar{\varphi})$, then $E(F, \omega)$ contains chains from φmm to φmm and from $\bar{\varphi} mm$ to $\bar{\varphi} mm$ for all sufficiently large m; in this case $E(F, \omega)$ is inconsistent. Conversely, if $E(F, \omega)$ is inconsistent, then by Lemma 2 it contains chains from φmm to φmm and from $\bar{\varphi} mm$ to $\bar{\varphi} mm$ for some atom φmm, whence $\langle 0, 0 \rangle \in \Theta(\varphi, \varphi)$ and $\langle 0, 0 \rangle \in \Theta (\bar{\varphi}, \bar{\varphi})$.

The solvability of Class 5.6 then follows from Lemma 6, since there is a uniform and effective procedure for deciding, given any presentation of a semilinear set, whether or not $\langle 0, 0 \rangle$ is in the set.

Our docility proof for the ∀∃∀-Krom Class builds on the solvability proof. Let F be a schema in Class 5.6; for each $r > 0$, let N_r be the set of biconditionals $(Pmn \Leftrightarrow Pm + r\ n)$ and $(Pmn \Leftrightarrow Pm\ n + r)$ for each predicate letter P of F and all terms m and n. Note that N_r is a set of Krom formulas.

Fact 1. If F has a finite model, then $E(F, \omega) \cup N_r$ is consistent for some r. If $E(F, \omega) \cup N_r$ is consistent, then F has a finite model of cardinality r.

Proof. If $\mathfrak{A}$ is a truth-assignment that verifies $E(F, \omega) \cup N_r$, then $\mathfrak{A}$ also verifies $\Delta_r[E(F, \omega)]$; and if $\mathfrak{A}$ verifies $\Delta_r[E(F, \omega)]$, then the induced truth-assignment $\mathfrak{A}(\Delta_r)$ verifies $E(F, \omega) \cup N_r$. (The formula-mapping Δ_r is defined on page 130.) Hence the result follows by the Special Finite Model Lemma. □

In the solvability proof we reduced Class 5.6 to the subclass of ∀∃∀-Krom schemata whose matrices contain elementary and monadic conjuncts only. From Fact 1 and the Chain Criterion it quickly follows that this reduction is conservative. For the rest of this proof, let F be a fixed schema in this subclass. Again reasoning as in the solvability proof, we have

Fact 2. $E(F, \omega) \cup N_r$ is inconsistent iff it contains chains from φjj to φjj and from $\bar{\varphi} jj$ to $\bar{\varphi} jj$ for some predicate letter φ and some j. □

Let s be twice the number of predicate letters in F, and let $t = (2s^5)!(2s^2)!$. We shall prove that if $E(F, \omega) \cup N_t$ is inconsistent, then $E(F, \omega) \cup N_r$ is inconsistent for each $r > 0$. Consequently, by Fact 1, F has a finite model only if F has a model of cardinality t.

A chain K is *r-elementary*, $r > 0$, iff each conjunct in K either is an elementary instance for F or is a conjunct in a member of N_r. For all signed predicate letters φ and ψ let $\Xi_r(\varphi, \psi)$ be the set of pairs $\langle p, q \rangle$ such that there is an r-elementary chain from φmn to $\psi m + p\ n + q$ for some m and n. The following is immediate from the definition.

Lemma 7. For all φ and ψ, and each $r > 0$, $\Xi_r(\varphi, \psi) = \{\mathbf{p} + \langle ir, jr \rangle \mid \mathbf{p} \in \Xi(\varphi, \psi)$ and $i, j \in \mathbb{Z}\}$. □

Let $\mathbf{p}_1, \ldots, \mathbf{p}_k \in \mathbb{Z}^2$. A $\mathbb{Z}$-*linear combination* of $\mathbf{p}_1, \ldots, \mathbf{p}_k$ is any pair of the form $i_1\mathbf{p}_1 + \cdots + i_k\mathbf{p}_k$ for $i_1, \ldots, i_k \in \mathbb{Z}$.

Fact 3. Let $L(\mathbf{p}_0, S)$ be a linear set. Then for each $r > 0$ the set $\{\mathbf{q} +$

ISBN 0-201-02540-X

$\langle ir, jr \rangle \, | \, \mathbf{q} \in L(p_0, S)$ and $i, j \in Z\}$ contains every pair $\mathbf{p}_0 + \mathbf{a}$, where $\mathbf{a}$ is a Z-linear combination of members of S.

Proof. It suffices to show that if $\langle q_1, q_2 \rangle \in S$, then $\langle -q_1, -q_2 \rangle = n(\langle q_1, q_2 \rangle) + \langle ir, jr \rangle$ for some nonnegative n and some $i, j \in Z$. This is obvious: let $n = r - 1$, $i = -q_1$, and $j = -q_2$. $\square$

For all signed predicate letters φ and ψ and each $r > 0$ let $X_r(\varphi, \psi) = \{\langle p, p \rangle \, | \, \langle p, p \rangle \in \Xi_r(\varphi, \psi)\}$. We now depict some relationships between $X_t(\varphi, \psi)$ and sets $X_r(\varphi, \psi)$ for arbitrary r. We use $|q|$ for the absolute value of q.

Lemma 8. Suppose $\langle p, p \rangle \in X_t(\varphi, \psi)$. Then either

(1) there is an integer a, $0 \neq |a| \leq 2s^4$, such that $\langle p + ka, p + ka \rangle \in X_r(\varphi, \psi)$ for every $k \in Z$ and every $r > 0$; or

(2) there is an integer b, $|b| \leq 2s^4$, such that $p - b \equiv 0 \pmod{2s^5!}$ and $\langle b, b \rangle \in X_r(\varphi, \omega)$ for every $r > 0$.

Proof. By Lemma 7, $\langle p, p \rangle = \langle p_1 + it, q_1 + jt \rangle$ for some $i, j \in Z$ and some $\langle p_1, q_1 \rangle$ in $\Xi(\varphi, \psi)$. By Lemma 4, $\langle p_1, q_1 \rangle \in L(\langle p_0, q_0 \rangle, S) \subseteq \Xi(\varphi, \psi)$, where p_0, q_0, and each integer occurring in a member of S has absolute value $\leq s^2$. Note that, by Lemma 7 and Fact 3, to show that a pair $\langle p + l, p + l \rangle$ is in $X_r(\varphi, \psi)$ for every $r > 0$, it suffices to show that $\langle p - p_0 + l, p - q_0 + l \rangle$ is a Z-linear combination of members of S, which in turn holds iff $\langle it + l, jt + l \rangle$ is a Z-linear combination of members of S.

Case 1. S contains pairs $\langle m_1, n_1 \rangle$ and $\langle m_2, n_2 \rangle$ such that $m_1 n_2 - m_2 n_1 \neq 0$. Let $a = m_1 n_2 - m_2 n_1$; thus $0 \neq |a| \leq 2s^4$. To show that $\langle p + ka, p + ka \rangle \in X_r(\varphi, \psi)$ for each $r > 0$ and each $k \in Z$, it suffices to show that $\langle it + ka, jt + ka \rangle$ is a Z-linear combination of $\langle m_1, n_1 \rangle$ and $\langle m_2, n_2 \rangle$ for each $k \in Z$. In fact, $\langle it + ka, jt + ka \rangle = i_1(\langle m_1, n_1 \rangle) + i_2(\langle m_2, n_2 \rangle)$, where $i_1 = k(m_1 - n_1) + ((in_2 - jm_2)t/a)$ and $i_2 = k(m_1 - n_1) + ((jm_1 - in_1)t/a)$. Note that i_1 and i_2 are integral: a divides t since $|a| \leq 2s^4$.

Case 2. S is nonempty, but contains no pairs as in Case 1. Then there are m and n such that each member of S is $\langle mc, nc \rangle$ for some c. Consequently, every member of $L(\langle p_0, q_0 \rangle, S)$ is $\langle p_0, q_0 \rangle + \langle mc', nc' \rangle$ for some c'.

Subcase (a). Suppose $m = n$. Since $p = p_1 + it = q_1 + jt$, and $\langle p_1, q_1 \rangle = \langle p_0, q_0 \rangle + \langle mc', nc' \rangle$ for some c', we have $0 = q_0 - p_0 + (j - i)t$. But $|q_0 - p_0| \leq 2s^2 < t$. Hence $i = j$. Let $\langle a, a \rangle$ be any member of S; thus $|a| \leq 2s^2$, whence a divides t. It follows that $\langle it + ka, jt + ka \rangle$ is a Z-linear combination of $\langle a, a \rangle$ for each k in Z, whence $\langle p + ka, p + ka \rangle \in X_r(\varphi, \psi)$ for each $r > 0$.

Subcase (b). Suppose $m \neq n$. Let $\langle d, e \rangle$ be any member of S; thus $d \neq e$ and $|d - e| \leq 2s^2$. Let $k = (i - j)t/(d - e)$. Since $t = (2s^5)!(2s^2)!$,

$k \equiv 0 \pmod{(2s^5)!}$. Let $b = p - kd - it$; then $b = p - ke - jt$, so that $\langle b, b \rangle = \langle p_1, q_1 \rangle - k(\langle d, e \rangle)$. Thus, by Fact 3, $\langle b, b \rangle \in X_r(\varphi, \psi)$ for each $r > 0$. Moreover, $p - b = kd + it \equiv 0 \pmod{(2s^5)!}$. It remains only to show that $|b| \leq 2s^4$. But $\langle b, b \rangle = \langle p_0, q_0 \rangle + \langle mc', nc' \rangle$ for some c'. Indeed, we must have $c' = (q_0 - p_0)/(m - n)$. Hence $b = q_1 + ((q_0 - p_0)m/(m - n)) = (q_0 m - p_0 n)/(m - n)$. Since p_0, q_0, m, and n have absolute values $\leq s^2$, $|b| \leq 2s^4$.

Case 3. S is empty. Then $\langle p, p \rangle = \langle p_0 + it, q_0 + jt \rangle$. Since $|q_0 - p_0| < t$, $i = j$ and $p_0 = q_0$. Let $b = p_0$. Then $\langle b, b \rangle \in L(\langle p_0, q_0 \rangle, S)$, whence $\langle b, b \rangle \in X_r(\varphi, \omega)$ for each $r > 0$; $|b| \leq s^2$; and $p - b = it \equiv 0 \pmod{(2s^5)!}$. $\square$

For all signed predicate letters φ and ψ, and each $r > 0$, let $Y_r(\varphi, \psi)$ be $X_r(\varphi, \psi) \cup \{\langle 0, 0 \rangle\}$ if $\varphi mm \vee \psi mm$ is a monadic instance for F (for all m); and let $Y_r(\varphi, \psi)$ be $X_r(\varphi, \psi)$ otherwise. Then, just as in the solvability proof, if, for some m, $E(F, \omega) \cup N_r$ contains a chain from φmm to $\psi m + p \; m + p$, then $\langle p, p \rangle \in \Sigma_{Y_r}(\Phi)$ for some $\langle \varphi, \psi \rangle$-sequence Φ. And if $\langle p, p \rangle \in \Sigma_{Y_r}(\Phi)$ for some $\langle \varphi, \psi \rangle$-sequence Φ, then for all sufficiently large m $E(F, \omega) \cup N_r$ contains a chain from φmm to $\psi m + p \; m + p$. Hence $E(F, \omega) \cup N_r$ is inconsistent iff for some predicate letter φ there exist a $\langle \varphi, \varphi \rangle$-sequence Φ_1 and a $\langle \bar{\varphi}, \bar{\varphi} \rangle$-sequence Φ_2 such that $\langle 0, 0 \rangle \in \Sigma_{Y_r}(\Phi_i)$, $i = 1, 2$. The following lemma therefore shows that if $E(F, \omega) \cup N_t$ is inconsistent, then so is $E(F, \omega) \cup N_r$ for each $r > 0$.

Lemma 9. For all signed predicate letters φ and ψ, if there is a $\langle \varphi, \psi \rangle$-sequence Φ such that $\langle 0, 0 \rangle \in \Sigma_{Y_t}(\Phi)$, then for each $r > 0$ there is a $\langle \varphi, \psi \rangle$-sequence Ψ such that $\langle 0, 0 \rangle \in \Sigma_{Y_r}(\Psi)$.

Proof. Call a sequence Φ *good* iff $\Sigma_{Y_t}(\Phi)$ contains a pair $\langle p, p \rangle$ for some $p \equiv 0 \pmod{(2s^5)!}$. We show that if there is a good $\langle \varphi, \psi \rangle$-sequence, then for each $r > 0$ there is a $\langle \varphi, \psi \rangle$-sequence Ψ such that $\langle 0, 0 \rangle \in \Sigma_{Y_r}(\Psi)$.

Let $\Phi = \langle \varphi_1, \ldots, \varphi_m \rangle$ be a good $\langle \varphi, \psi \rangle$-sequence minimal in length. Thus there are integers $p_1, \ldots, p_{m-1}$ such that $p_1 + \cdots + p_{m-1} \equiv 0 \pmod{(2s^5)!}$ and $\langle p_i, p_i \rangle \in Y_t(\bar{\varphi}_i, \varphi_{i+1})$, $1 \leq i < m$. For each $i < m$ let q_i and c_i be the integers determined by the first applicable clause of the following three:

(i) if $\langle p_i, p_i \rangle \in X_t(\bar{\varphi}_i, \varphi_{i+1})$ and there is an integer a fulfilling property (1) of Lemma 8, then let $q_i = p_i$ and $c_i = a$;

(ii) if $\langle p_i, p_i \rangle \in X_t(\bar{\varphi}_i, \varphi_{i+1})$ and there is an integer b fulfilling property (2) of Lemma 8, then let $q_i = b$ and $c_i = 0$;

(iii) if $\langle p_i, p_i \rangle \notin X_t(\bar{\varphi}_i, \varphi_{i+1})$, in which case $p_i = 0$ and $\bar{\varphi}_i mm \vee \varphi_{i+1} mm$ is a monadic instance for F, then let $q_i = c_i = 0$.

It follows that $\langle q_i + kc_i, q_i + kc_i \rangle \in Y_r(\bar{\varphi}_i, \varphi_{i+1})$ for each $i < m$, all k in Z, and all $r > 0$. Moreover, $p_i - q_i \equiv 0 \pmod{(2s^5)!}$ for each $i < m$,

ISBN 0-201-02540-X

whence $q_1 + \cdots + q_{m-1} \equiv 0 \pmod{(2s^5)!}$. Suppose some c_i is nonzero. Since $|c_i| \leq 2s^4$, c_i divides $(2s^5)!$. Hence $q_1 + \cdots + q_{m-1} + kc_i = 0$ for some k. But $q_1 + \cdots + q_{m-1} + kc_i$ is in $\Sigma_{Y_r}(\Phi)$ for each r, and we are done. Suppose each c_i is zero. Then $|q_i| \leq 2s^4$ for each $i < m$. If $m \leq s$, then $|q_1 + \cdots + q_{m-1}| < s(2s^4) < (2s^5)!$. Hence $q_1 + \cdots + q_{m-1} = 0$, so that $\langle 0; 0 \rangle \in \Sigma_{Y_r}(\Phi)$ for every $r > 0$ and again we are done. If $m > s$, then by the pigeonhole argument, there are integers i and k, $1 \leq i < i + k \leq i + s < m$, such that $\varphi_i = \varphi_{i+k}$. Let $q = q_i + q_{i+1} + \cdots + q_{i+k-1}$. If $q \equiv 0 \pmod{(2s^5)!}$, then $(q_1 + \cdots + q_{m-1}) - q \equiv 0 \pmod{(2s^5)!}$. But $(q_1 + \cdots + q_{m-1}) - q \in \Sigma_{Y_t}(\Phi')$, where $\Phi' = \langle \varphi_1, \ldots, \varphi_{i-1}, \varphi_{i+k}, \ldots, \varphi_m \rangle$. Hence in this case Φ' is a good $\langle \varphi, \psi \rangle$-sequence shorter than Φ, which contradicts the choice of Φ. We may conclude that $q \not\equiv 0 \pmod{2s^5)!}$. Since $|q| \leq s(2s^4)$, q divides $(2s^5)!$. Hence there is an integer j such that $jq + q_1 + \cdots + q_{m-1} = 0$. Let $r > 0$. Then there is a nonnegative integer j' such that $j'q + q_1 + \cdots + q_{m-1} = nr$ for some integer n. Let $\Psi = \langle \varphi_1, \ldots, \varphi_{i+k}, \varphi_{i+1}, \ldots, \varphi_{i+k}, \ldots, \varphi_{i+1}, \ldots, \varphi_{i+k}, \varphi_{i+k+1}, \ldots, \varphi_m \rangle$ be the sequence of length $m + j'k$ obtained from Φ by introducing j' new occurrences of the sequence $\langle \varphi_{i+1}, \ldots, \varphi_{i+k} \rangle$. Then $\Sigma_{Y_r}(\Psi)$ contains $\langle nr, nr \rangle$. And then, since if a set $X_r(\lambda, \mu)$ contains $\langle l, l \rangle$ it also contains $\langle l + r, l + r \rangle$ and $\langle l - r, l - r \rangle$, it follows that $\langle 0, 0 \rangle \in \Sigma_{Y_r}(\Psi)$. $\square$

4. Finitely Controllable and Docile ∀∃∀-Classes with Identity

We claimed in §2 of Chapter 8 that Classes 5.3 and 5.4 with Identity are finitely controllable, and that Classes 5.1, 5.5, and 5.6 with Identity are docile. Here we substantiate those claims, by sketching the way in which the proofs for the corresponding classes without identity may be extended. Throughout this section we use the numerical notation for terms in the domains of ∀∃∀-schemata, first introduced in Chapter 5.

Let F be an ∀∃∀-schema containing identity. A truth-assignment $\mathfrak{A}$ on $\tilde{L}(F, \omega)$ is *p-strict*, $p > 0$, iff, for all m and $n < p$, $\mathfrak{A}$ verifies $m \equiv n$ iff $m = n$. The proofs in this section differ from those for the ∀∃∀-classes without identity principally in the attention paid to *p*-strictness.

Classes 5.3 and 5.4 with Identity. Let F be an ∀∃∀-schema containing identity such that either each atomic subformula of F has one of the forms Py_1y_2 and Py_2x, or $b(F)$ does not contain $\{y_1, y_2\}$, or $b(F)$ does not contain $\{x, y_2\}$. Then F has a model only if F has a finite model.

Proof. It suffices to construct, for each F in one of these classes and each truth-assignment $\mathfrak{A}$, a term-mapping δ, a formula-mapping Γ, and an

ISBN 0-201-02540-X

instance-mapping Ψ such that

(a) δ has finite range;

(b) $\mathfrak{A}(\Gamma(\Delta(\langle H\rangle_i))) = \mathfrak{A}(\langle\Psi(H)\rangle_i)$ for each Herbrand instance H and each i;

(c) Γ carries each equation $m \equiv n$ to an equation $m' \equiv n'$, where $m' = n'$ iff $m = n$.

For suppose F has a model, and let $\mathfrak{A}$ be a standard truth-assignment verifying $E(F, \omega)$. If for some p $\mathfrak{A}$ is not p-strict, then, as we saw in the solvability proof for Classes 5.1, 5.5, and 5.6 with Identity, $\mathfrak{A}$ provides a finite model for F. Hence we may suppose that $\mathfrak{A}$ is p-strict for each p. By (c), then, $\mathfrak{A}(\Gamma)$ is p-strict for each p. Moreover, by (b), $\mathfrak{A}(\Gamma)$ verifies $\Delta[E(F, \omega)]$. Hence $\mathfrak{A}(\Gamma)$ provides a model for F over $\delta[D(F, \omega)]$, which, by (a), is finite.

The finite controllability of Class 5.4 with Identity is immediate: the construction in the amiability proof for Class 5.4 without identity fulfills the three conditions.

The finite controllability of Class 5.3 with Identity requires a modification of the argument of Chapter 5. Given $\mathfrak{A}$, we first choose q and r such that $0 \leq q < r \leq 2d(F, 2)$, q is congruent to r (mod 2), and $\mathfrak{p}_\mathfrak{A}(q, q+1) = \mathfrak{p}_\mathfrak{A}(r, r+1)$. We then let Γ carry each m to $q + i$, where i is the least nonnegative remainder of m (mod $r - q$). Clearly (a) is fulfilled. We define Γ thus: $\Gamma(Pmn) = Pnn$ if $m = n$; $\Gamma(Pmn) = Pm\,m+1$ if $n = m + 1$ or if $n = q$ and $m = r - 1$; $\Gamma(Pmn) = Pn+1\,n$ if m is congruent to n (mod 2) and $m \neq n$; and $\Gamma(Pmn) = Pm\,m+2$ if m is incongruent to n (mod 2), $n \neq m + 1$, and either $n \neq q$ or $m \neq r - 1$. Clearly (c) is fulfilled. The instance-mapping Ψ is defined thus: $\Psi(F^*(y_1/m, y_2/n)) = F^*(y_1/p, y_2/q)$, where $\Gamma(Pmn) = Ppq$. The verification of condition (b) is routine. $\square$

To show the docility of Classes 5.1, 5.5, and 5.6 with Identity, we shall apply the Special Finite Model Lemma to $\forall\exists\forall$-schemata that contain identity. Let F be such a schema. Recall that, for each $p > 0$, δ_p is the term-mapping that carries each m to the least nonnegative remainder of m (mod p); and Δ_p is the δ_p-canonical formula-mapping. The proof of the Special Finite Model Lemma yields the following: if F has a finite model of cardinality q, then for some $p \leq q$ there is a truth-assignment $\mathfrak{A}$ that verifies $\Delta_p[E(F, \omega)]$ and is p-strict. Conversely, any truth-assignment that verifies $\Delta_p[E(F, \omega)]$ and is p-strict provides a model for F of cardinality p.

Class 5.1 with Identity. Let F be an $\forall\exists\forall$-schema containing identity each of whose atomic subformulas has one of the seven forms Py_1y_2, Pxy_2, Py_1y_1, Pxx, Py_2y_2, Py_1x, Pxy_1. Then F has a finite model only if F has a finite model of cardinality $\leq 2^{2d(F,2)}$.

ISBN 0-201-02540-X

Proof.   Suppose that $p > 2^{2d(F,2)}$ and that $\mathfrak{A}$ is a p-strict truth-assignment that verifies $\Delta_p[E(F, \omega)]$. It suffices to construct, for some $r > 0$, a $(p - r)$-strict truth-assignment that verifies $\Delta_{p-r}[E(F, \omega)]$.

For each $m \geq 0$ let

$$\xi_{\mathfrak{A}}{}^L(m) = \{\mathfrak{p}_{\mathfrak{A}}(0, m, n) \mid n \leq p - 1 \text{ and } n < m\};$$
$$\xi_{\mathfrak{A}}{}^G(m) = \{\mathfrak{p}_{\mathfrak{A}}(0, m, n) \mid m < n \leq p - 1\}.$$

Since $p > 2^{2d(F,2)}$, by the pigeonhole argument there are numbers q and r, $0 < q < q + r \leq p - 1$, such that $\xi_{\mathfrak{A}}{}^L(q) = \xi_{\mathfrak{A}}{}^L(q + r)$ and $\xi_{\mathfrak{A}}{}^G(q) = \xi_{\mathfrak{A}}{}^G(q + r)$.

As in the docility proof for Class 5.1 without identity (Chapter 5, §3), we shall define a formula-mapping Γ, and hence an induced truth-assignment $\mathfrak{A}(\Gamma)$, on the basis of a term-mapping γ and a mapping η on pairs of terms. These mappings are to fulfill the following six conditions for all $m, n \leq p - r - 1$:

(1) $\gamma m \leq p - 1$; if $m < p - r - 1$, then $\gamma m < p - 1$;
(2) $\eta(m, n) \leq p - 1$;
(3) $\mathfrak{p}_{\mathfrak{A}}(0, \gamma(m + 1), \eta(m + 1, n)) = \mathfrak{p}_{\mathfrak{A}}(0, \gamma m + 1, \eta(m, n))$;
(4) $\eta(m, m) = \gamma m$ and $\eta(m, m + 1) = \gamma m + 1$;
(5) $\eta(m, 0) = \gamma 0 = 0$;
(6)   if $m < n$, then $\gamma m < \eta(m, n)$; and if $m > n$, then $\gamma m > \eta(m, n)$.

Note that the first five conditions were obtained in the earlier docility proof. The sixth is new, and is needed to ensure that the induced truth-assignment $\mathfrak{A}(\Gamma)$ is $(p - r)$-strict. It was to obtain this new condition that we replaced the function $\xi_{\mathfrak{A}}$ used in the earlier proof by the two functions $\xi_{\mathfrak{A}}{}^L$ and $\xi_{\mathfrak{A}}{}^G$.

We define γ thus: if $m < q$, then $\gamma m = m$, and if $m \geq q$, then $\gamma m = m + r$. From the choice of q and r we have $\xi_{\mathfrak{A}}{}^L(\gamma(m+1)) = \xi_{\mathfrak{A}}{}^L(\gamma m + 1)$ and $\xi_{\mathfrak{A}}{}^G(\gamma(m + 1)) = \xi_{\mathfrak{A}}{}^G(\gamma m + 1)$ for each m. We define η thus:

(η-1)   For each m, $\eta(m, m) = \gamma m$ and $\eta(m, m + 1) = \gamma m + 1$.

(η-2)   If $m + 1 < n \leq p - r - 1$, then $\eta(m, n)$ is the least k such that $\gamma m < k \leq p - 1$ and $\mathfrak{p}_{\mathfrak{A}}(0, \gamma m + 1, k) = \mathfrak{p}_{\mathfrak{A}}(0, \gamma(m + 1), \eta(m + 1, n))$. On the assumption that $\gamma m < \gamma(m + 1) < \eta(m + 1, n) \leq p - 1$, there will be such a k, since $\xi_{\mathfrak{A}}{}^G(\gamma m + 1) = \xi_{\mathfrak{A}}{}^G(\gamma(m + 1))$.

(η-3)   If $n \leq m \leq p - r - 1$, then $\eta(m + 1, n)$ is the least k such that $k < \gamma(m + 1) \leq p - 1$ and $\mathfrak{p}_{\mathfrak{A}}(0, \gamma(m + 1), k) = \mathfrak{p}_{\mathfrak{A}}(0, \gamma m + 1, \eta(m, n))$. On the assumption that $\eta(m, n) \leq \gamma m < \gamma m + 1$, there will be such a k, since $\xi_{\mathfrak{A}}{}^L(\gamma(m + 1)) = \xi_{\mathfrak{A}}{}^L(\gamma m + 1)$.

(η-4)   If $\eta(m, n)$ is not determined by (η-1)–(η-3), then $\eta(m, n) = 0$.

ISBN 0-201-02540-X

Condition (6) is then immediate and conditions (1)–(5) follow just as in the docility proof for Class 5.1 without identity.

We define Γ thus: for each atom Pmn in $L(F, \omega)$, $\Gamma(Pmn) = P\gamma m\, \eta(m, n)$. Again as in the earlier docility proof, conditions (1)–(5) imply that the induced truth-assignment $\mathfrak{A}(\Gamma)$ verifies $\Delta_{p-r}[E(F, \omega)]$. Moreover, $\mathfrak{A}(\Gamma)$ is $(p - r)$-strict. For let $m, n \le p - r - 1$; thus γm and $\eta(m, n)$ are no greater than $p - 1$. By definition, $\mathfrak{A}(\Gamma)$ verifies $m \equiv n$ iff $\mathfrak{A}$ verifies $\gamma m \equiv \eta(m, n)$. Since $\mathfrak{A}$ is p-strict, $\mathfrak{A}$ verifies $\gamma m \equiv \eta(m, n)$ iff $\gamma m = \eta(m, n)$. And by conditions (4) and (6), $\gamma m = \eta(m, n)$ iff $m = n$. $\square$

Class 5.5 is the $\forall\exists\forall$-Horn Class. In §5 of Chapter 6 we gave a reduction that eliminates from any $\forall\exists\forall$-schema all k-place predicate letters for $k \ne 2$. This reduction preserves Horn (and Krom) matrices, and is conservative. Hence it sufficed to prove the solvability and docility of the class of $\forall\exists\forall$-Horn schemata in which each predicate letter is dyadic. This remains true when identity is allowed, since the reduction in Chapter 6 can easily be seen to apply to $\forall\exists\forall$-schemata F containing identity. Indeed, if G is obtained from F by means of this reduction and if $q > 0$, then there is a q-strict truth-assignment verifying $\Delta_q[E(F, \omega)]$ iff there is a q-strict truth-assignment verifying $\Delta_q[E(G, \omega)]$.

Class 5.5. with Identity.   Let F be an $\forall\exists\forall$-Horn schema containing identity in which each predicate letter is dyadic. Then F has a finite model only if F has a model of cardinality $\le 2d(F, 2) + 3$.

Proof.   Suppose that $p > 2d(F, 2) + 3$ and that there is a p-strict truth-assignment that verifies $\Delta_p[E(F, \omega)]$. It suffices to show the existence, for some $r > 0$, of a $(p - 2r)$-strict truth-assignment that verifies $\Delta_{p-2r}[E(F, \omega)]$.

We may assume without loss of generality that $y_1 \equiv y_1$ is a conjunct in the matrix of F. Let $\mathfrak{A}$ be the truth-assignment that verifies an atom A iff $\Delta_p[E(F, \omega)]$ truth-functionally implies A. By the Truth-Functional Lemma for Horn formulas (Chapter 5, §2), $\mathfrak{A}$ verifies $\Delta_p[E(F, \omega)]$. Moreover, $\mathfrak{A}$ is p-strict. For if $m < p$, then $m \equiv m$ is a conjunct in some member of $\Delta_p[E(F, \omega)]$, whence $\mathfrak{A}$ verifies $m \equiv m$. And if $\mathfrak{A}$ verified $m \equiv n$ for some distinct $m, n < p$, then $\Delta_p[E(F, \omega)]$ would truth-functionally imply $m \equiv n$, whence no p-strict truth-assignment verifying $\Delta_p[E(F, \omega)]$ could exist, contrary to hypothesis.

We may then proceed exactly as in the docility proof for Class 5.5 without identity. We first pick a suitable r, and then construct a term-mapping γ and a formula-mapping Γ, with the result that $\mathfrak{A}(\Gamma)$ verifies $\Delta_{p-2r}[E(F, \omega)]$. It also follows from that construction that $\mathfrak{A}(\Gamma)$ is $(p - 2r)$-strict. For suppose $0 \le n \le m < p - 2r$. Then $\Gamma(m \equiv n) = (\gamma(m - n) \equiv 0)$; since $\mathfrak{A}$ is p-strict, $\gamma 0 = 0$, and $0 < \gamma(m - n) < p$ whenever $0 < m - n < p - 2r$, $\mathfrak{A}(\Gamma)$ verifies $m \equiv n$ iff $m = n$. Now suppose $0 \le m < n < p - 2r$.

ISBN 0-201-02540-X

Then $\Gamma(m \equiv n) = (0 \equiv \gamma(n - m))$ and $0 < \gamma(n - m) < p$. Since $\mathfrak{A}$ is p-strict, $\mathfrak{A}(\Gamma)$ falsifies $m \equiv n$. Thus, for all m, $n < p - 2r$, $\mathfrak{A}(\Gamma)$ verifies $m \equiv n$ iff $m = n$. □

We come finally to Class 5.6 with Identity; Class 5.6 is the ∀∃∀-Krom Class. As in the case of Class 5.5, we may restrict our attention to schemata in which each predicate letter is dyadic. If F is such a schema, then for each $r > 0$ let N_r be the set of Krom formulas defined on page 241, and let M_r be

$$\{-(m \equiv n) \vee -(m \equiv n) \,|\, m, n \geq 0, \, m \text{ and } n \text{ incongruent (mod } r)\}.$$

We also assume throughout that $y_1 \equiv y_1 \vee y_1 \equiv y_1$ is a conjunct in the matrix F^M of F.

Suppose a truth-assignment $\mathfrak{A}$ verifies $E(F, \omega) \cup N_r \cup M_r$. Since $\mathfrak{A}$ verifies N_r, $\mathfrak{A}(A) = \mathfrak{A}(\Delta_r(A))$ for each atom A; since $\mathfrak{A}$ also verifies $E(F, \omega)$, $\mathfrak{A}$ verifies $\Delta_r[E(F, \omega)]$. Moreover, $\mathfrak{A}$ verifies $m \equiv m \vee m \equiv m$ for each $m \geq 0$; and since $\mathfrak{A}$ verifies M_r, $\mathfrak{A}$ falsifies $m \equiv n$ for all distinct m, $n < r$. Hence $\mathfrak{A}$ is r-strict.

Conversely, suppose $\mathfrak{A}$ is an r-strict truth-assignment verifying $\Delta_r[E(F, \omega)]$. Then the induced truth-assignment $\mathfrak{A}(\Delta_r)$ verifies $E(F, \omega) \cup N_r \cup M_r$.

We have shown that if $E(F, \omega) \cup N_r \cup M_r$ is consistent, then F has a model of cardinality r; and if F has a model of cardinality q, then $E(F, \omega) \cup N_r \cup M_r$ is consistent for some $r \leq q$.

Given an ∀∃∀-Krom schema G_0 containing identity in which each predicate letter is dyadic, we wish to reduce G_0 conservatively to an ∀∃∀-Krom schema containing identity whose matrix contains elementary and monadic conjuncts only. The procedures used for Class 5.6 without identity may be applied here too, with only one minor change. In the last step of the reduction (page 236), we introduce a new predicate letter $\check{P}$ for all the predicate letters P of F except $\equiv$. We then define Π as follows: $\Pi(Pu_1u_2) = \check{P}u_2u_1$ and $\Pi(\check{P}u_1u_2) = Pu_2u_1$ for each predicate letter P of F except $\equiv$; $\Pi(u_1 \equiv u_2) = (u_2 \equiv u_1)$. (We need not introduce a predicate letter $\check{P}$ when P is $\equiv$, since $\equiv$ is always interpreted as a symmetrical relation.) It is a routine matter to check that, with this change, the reduction is successful.

Thus we consider an arbitrary ∀∃∀-Krom schema F with identity whose matrix contains elementary and monadic conjuncts only. We may carry out the analysis of chains given in the preceding section, treating $\equiv$ as an ordinary predicate letter. (Hence $\equiv$ and $-\equiv$ are signed predicate letters.) In particular we can effectively decide whether $\Xi(\equiv, \equiv)$ is nonempty, that is, whether there is an elementary chain from $m \equiv n$ to $p \equiv q$ for some m, n, p, q. If $\Xi(\equiv, \equiv)$ is nonempty, then $E(F, \omega) \cup N_r \cup M_r$ is inconsistent for each $r > 2$. For suppose $\langle i, j \rangle \in \Xi(\equiv, \equiv)$. Then for all sufficiently large

ISBN 0-201-02540-X

m and n there are elementary chains from $m \equiv n$ to $m + i \equiv n + j$. Since $r > 2$, we may pick m and n so that m and n are incongruent (mod r) and $m + i$ and $n + j$ are incongruent (mod r). But then M_r truth-functionally implies $- (m \equiv n) \wedge -(m + i \equiv n + j)$, and $E(F, \omega)$ truth-functionally implies $m \equiv n \vee m + i \equiv n + j$. Hence $E(F, \omega) \cup M_r$ is inconsistent. Thus if $\Xi(\equiv, \equiv)$ is nonempty, F either has no finite models or else has a model of cardinality ≤ 2.

We now assume that $\Xi(\equiv, \equiv)$ is empty. Let S be the set of signed predicate letters φ such that $\langle p, q \rangle \in \Xi(\equiv, \varphi)$ for some distinct p, q. We can effectively decide membership in S. Indeed, Lemma 4 of the preceding section assures us that if $\langle p, q \rangle \in \Xi(\equiv, \varphi)$ for some distinct p, q, then $\Xi(\equiv, \varphi)$ contains a pair $\langle a, b \rangle$ such that $0 < |a - b| \leq 2s^2$, where s is twice the number of predicate letters in F. For $\langle p, q \rangle \in L(\langle p_0, q_0 \rangle, S) \subseteq \Xi(\equiv, \varphi)$, where $L(\langle p_0, q_0 \rangle, S)$ is a linear set such that p_0, q_0, and each integer occurring in S has absolute value at most s^2. Since $p \neq q$, either $p_0 \neq q_0$, so we may take $\langle a, b \rangle = \langle p_0, q_0 \rangle$, or else $p_0 = q_0$ and there is a pair $\langle p_1, q_1 \rangle$ in S such that $p_1 \neq q_1$, so we may take $\langle a, b \rangle = \langle p_0 + p_1, q_0 + q_1 \rangle$.

Let $G = \forall y_1 \exists x \forall y_2 [F^M \wedge \bigwedge_{\varphi \in S} (\varphi y_1 y_1 \vee \varphi y_1 y_1)]$. We shall show that, for each $r > 2s^2$, $E(G, \omega) \cup E_r$ is consistent iff $E(F, \omega) \cup N_r \cup M_r$ is consistent. This suffices for docility: by the results of the previous section, $E(G, \omega) \cup N_r$ is consistent for some r iff $E(G, \omega) \cup N_t$ is consistent, where $t = (2s^5)!(2s^2)!$. Hence F has a finite model iff it has a finite model either of cardinality $\leq 2s^2$ or of cardinality t.

Lemma 1. For each $r > 2s^2$, $E(F, \omega) \cup N_r \cup M_r$ truth-functionally implies $E(G, \omega) \cup N_r$.

Proof. Let $m \geq 0$ and let $\varphi \in S$. It suffices to show that $E(F, \omega) \cup N_r \cup M_r$ truth-functionally implies $\varphi m m \vee \varphi m m$. As we have seen, $\Xi(\equiv, \varphi)$ contains a pair $\langle a, b \rangle$ such that $0 < |a - b| \leq 2s^2$. Hence, for all sufficiently large q, $E(F, \omega)$ contains a chain from $q - a \equiv q - b$ to $\varphi q q$, whence $E(F, \omega)$ truth-functionally implies $q - a \equiv q - b \vee \varphi q q$. Since $r > 2s^2$, $q - a$ and $q - b$ are incongruent (mod r), so that M_r truth-functionally implies $-(q - a \equiv q - b)$. Hence $E(F, \omega) \cup M_r$ truth-functionally implies $\varphi q q$ for all sufficiently large q. In particular, for sufficiently large n, $E(F, \omega) \cup M_r$ truth-functionally implies $\varphi m + nr\ m + nr$. But N_r truth-functionally implies $(\varphi m + nr\ m + nr \Leftrightarrow \varphi m m)$. Hence $E(F, \omega) \cup N_r \cup M_r$ truth-functionally implies $\varphi m m$. $\square$

Lemma 2. For each $r > 0$, if $E(G, \omega) \cup N_r$ is consistent, then $E(F, \omega) \cup N_r \cup M_r$ is consistent.

Proof. Suppose $E(F, \omega) \cup N_r \cup M_r$ inconsistent. By the Chain Criterion, $E(F, \omega) \cup N_r \cup M_r$ contains chains J_1 from A to A and J_2 from $\bar{A}$

ISBN 0-201-02540-X

to $\bar{A}$ for some atom A. If J_1 and J_2 are contained in $E(F, \omega) \cup N_r$, then they are contained in $E(G, \omega) \cup N_r$, whence $E(G, \omega) \cup N_r$ is inconsistent. Otherwise, some member of M_r is a conjunct in J_1 or in J_2, so that some equation $m \equiv n$, where m and n are incongruent (mod r), occurs positively in J_1 or in J_2. By Lemma 1 of the preceding section, $E(F, \omega) \cup N_r \cup M_r$ contains a chain K from $m \equiv n$ to $m \equiv n$. And $K = K_1 \wedge K_2$, where $E(F, \omega) \cup N_r$ contains K_1 and either K_2 is empty (so that $K_1 = K$) or else the first conjunct in K_2 is a member of M_r. Thus K_1 is from $m \equiv n$ to $p \equiv q$, where p and q are incongruent (mod r). Moreover, K_1 is not r-elementary. For if it were then, by Lemma 7 of the previous section, $\Xi(\equiv, \equiv)$ would be nonempty, contrary to hypothesis. Hence $K_1 = K_3 \wedge K_4 \wedge K_5$ where, for some signed predicate letters φ and ψ, and some integers i and j, K_3 is an r-elementary chain from $m \equiv n$ to φii, K_4 is a chain contained in $E(F, \omega) \cup N_r$ from $\bar{\varphi} ii$ to ψjj, and K_5 is an r-elementary chain from $\bar{\psi} jj$ to $p \equiv q$. Hence $\langle i - m, i - n \rangle \in \Xi_r(\equiv, \varphi)$ and $\langle p - j, q - j \rangle \in \Xi_r(\psi, \equiv)$. Since $i - m$ and $i - n$ are incongruent (mod r), by Lemma 7 of the preceding section $\Xi(\equiv, \varphi)$ contains a pair $\langle m', n' \rangle$ such that $m' \neq n'$. Similarly, $\Xi(\equiv, \bar{\psi})$ contains a pair $\langle m', n' \rangle$ such that $m' \neq n'$. But then $\varphi \in S$ and $\psi \in S$. Hence $E(G, \omega)$ truth-functionally implies $\varphi pp \wedge \psi qq$. Since $E(G, \omega) \cup N_r$ contains K_4, $E(G, \omega) \cup N_r$ truth-functionally implies $\bar{\varphi} pp \vee \bar{\psi} qq$. Thus $E(G, \omega) \cup N_r$ is inconsistent. $\square$

5. An Extension of Class 5.1

The class that we here prove solvable, Class A.3, presents certain anomalies. Although the argument that establishes its amenability is in essence no different from that for Class 5.1, the docility of Class A.3—unlike that of Class 5.1—is open. Moreover, again unlike Class 5.1, the extension of Class A.3 obtained by allowing the identity sign is unsolvable.

Class A.3 differs from Class 5.1 by allowing prenex schemata with prefixes $\forall y_1 \exists x_1 \cdots \exists x_m \forall y_2 \cdots \forall y_n$, rather than just $\forall \exists \forall$-schemata. (Hence, of course, the numerical notation for terms used for Class 5.1 is no longer applicable.) The restrictions on forms of atomic subformulas are similar to those in the specification of Class 5.1.

Class A.3. Let F be any prenex schema with prefix $\forall y_1 \exists x_1 \cdots \exists x_m \forall y_2 \cdots \forall y_n$ each atomic subformula of which has one of the following forms: Pvv, where v is any variable of F, $Py_1 x_j$, $Px_j y_1$, $Px_j x_k$, $Py_1 y_l$, $Px_j y_l$, where $1 \leq j, k \leq m$ and $2 \leq l \leq n$. Then $E(F, 2^{d(F,2)})$ is adequate.

Proof. For notational convenience we use f_i rather than f_{x_i} for the 1-place indicial correlate of x_i, $1 \leq i \leq m$. Thus the functional form F^* of F is $F^M(x_1/f_1(y_1), \ldots, x_m/f_m(y_1))$. Let $\mathfrak{A}$ be any truth-assignment on $\bar{L}(F, \omega)$.

ISBN 0-201-02540-X

We show first that it suffices to have a term-mapping γ and a mapping η on pairs of terms in $D(F, \omega)$ satisfying the following conditions:

(1) $\gamma[D(F, \omega)] \subseteq D(F, 2^{d(F,2)})$;

(2) $\eta[D(F, \omega) \times D(F, \omega)] \subseteq D(F, 2^{d(F,2)})$;

(3) $\mathfrak{p}_{\mathfrak{A}}(\gamma f_k(t), \eta(f_k(t), s)) = \mathfrak{p}_{\mathfrak{A}}(f_k(\gamma t), \eta(t, s))$ for all terms t and s in $D(F, \omega)$ and every number $k \le m$;

(4) $\eta(t, t) = \gamma t$ and $\eta(t, f_k(t)) = f_k(\gamma t)$ for every term t in $D(F, \omega)$ and every number $k \le m$.

Given such γ and η, we define a formula-mapping Γ and an instance-mapping Ψ thus:

(A) For each atom Pts in $L(F, \omega)$ let $\Gamma(Pts) = P\gamma t\, \eta(t, s)$.

(B) For each Herbrand instance $H = F^*(y_1/t_1, y_2/t_2, \ldots, y_n/t_n)$ of F, let $\Psi(H) = F^*(y_1/\gamma t_1, y_2/\eta(t_1, t_2), \ldots, y_n/\eta(t_1, t_n))$.

Conditions (1) and (2) imply the boundedness condition

$$\Psi[E(F, \omega)] \subseteq E(F, 2^{d(F,2)}).$$

We must show the central condition

$$\mathfrak{A}(\Gamma(\langle H \rangle_i)) = \mathfrak{A}(\langle \Psi(H) \rangle_i)$$

for each Herbrand instance H of F and each member i.

Let $H = F^*(y_1/t_1, \ldots, y_n/t_n)$. Just as in the proof for Class 5.1, if $\langle F \rangle_i = Py_1 y_l$, $2 \le l \le n$, then $\Gamma(\langle H \rangle_i) = \langle \Psi(H) \rangle_i$. If $\langle F \rangle_i = Py_1 y_1$ or $Py_1 x_j$, then by condition (4), once again $\Gamma(\langle H \rangle_i) = \langle \Psi(H) \rangle_i$. Again, just as in the proof for Class 5.1, the central condition follows from condition (3) if $\langle F \rangle_i = Px_k y_l$ and from conditions (3) and (4) if $\langle F \rangle_i = Px_k y_1$, $1 \le k \le m$ and $2 \le l \le n$. We are left with the cases $\langle F \rangle_i = Px_j x_k$ and $\langle F \rangle_i = Py_l y_l$, $1 \le j, k \le m$ and $2 \le l \le n$.

If $\langle F \rangle_i = Px_j x_k$, then $\Gamma(\langle H \rangle_i) = P\gamma f_j(t_1)\, \eta(f_j(t_1), f_k(t_1))$, which by condition (3) is equivalent under $\mathfrak{A}$ to $Pf_j(\gamma t_1)\, \eta(t_1, f_k(t_1))$, which in turn by condition (4) is identical with $Pf_j(\gamma t_1)\, f_k(\gamma t_1)$, which is just $\langle \Psi(H) \rangle_i$.

If $\langle F \rangle_i = Py_l y_l$, then $\Gamma(\langle H \rangle_i) = P\gamma t_l\, \eta(t_l, t_l) = P\gamma t_l\, \gamma t_l$, while $\langle \Psi(H) \rangle_i = P\eta(t_1, t_l)\, \eta(t_1, t_l)$. But, continuing the analogy with Class 5.1, it suffices to show that for each term s the profile $\mathfrak{p}_{\mathfrak{A}}(\eta(t, s))$ is fixed as t varies. Now for any term t in $D(F, \omega)$ let the *ancestors* of t be the terms t, $\hat{y}_1 t$, $\hat{y}_1 \hat{y}_1 t, \ldots, \mathfrak{1}$. Condition (3) implies that if t is an ancestor of t' then $\mathfrak{p}_{\mathfrak{A}}(\eta(t, s)) = \mathfrak{p}_{\mathfrak{A}}(\eta(t', s))$. This alone was sufficient for Class 5.1, since there the domain is generated by one 1-place function sign, so that of any two terms t and t' either t is an ancestor of t' or t' is an ancestor of t. Here, however, since the domain is generated by m 1-place function signs, we need the additional fact that every two terms have a common ancestor. This is obvious, since $\bot$ is an ancestor of every term.

Thus the adequacy of $E(F, 2^{d(F,2)})$ would be established if we could find a term-mapping γ and a mapping η fulfilling conditions (1)–(4). By using the pigeonhole argument just as we did for Class 5.1, we construct a term-mapping γ such that

(γ-1) $\gamma[D(F, \omega)] \subseteq D(F, 2^{d(F,2)} - 1)$;

(γ-2) $\xi_{\mathfrak{A}}(\gamma f_k(t)) = \xi_{\mathfrak{A}}(f_k(\gamma t))$ for every term t in $D(F, \omega)$ and every number $k \leq m$, where the function $\xi_{\mathfrak{A}}$ is defined, as before, by

$$\xi_{\mathfrak{A}}(t) = \{\mathfrak{p}_{\mathfrak{A}}(t, s) \mid s \in D(F, 2^{d(F,2)})\}.$$

We define the mapping η as follows:

(η-1) $\eta(t, t) = \gamma t$ and $\eta(t, f_k(t)) = f_k(\gamma t)$ for every term t in $D(F, \omega)$ and every $k \leq m$.

(η-2)   If $f_k(t)$ is an ancestor of s but not identical with s, $1 \leq k \leq m$, then $\eta(t, s)$ is the earliest term in $D(F, 2^{d(F,2)})$ such that

$$\mathfrak{p}_{\mathfrak{A}}(f_k(\gamma t), \eta(t, s)) = \mathfrak{p}_{\mathfrak{A}}(\gamma f_k(t), \eta(f_k(t), s)).$$

(η-3)   If either t is not an ancestor of s or $t = s$, then for each $k \leq m$, $\eta(f_k(t), s)$ is the earliest term in $D(F, 2^{d(F,2)})$ such that

$$\mathfrak{p}_{\mathfrak{A}}(\gamma f_k(t), \eta(f_k(t), s)) = \mathfrak{p}_{\mathfrak{A}}(f_k(\gamma t), \eta(t, s)).$$

Note that the values of η fixed by (η-1) are in $D(F, 2^{d(F,2)})$. Hence, by (η-2) and induction, there always exist a value for $\eta(t, s)$ in clause (η-2) and a value for $\eta(f_k(t), s)$ in clause (η-3). To find $\eta(t, s)$ where t is not an ancestor of s it suffices to start with $\eta(t', s)$, where t' is the common ancestor of t and s of greatest height, and apply clause (η-3) repeatedly. Hence η is determined on all pairs. Conditions (2), (3), and (4) then follow directly. $\square$

The docility of Class A.3 is open. The difficulty in extending the argument for Class 5.1 comes from the fact that the schemata in Class A.3 have prefixes $\forall y_1 \exists x_1 \cdots \exists x_m \forall y_2 \cdots \forall y_n$, so that the Special Finite Model Lemma of Chapter 5, §3, does not apply. However, if we restrict m to 1, that is, allow only one x-variable, then the Special Finite Model Lemma does apply. (Although we stated the Lemma solely for $\forall\exists\forall$-schemata, the Lemma requires only that the functional form of F^* contain but one function sign and that the function sign have but one argument). Indeed, for this subclass of A.3 the docility proof for Class 5.1 goes through almost word for word.

Like the docility proof for Class 5.1, the solvability proof for Class 5.1 with Identity turns on there being exactly one term of height r in $D(F, \omega)$

ISBN 0-201-02540-X

for each r. Hence that proof is inapplicable to Class A.3 with Identity. Indeed, the addition of identity to Class A.3 yields unsolvability, even if we restrict the number of x-variables to two. The proof of this, just as in the case of the mirror-image of Class 1.3 with Identity (see Chapter 8, §2), involves an encoding of an unsolvable word problem (see Lewis, 1974, page 4–11).

6. More on the Gödel–Kalmár–Schütte Class with Identity

Let $F = \forall y_1 \forall y_2 \exists x_1 \cdots \exists x_n F^M$ be a schema in Class 2.5 with Identity, and suppose F is k-strongly satisfiable for some k (see page 220). We show that F has a finite model. For expository convenience we limit ourselves to the case in which F contains only dyadic predicate letters. Let $\mathfrak{A}$ be a standard truth-assignment verifying $E(F, \omega)$ under which there are no distinguished terms of level $k + 1$, and let $u_1,\ldots,u_q$ be the distinguished terms of levels $\leq k$. A term t in $D(F, \omega)$ is *proper* iff it is $\mathfrak{A}$-distinct from each of $u_1,\ldots,u_q$.

Let $G = \forall y_1 \forall y_2 \exists x_1 \cdots \exists x_n [\bigvee F^M(x_1/v_1,\ldots,x_n/v_n)]$, where the disjunction is over all n-tuples $\langle v_1,\ldots,v_n \rangle$ of variables of F. Since G and F are logically equivalent, it suffices to show that G has a finite model. We start by constructing a truth-assignment $\mathfrak{B}$ verifying $E(G, \omega)$ that has certain special properties. To do this we inductively define a term-mapping γ and a mapping γ^* on pairs of terms.

Let $\gamma\bot = \bot$ and $\gamma^*(\bot, \bot) = \langle \bot, \bot \rangle$. Now suppose γ and γ^* defined on $D(G, p)$ $(=D(F, p))$, $p \geq 1$, and suppose $\gamma^*(s, s) = \langle \gamma s, \gamma s \rangle$ for each s of height $\leq p$. Then for each term $f_{x_j}(s\, t)$ of height $p + 1$, $1 \leq j \leq n$, let $\gamma f_{x_j}(s\, t) = f_{x_j}(\gamma^*(s, t))$. If s_1, s_2 are terms from $D(G, p + 1)$ at least one of which has height $p + 1$, we define $\gamma^*(s_1, s_2)$ thus.

(1) If $\langle s_1, s_2 \rangle = \langle \hat{v}t, \hat{w}t \rangle$ for some x-focused pair $\langle v, w \rangle$ and some x_n-term t of height $p + 1$ such that $\hat{v}\gamma t$ and $\hat{w}\gamma t$ are $\mathfrak{A}$-distinct and proper, then let $\gamma^*(s_1, s_2) = \langle \hat{v}\gamma t, \hat{w}\gamma t \rangle$.

(2) If $\gamma^*(s_1, s_2)$ is not determined by (1) and either $s_1 = s_2$, or γs_1 and γs_2 are $\mathfrak{A}$-distinct, or at least one of γs_1 and γs_2 is not proper, then let $\gamma^*(s_1, s_2) = \langle \gamma s_1, \gamma s_2 \rangle$.

(3) If $\gamma^*(s_1, s_2)$ is determined by neither (1) nor (2), then let $\gamma^*(s_1, s_2) = \langle t_1, t_2 \rangle$, where $\langle t_1, t_2 \rangle$ is the earliest pair such that $\mathfrak{p}_{\mathfrak{A}}(u_1,\ldots,u_q, t_i) = \mathfrak{p}_{\mathfrak{A}}(u_1,\ldots,u_q, \gamma s_i)$, $i = 1$ and 2, t_1 and t_2 are $\mathfrak{A}$-distinct, and t_1 is earlier than t_2 iff s_1 is earlier than s_2. Such a pair $\langle t_1, t_2 \rangle$ exists since there are no distinguished terms of level $k + 1$.

This completes the definition of γ and γ^*. The following properties are easily verified: for each s of height >1 and each variable v of F,

ISBN 0-201-02540-X

$\mathfrak{p}_{\mathfrak{A}}(u_1,\ldots,u_q, \gamma \hat{v}s) = \mathfrak{p}_{\mathfrak{A}}(u_1,\ldots,u_q, \hat{v}\gamma s)$; for all s_1 and s_2, if $\gamma^*(s_1, s_2) = \langle t_1, t_2 \rangle$, then $\gamma^*(s_2, s_1) = \langle t_2, t_1 \rangle$ and $\mathfrak{p}_{\mathfrak{A}}(u_1,\ldots,u_q, t_i) = \mathfrak{p}_{\mathfrak{A}}(u_1,\ldots,u_q, \gamma s_i)$, $i = 1, 2$.

Let $\Gamma(Ps_1s_2) = P\gamma^*(s_1, s_2)$ for each atom Ps_1s_2 in $L(G, \omega)$. We show that $\mathfrak{A}(\Gamma)$ verifies $E(G, \omega)$. Let s be any x_n-term and let $H = G^*(y_1/\hat{y}_1s, y_2/\hat{y}_2s)$. Let $s_1,\ldots,s_n$ and $t_1,\ldots,t_n$ be defined thus: for $1 \leq j \leq n$, if $\mathfrak{A}$ verifies $\hat{x}_j\gamma s \equiv \hat{v}\gamma s$ for v among $y_1, y_2, x_1,\ldots,x_{j-1}$, then let $s_j = \hat{v}s$ and $t_j = \hat{v}\gamma s$ for the earliest such v; otherwise, let $s_j = \hat{x}_js$ and $t_j = \hat{x}_j\gamma s$. Now let $I = F^M(y_1/\hat{y}_1s, y_2/\hat{y}_2s, x_1/s_1,\ldots,x_n/s_n)$, and $I' = F^M(y_1/\hat{y}_1\gamma s, y_2/\hat{y}_2\gamma s, x_1/t_1,\ldots,x_n/t_n)$. By the construction of G, I is a disjunct in H. Since $\mathfrak{A}$ is standard and verifies $\hat{x}_j\gamma s \equiv t_j$ for each $j \leq n$, $\mathfrak{A}(I') = \mathfrak{A}(F^*(y_1/\hat{y}_1\gamma s, y_2/\hat{y}_2\gamma s))$. Hence it suffices to show that $\mathfrak{A}(\Gamma(\langle I \rangle_i)) = \mathfrak{A}(\langle I' \rangle_i)$ for each i. But this follows straightforwardly from the properties of γ and γ^*.

Moreover, $\mathfrak{A}(\Gamma)$ is standard. For by the definition of γ^* an equation $s \equiv t$ is verified by $\mathfrak{A}(\Gamma)$ only if either $s = t$ or else $\mathfrak{A}$ verifies $\gamma s \equiv u_i \wedge \gamma t \equiv u_i$ for some $i \leq q$; standardness then follows quickly.

Let $\mathfrak{B} = \mathfrak{A}(\Gamma)$. The argument of the preceding paragraph shows that there are terms $u_1',\ldots,u_r'$, where $r \leq q$, such that $\mathfrak{B}$ verifies an equation $s \equiv t$ only if either $s = t$ or else $\mathfrak{B}$ verifies $s \equiv u_i' \wedge t \equiv u_i'$ for some $i \leq r$. Given $\mathfrak{B}$, we may construct a finite model for F by applying the method used for the finite controllability of Class 2.7 (pages 96–101). By Combinatorial Lemmas 1 and 2 there is an equivalence relation $\simeq$ of finite order on $D(F, \omega)$ with the following properties: $\simeq$ is faithful; if $s \simeq t$, then either $s = t = \dagger$ or else s and t have height >1 and

$$\mathfrak{p}_{\mathfrak{B}}(u_1',\ldots,u_r', \hat{y}_1s, \hat{y}_2s, \hat{x}_1s,\ldots,\hat{x}_ns)$$

$$= \mathfrak{p}_{\mathfrak{B}}(u_1',\ldots,u_r', \hat{y}_1t, \hat{y}_2t, \hat{x}_1t,\ldots,\hat{x}_nt);$$

if $s \simeq t$ and $\hat{y}_is \simeq \hat{y}_jt$, then $\hat{y}_is \simeq \hat{y}_js \simeq \hat{y}_it \simeq \hat{y}_jt$. A term t is tight iff $\hat{y}_1t \simeq \hat{y}_2t$ only if $\hat{y}_1t = \hat{y}_2t$. Let η carry each term to its earliest $\simeq$-equivalent. Define a mapping β^* on pairs of terms thus: if there is a tight x_n-term t and an x-focused pair $\langle v, w \rangle$ such that $\langle s_1, s_2 \rangle \simeq \langle \hat{v}t, \hat{w}t \rangle$, then let $\beta^*(s_1, s_2) = \langle \hat{v}t, \hat{w}t \rangle$ for the earliest such t and $\langle v, w \rangle$; otherwise let $\beta^*(s_1, s_2) = \langle \eta s_1, \eta s_2 \rangle$. Define a term-mapping δ by: $\delta\dagger = \dagger$, and $\delta f_{x_j}(s_1s_2) = \eta f_{x_j}(\beta^*(\delta s_1, \delta s_2))$. By reasoning as in the finite controllability proof for Class 2.7, we may infer that $\mathfrak{p}_{\mathfrak{B}}(\beta^*(\delta\hat{v}t, \delta\hat{w}t)) = \mathfrak{p}_{\mathfrak{B}}(\hat{v}\delta t, \hat{w}\delta t)$ for each x_n-term t and all variables v and w of G. Consequently, if $\Gamma'(Ps_1s_2) = P\beta^*(s_1, s_2)$ for each atom Ps_1s_2 in $L(G, \omega)$ and if Δ is the δ-canonical formula-mapping, then $\mathfrak{B}(\Gamma')$ verifies $\Delta[E(G, \omega)]$. We show now that $\mathfrak{B}(\Gamma')$ is standard on $\Delta[E(G, \omega)]$. Note first that, by the choice of $\simeq$, if $s \simeq t$ and $\mathfrak{B}$ verifies $s \equiv u_j'$ for some $j \leq r$, then $\mathfrak{B}$ verifies $t \equiv u_j'$. Moreover, $\beta^*(s_1, s_2) \simeq \langle s_1, s_2 \rangle$ for all s_1 and s_2. Now suppose $\mathfrak{B}(\Gamma')$ verifies $s_1 \equiv s_2$,

ISBN 0-201-02540-X

where s_1 and s_2 are distinct members of the range of δ. By the facts just mentioned and the special property of $\mathfrak{B}$, $\mathfrak{B}$ verifies $s_1 \equiv u_j{}' \wedge s_2 \equiv u_j{}'$ for some $j \leq r$. Consequently, since $\mathfrak{B}$ is standard, for each term t and $i = 1, 2$ $\mathfrak{p}_\mathfrak{B}(\beta^*(s_i, t)) = \mathfrak{p}_\mathfrak{B}(u_j{}', t')$, where t' is some $\simeq$-equivalent of t. By the definition of $\simeq$, $\mathfrak{p}_\mathfrak{B}(u_j{}', t') = \mathfrak{p}_\mathfrak{B}(u_j{}', t)$. Hence $\mathfrak{p}_\mathfrak{B}(\beta^*(s_1, t)) = \mathfrak{p}_\mathfrak{B}(\beta^*(s_2, t))$, whence $\mathfrak{B}(\Gamma')$ is standard.

It follows that $\mathfrak{B}(\Gamma')$ provides a model for G of cardinality at most that of $\delta[D(G, \omega)]$, that is, of cardinality at most the number of $\simeq$-equivalence classes. This number is primitive recursively calculable from G and r. Moreover, r is at most q, the number of distinguished terms under the original truth-assignment $\mathfrak{A}$ that verifies $E(F, \omega)$. Since these distinguished terms are of levels $\leq k$, q is at most $\lambda(k)$, where λ is the function defined thus: $\lambda(1) = d(F, 1)$, $\lambda(i + 1) = \lambda(i) + d(F, \lambda(i))$. In sum, we may calculate primitive recursively from F and k the size of a finite model for F.

ISBN 0-201-02540-X

References

In this list *J.S.L.* stands for *The Journal of Symbolic Logic* and *M.A.* for *Mathematische Annalen*.

Aanderaa, Stål O.
- 1966   A new undecidable problem with applications in logic. Ph.D. thesis, Harvard University, Cambridge, Mass.
- 1966a   ''A reduction method for special cases of the decision problem,'' *Notices Amer. Math. Soc.* **13**, 261.
- 1971   ''On the decision problem for formulas in which all disjunctions are binary,'' *Proc. Second Scandinavian Logic Symposium* (J. E. Fenstad, ed.), pp. 1–18. North-Holland, Amsterdam.
——, and Warren D. Goldfarb
- 1974   ''The finite controllability of the Maslov case,'' *J.S.L.* **39**, 509–518.
——, and Harry R. Lewis
- 1973   ''Prefix classes of Krom formulas,'' *J.S.L.* **38**, 628–642.
- 1974   ''Linear sampling and the ∀∃∀ case of the decision problem,'' *J.S.L.* **39**, 519–547.

Ackermann, Wilhelm
- 1928   ''Über die Erfüllbarkeit gewisser Zählausdrücke,'' *M.A.* **100**, 638–649.
- 1936   ''Beiträge zum Entscheidungsproblem der mathematischen Logik,'' *M.A.* **112**, 419–432.
- 1954   *Solvable cases of the decision problem*. North-Holland, Amsterdam.

Behmann, Heinrich
- 1922   ''Beiträge zur Algebra der Logik, insbesondere zum Entscheidungsproblem,'' *M.A.* **86**, 163–229.

Bernays, Paul, and Moses Schönfinkel
- 1928   ''Zum Entscheidungsproblem der mathematischen Logik,'' *M.A.* **99**, 342–372.

Chang, Chen Chung, and H. Jerome Keisler
- 1962   ''An improved prenex normal form,'' *J.S.L.* **27**, 317–326.

Church, Alonzo
- 1936   ''A note on the Entscheidungsproblem,'' *J.S.L.* **1**, 40–41; correction, *J.S.L.* **1**, 101–102.
- 1951   ''Special cases of the decision problem,'' *Rev. phil. Louvain* **49**, 203–221; correction, *Rev. phil. Louvain* **50**, 270–272.
- 1956   *Introduction to mathematical logic*, Vol. 1. Princeton Univ. Press, Princeton, N.J.

Denton, John
- 1963   Applications of the Herbrand theorem. Ph.D. thesis, Harvard University, Cambridge, Mass.

Dreben, Burton
  1959   A systematic approach to the decision problem. Unpublished monograph.
  1961   "Solvable Surányi subclasses: An introduction to the Herbrand theory,"
          Proc. Harvard symposium on digital computers and their applications, 3–
          6 April 1961. (*Annals of the Computation Laboratory of Harvard Univer-
          sity* **31**, 32–47. Harvard Univ. Press, Cambridge, Mass., 1962.)
  ——, and John Denton
  1963   "Three solvable cases," *Notices Amer. Math. Soc.* **10**, 590.
  ——, Andrew S. Kahr, and Hao Wang
  1962   "Classification of AEA formulas by letter atoms," *Bull. Amer. Math.
          Soc.* **68**, 528–532.
Erdös, Paul
  1963   "On a problem in graph theory," *Math. Gazette* **47**, 220–223.
Friedman, Joyce
  1957   "Extensions of two solvable cases of the decision problem," *J.S.L.* **22**,
          108.
Gödel, Kurt
  1930   "Die Vollständigkeit der Axiome des logischen Funktionenkalküls," *Mo-
          natshefte Math. Phys.* **37**, 349–360.
  1932   "Ein Spezialfall des Entscheidungsproblems der theoretischen Logik,"
          Ergebn. math. Kolloq. **2**, 27–28.
  1933   "Zum Entscheidungsproblem des logischen Funktionenkalküls," *Monats-
          hefte Math. Phys.* **40**, 433–443.
Goldberg, Richard
  1963   "On the solvability of a subclass of the Surányi reduction class," *J.S.L.*
          28, 237–244.
Goldfarb, Warren D.
  1971   "Review of Thoralf Skolem, *Selected Works in Logic*," *J. Philos.* **68**,
          520–530.
  1975   Decision problems for quantification theory. Ph.D. thesis, Harvard Uni-
          versity, Cambridge, Mass.
  1979   "Logic in the twenties: the nature of the quantifier," *J.S.L.* **44.**
  1979a   "The Gödel Class with Identity is not primitive recursively solvable,"
          Notices Amer. Math. Soc. **26**, A-390.
  1980   "On the Gödel Class with Identity" (to appear, *J.S.L.*).
  ——, and Harry R. Lewis
  1975   "Skolem reduction classes, *J.S.L.* **40**, 62–68.
Herbrand, Jacques
  1930   *Recherches sur la théorie de la démonstration*, Thesis, University of
          Paris; also *Prace Towarzystwa Naukowego Warszawskiego, Wydział III*,
          no. 33. English translation in Herbrand (1971).
  1931   "Sur le problème fondamental de la logique mathématique," *Sprawoz-
          dania z posiedzeń Towarzystwa Naukowego Warszawskiego, Wydział III*,
          24, 12–56. English translation in Herbrand (1971).
  1971   *Logical writings* (Warren D. Goldfarb, ed.). Reidel, Dordrecht, and Har-
          vard Univ. Press, Cambridge, Mass.
Hilbert, David, and Wilhelm Ackermann
  1928   *Grundzüge der theoretischen Logik*. Springer, Berlin.

ISBN 0-201-02540-X

Horn, Alfred
  1951   "On sentences which are true of direct unions of algebras," *J.S.L.* **16,**
         14–21.
Kahr, Andrew S., Edward F. Moore, and Hao Wang
  1961   "Entscheidungsproblem reduced to the ∀∃∀ case," *Proc. Nat. Acad. Sci.
         U.S.A.* **48,** 365–377 (1962).
Kalmár, László
  1933   "Über die Erfüllbarkeit derjenigen Zählausdrücke, welche in der Nor-
         malform zwei benachbarte Allzeichen enthalten," *M.A.* **108,** 466–484.
Kostyrko, V. F.
  1962   "On a case of the decision problem in the restricted predicate calculus,"
         Uspekhi Mat. Nauk **17** (106), 213.
Krom, Melven R.
  1967   "The decision problem for segregated formulas in first-order logic," *Math-
         ematica scandinavica* **21,** 233–240.
  1967a "The decision problem for a class of first-order formulas in which all
         disjunctions are binary," *Z. math. Logik und Grundlagen der Math.* **13,**
         15–20.
  1970   "The decision problem for formulas in prenex conjunctive normal form
         with binary disjunctions," *J.S.L.* **35,** 210–216.
Lewis, Harry R.
  1974   Herbrand expansions and reductions of the decision problem. Ph.D. the-
         sis, Harvard University, Cambridge, Mass.
  1979   *Unsolvable classes of quantificational formulas.* Addison-Wesley, Ad-
         vanced Book Program, Reading, Mass.
———, and Warren D. Goldfarb
  1973   "The decision problem for formulas with a small number of atomic subfor-
         mulas," *J.S.L.* **38,** 471–480.
Löwenheim, Leopold
  1915   "Über Möglichkeiten im Relativkalkül," *M.A.* **68,** 169–207.
Maslov, S. Yu.
  1964   "An inverse method of establishing deducibilities in the classical predicate
         calculus," *Dokl. Akad. Nauk SSSR* **159,** 1420–1424.
Orevkov, V. P.
  1969   "Two undecidable classes of formulas in classical predicate calculus,"
         Seminars in Math., V.A. Steklov Math. Inst., Leningrad **8,** 98–102.
Péter, Rózsa
  1967   *Recursive functions.* Academic Press, New York.
Ramsey, F. P.
  1928   On a problem of formal logic, *Proc. London Math. Soc. (Ser. 2)* **30** (Part
         4), 338–384.
Schütte, Kurt
  1934   "Untersuchungen zum Entscheidungsproblem der mathematischen
         Logik," *M.A.* **109,** 572–603.
  1934a "Über die Erfüllbarkeit einer Klasse von logischen Formeln," *M.A.* **110,**
         161–194.
Skolem, Thoralf
  1919   "Untersuchungen über die Axiome des Klassenkalküls und über Produk-

ISBN 0-201-02540-X

tations- und Summationsprobleme, welche gewisse Klassen von Aussagen betreffen," *Skr. Vidensk. Kristiania, I. Mat.-naturvidenskabelig klasse* **3**.

1920   "Logisch-kombinatorische Untersuchungen über die Erfüllbarkeit oder Beweisbarkeit mathematischer Sätze nebst einem Theorem über dichte Mengen," *Skr. Vidensk. Kristiania, I. Mat.-naturvidenskabelig klasse* **4**.

1922   "Einige Bemerkungen zur axiomatischen Begründung der Mengenlehre," *Proc. Fifth Scandinavian Math. Congr., Helsinki 4–7 July 1922*, pp. 217–232, Akademiska Bokhandeln, Helsinki, 1923.

1928   "Über die mathematische Logik," *Norsk mat. tidsskr.* **10**, 125–142.

1929   "Über einige Grundlagenfragen der Mathematik," *Skr. Norske Videnskaps-Akademi i Oslo, I. Mat.-naturvidenskabelig klasse* **4**, 1–49.

1935   "Über die Erfüllbarkeit gewisser Zählausdrücke," *Skr. Norske Videnskaps-Akademi i Oslo, I. Mat.-naturvidenskabelig klasse* **6**, 1–14.

Surányi, Janos

1950   "Contributions to the reduction theory of the decision problem, second paper," *Acta Math. Acad. Sci. Hungaricae* **1**, 261–270.

Trakhtenbrot, B. A.

1950   "Impossibility of an algorithm for the decision problem in finite classes," *Dokl. Akad. Nauk SSSR* **70**, 569–572.

Turing, Alan M.

1937   "On computable numbers, with an application to the Entscheidungsproblem," *Proc. London Math. Soc.* **42**, 230–265; correction, **43**, 544–546.

van Heijenoort, Jean

1967   *From Frege to Gödel: A source book in mathematical logic 1879–1931.* Harvard Univ. Press, Cambridge, Mass.

Wang, Hao

1962   "Dominoes and the AEA case of the decision problem," *Proc. symposium on the mathematical theory of automata, New York, 1962*, pp. 23–55. Polytechnic Press, Brooklyn, N.Y., 1963.

ISBN 0-201-02540-X

Indexes

Index of Solvable Classes

(1) for each x_n-term t, if $\hat{Y}_1[t]$ and $\hat{Y}_2[t]$ are distinct sets of cardinality >1, then no member of Y_1 is hereditarily t-united to any member of Y_2; or

(2) there is a y-variable that cleaves $b(F)$ into sets S_1 and S_2 such that $Y_1 \in S_1$ and $Y_2 \in S_2$.   142

Class 6.2.   Schemata obtained from Class 6.1 by closing under the following operation: if a schema F is in the class, then so is every schema F' whose prefix is like that of F but for containing additional final variables $w_1,\ldots,w_q$, where either all the w_i are y-variables or else all the w_i are x-variables, and whose basis is like that of F but for containing, for some variable v of F, the sets $\{v, w_1,\ldots,w_q\}$, $\{w_1\},\ldots,\{w_q\}$.   145

Class 6.3.   Skolem schemata F such that, for some $k > 0$, each y-set in $b(F)$ contains at most k variables, and each x-focused set in $b(F)$ contains at least $k + 1$ variables, of which at least k are x-variables.   154

Class 7.1.   Prenex schemata containing two distinct atomic subformulas.   192

Class 7.2.   For each m and n, schemata containing at most m distinct atomic subformulas and n distinct y-variables.   195

Initially-extended Ackermann Class with Identity.   Prenex schemata containing identity with prefixes $\exists z_1 \cdots \exists z_l \forall y \exists x_1 \cdots \exists x_n$.   207

Class A.1.   Schemata with Horn matrices in which the number of negative occurrences of each predicate letter is at most one.   226

Class A.2.   Schemata with Horn matrices in which the number of positive occurrences of each predicate letter is at most one.   226

Class A.3.   Prenex schemata F with prefixes $\forall y_1 \exists x_1 \cdots \exists x_m \forall y_2 \cdots \forall y_n$ each atomic subformula of which has one of the following forms: Pvv, where v is any variable of F, $Py_1 x_j$, $Px_j y_1$, $Px_j x_k$, $Py_1 y_l$, $Px_j y_l$, where $1 \leq j,\ k \leq m$ and $2 \leq l \leq n$.   250

Index of Special Notation

General Index

Numerals set in *italics* indicate pages on which complete literature citations are given.

Aanderaa, S., 108, 124, 129, 135, 146, 215, 234, 239, *257*
Ackermann, W., 36, 40, 135, 137, 185, 206, 215, *257, 258*
Ackermann Class, 41–42, 83–84
   with Identity, 206
𝔄-cycle, 136
adequacy, 35
Adequacy Lemma, 48
𝔄-distinct, 211
advanced, 114
∀∃∀-Horn Class, 126–129, 133–135
   with Identity, 204–205, 247–248
∀∃∀-Krom Class, 129, 234–244
   with Identity, 204–205, 248–250
∀∃∀-schema, 119
amenability, 35
𝔄-match, 176
amiability, 85
Amiability Lemma, 88
Amplified Combinatorial Lemma 2, 112, 233
ancestor, 251
argument, 9, 10
arrangement, 47
atom, 11
atomic formula, 10

Basic Unsolvable Class, 150
basis, 47
   , pruned, 65
basis-set, 146
Behmann, H., 2, 36, *257*
Bernays, P., 2, 19, *257*
Bernays-Schönfinkel Class, 19
   with Identity, 211–214
boundedness condition, 34

canonical formula-mapping, 29
central condition, 34
chain, 234
Chang, C., 214, *257*
Church, A., 2, 4, *257*
Classes A–D, 160–163
cleaves, 142

coinstantiable, 25, 191
coinstantiated, 25
coinstantiation, 25
collapsing, 148
Collapsing Reduction 1, 151
   , Generalized, 156
Collapsing Reduction 2, 151
   , Generalized, 157
Combinatorial Lemma 1, 90, 231
   , Generalized, 102, 232
Combinatorial Lemma 2, 97, 232
   , Amplified, 112, 233
   , Generalized, 102, 233
companion, 23
component, 11
Compression Lemma, 30
condensable, 20
conjunct in, 10
conjunctive normal form, 10
connected, 4
Consistency Lemma, 28
consistent, 16
contraction, 164
covers, 174, 175

deleting, 64
Denton, J., 4, 135, *257, 258*
disjunct in, 10
disjunctive normal form, 10
distinguished term, 219
docility, 7, 119, 130
domain, 13
dominates, 67
Dreben, B., 4, 54, 119, 135, 146, *258*

elementary chain, 236
elementary conjunct, 235
elementary instance, 236
engender, 25
Equivalence Lemma, 80
Erdös, P., 231, *258*
essentially monadic, 40
Essentially Monadic Class, 40
exhibits, 174

Addison-Wesley
Advanced Book Program

UNSOLVABLE CLASSES OF QUANTIFICATIONAL FORMULAS

Harry R. Lewis, *Aiken Computation Laboratory, Harvard University*

This volume is the first major published survey of unsolvability results for the predicate calculus since Surányi's treatment in 1959. The book deals with the theoretical possibility of determining by mechanical means whether particular statements of mathematical logic can be true; this is the oldest problem in the theory of computability. Logicians, mathematicians, and computer scientists will welcome this publication.

1979, 214 pp., illus.
Paperbound   0-201-04069-7